OR RETURN

OR RETURN

Western Diseases

WESTERN DISEASES

❖❖

Their Dietary Prevention and Reversibility

❖❖

Edited by

Norman J. Temple, PhD

Faculty of Science, Athabasca University, Athabasca, Alberta, Canada

and

Denis P. Burkitt, MD

Foreword by

Sir Richard Doll, FRS, FRCP

HUMANA PRESS 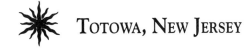 TOTOWA, NEW JERSEY

*Dedicated to Dr. Hugh Trowell,
a pioneer in the concept of Western diseases*

and

*To the memory of Denis P. Burkitt, who died having
just completed his work on this book*

© 1994 Humana Press Inc.
999 Riverview Dr., Suite 208
Totowa, NJ 07512

Printed in the United States of America. 9 8 7 6 5 4 3 2 1

Library of Congress Cataloging in Publication Data

Western diseases : their dietary prevention and reversibility / edited
  by Norman J. Temple and Denis P. Burkitt.
        p. cm.
    Includes index.
    ISBN 0-89603-264-7
    1. Nutritionally induced diseases—Epidemiology. 2. Chronic
diseases—Epidemiology. 3. Medical geography. 4. Diet in disease.
I. Temple, Norman J. II. Burkitt, D. P. (Denis P.) (d. 1993)
RA645.N87W469 1994
614.5939—dc20   94-14051
                                                        CIP

Foreword

Sir Richard Doll, FRS, FRCP

ICRF Cancer Research Studies Unit
Radcliffe Infirmary, Oxford, UK

The twentieth century has seen few changes more remarkable than the improvement in health that has occurred nearly everywhere, most spectacularly in the economically developed countries. In these countries improved nutrition, better housing, the control of infection, smaller family sizes, and higher standards of education have brought about a situation in which more than 97% of all liveborn children can expect to survive the first half of the three score years and ten that formerly was regarded as the allotted span of life. From then on, however, the position is less satisfactory. Some improvement has occurred; but the proportion of survivors who die prematurely, that is under 70 years of age, varies from 25% to over 50% in men and from 13% to 28% in women, the extremes in both sexes being recorded, respectively, in Japan and Hungary.

Most of these deaths under 70 years of age must now be called premature, even in Japan. For most of them are not the result of any inevitable aging process, but instead are the consequences of diseases (or types of trauma) that have lower—often much lower—age-specific incidence rates in many of the least developed countries. The causes of these diseases are, moreover, no longer shrouded in mystery, since most of them have been found to be characteristics of the economically developed societies that are, for the most part, capable of controlling these causes without sacrificing the many benefits that agricultural and industrial developments have brought.

One such characteristic is the use of tobacco, which is estimated to account for 20% of all premature deaths in middle age in the developed world. Another is traffic accidents and a third, far less important despite the prominence it is given, is the chemical pollution to which workers are exposed in industrial plants and to which the public is exposed outside from the dispersal of waste. There remains, however, a large block of premature deaths and of their associated morbidity that, in total, exceeds that arising from tobacco. The contributors to *Western Diseases* suggest that this nontobacco-caused mortality and morbidity may be rooted in the diet that is typical of the economically developed countries, most notably the excess of calories, the large proportions contributed by saturated fat and sugar, and the deficiency of dietary fiber. Much of the evidence in support of this belief has come from the work of the principal contributors to this book and the case they make is certainly strong enough to justify using the evidence for public education. A few of the proposed relationships may not survive the test of time; but the general message is clear. Action based on it would go far toward achieving the realistic objective of reducing the risk of premature death in middle age to around 10% within the next 30 years.

Preface

The majority of doctors, if asked what they considered to be the most important advances in the field of medicine and health that have occurred during the last 50 years would no doubt select some of the more dramatic technical achievements that have been witnessed throughout that time. And wonderful they are too. A selection might be made from joint prostheses, organ transplantation, and microsurgery. Another huge advance has been the opening up of the exciting field of molecular biology with all its amazing possibilities. Those with an eye on disease prevention might well opt for the worldwide eradication of smallpox, which has certainly been one of the greatest medical achievements of the past half century, comparable in value to the advent of immunization against poliomyelitis.

Few would make the choice of that highly respected British epidemiologist, the late Thomas McKeown. After carefully considering the advances in health care achieved during the last and present centuries he concluded that the major medical advance made during the 19th century was the recognition that infective disease, then the commonest cause of death worldwide, was the result of factors in the environment that could be reduced or even eliminated. The public health measures introduced consequent on this observation dramatically reduced infective disease in all the more affluent and industrialized populations. He went on to argue that the most important and potentially beneficial observation made in the 20th century is that precisely the same analysis applies to our chronic noninfective diseases, which are likewise attributable to factors in the environment that, once identified, can be reduced if not eliminated. The all-important discovery responsible for this change of attitude toward noninfective diseases was the momentous realization that the majority of the commonest disorders in Western populations were rare or even unknown in these countries before the First World War, are still rare throughout much of the Third World, and yet have comparable prevalences among black and white Americans today. Although this striking phenomenon had been hinted at by a few perceptive individuals in the past, whatever seeds were sown did not begin to bear fruit until after publication of two books: Hugh Trowell's *Non-infective Disease in Africa* (1960) and T. L. Cleave's *Diabetes, Coronary Thrombosis and the Saccharine Disease* (1966).

The enormous relevance of this discovery was slow to take root in the profession. Doctors are not trained to confront disease with the outlook of Louis Pasteur, who remarked, "Whenever I meditate on a disease, I never think of finding a remedy for it, but rather a means of preventing it." It will require a whole change of outlook for the question: "How can this be prevented?" to acquire anything like equal prominence with: "How can it be cured?"

This book is an attempt to draw attention not only to the fact that a huge proportion of diseases responsible for death and suffering in the Western world are potentially preventable, but also, in several instances, are actually reversible without recourse to drugs or surgery. Moreover, these diseases are the major consumers of health resources, and their large-scale reduction could have incalculable results in reducing health expenditure. In the following chapters this potential for disease reduction is viewed from many different, but complementary angles by a wide range of authorities. We hope that the great significance of the message will be disseminated not only throughout the medical profession, but also to the general public.

The major part of this book deals with the relationship between nutrition and Western disease. In addition, that knowledge is transplanted into concrete policy suggestions in the realm of health promotion. We consider the integration of these two areas—nutrition and health promotion—to be an important feature of this book. To focus on one and to neglect the other is to see only half the picture.

Norman J. Temple
Denis P. Burkitt

Contents

Contributors

ABAYOMI O. AKANJI • *VA Medical Center, University of Kentucky, Lexington, KY*

JAMES W. ANDERSON • *VA Medical Center, University of Kentucky, Lexington, KY*

DENIS P. BURKITT • *Deceased*

T. COLIN CAMPBELL • *Division of Nutritional Sciences, Martha Van Renselaer Hall, Cornell University, Ithaca, NY*

JUNSHI CHEN • *Institute of Nutrition and Food Hygiene, Chinese Academy of Preventive Medicine, Beijing, The People's Republic of China*

HANS DIEHL • *Lifestyle Medicine Institute, Loma Linda, CA*

INGRID I. GLATTHAAR • *Department of Human Nutrition, Medical University of Southern Africa, Medunsa, South Africa*

MARJORIE GOTT • *5 Fern Crescent, Groby, Leics, UK*

KEN W. HEATON • *Department of Medicine, University of Bristol, Bristol Royal Infirmary, Bristol, UK*

DEMETRE LABADARIOS • *Department of Human Nutrition, Faculty of Medicine, University of Stellenbosch, Tygerberg, South Africa*

KERIN O'DEA • *Deakin Institute of Human Nutrition, Deakin University, Geelong, Australia*

NORMAN J. TEMPLE • *Faculty of Science, Athabasca University, Athabasca, Alberta, Canada*

ALEXANDER R. P. WALKER • *South African Institute for Medical Research, Johannesburg, South Africa*

PART I

DISEASES CHARACTERISTIC OF MODERN WESTERN CULTURE

The Emergence of a Concept

Denis P. Burkitt

1. History of Dietary Fiber

A deficiency of dietary fiber was one of the first recognized characteristics of diets of more affluent populations in contrast to poorer ones. Consequently, the progressive understanding of this component of food has been closely tied to the emergence of the concept of Western disease. It is therefore appropriate to sketch the rise of dietary fiber from being viewed as a food contaminant to its honored role in nutrition today, analogous to the rise of Cinderella from kitchen maid to princess.

In the distant past (i.e., before 1970 or so) what is now known as dietary fiber was commonly referred to under the misleading term *roughage* because it was considered to be hard and consequently irritating to the lining of the gut, a property that was erroneously credited for its recognized laxative effect. Hugh Trowell aptly remarked that "softage" would have been a more appropriate name, since one of its properties is to render soft stools. Ken Heaton pointed out that soft food predisposes to constipation and vice versa. He coined the phrase "hard in and soft out" or "soft in and hard out."

So-called roughage was viewed as a food contaminant that called for means of extraction. These were provided

From: *Western Diseases: Their Dietary Prevention and Reversibility*
Edited by: N. J. Temple and D. P. Burkitt Copyright ©1994 Humana Press, Totowa, NJ

by developments in milling techniques that removed progressively coarser elements of the flour, notably the bran. It was generally used for animal fodder.

Early in the nineteenth century the term *crude fiber* was applied to what remained of plant food after heating first in dilute acid and then in dilute alkali. This was almost entirely composed of cellulose and lignin, and almost none of the components now designated noncellulose polysaccharides. This practice, developed for veterinary purposes, was quite inappropriate for human nutrition. Consequently, older tables giving the crude-fiber content of food can be highly misleading.

Not until the 1970s did Hugh Trowell, David Southgate, and others expose the deficiencies in the crude-fiber concept and suggest that fiber should be considered as that part of plant foods that was not degraded by human digestive enzymes during its passage through the stomach and small intestine and consequently entered the colon relatively unchanged. This fraction of food was eventually designated *dietary fiber (1)*.

It later became apparent that most of the fiber is degraded by bacterial action in the colon, but also that some of the starch component of food resists digestion in the small intestine and reaching the colon as *resistant starch* has similar effects there to dietary fiber *(2)*.

Fiber's history goes back to the time of Hippocrates, who wrote in the fourth century BC, "To the human body it makes a great difference whether the bread be made of fine flour or coarse, whether of wheat with the bran or without the bran." In 900 AD Hakim, a Persian physician, wrote, "Wheat is a beneficial cereal. Chapaties are made from wheat flour. The chapaties containing more bran come out of the digestive tract quicker, but are less nutritious."

Bran is referred to at least once in Shakespeare's plays. In *Coriolanus* (Act 1, Scene 1) the belly proudly announces to the other organs its own unselfishness, reminding them of how it generously gives up the food put into it to nourish other parts of the body, and ends its speech with the words, "Yet I can make my audit up that all from me do back receive the flour of all and leave me but the bran."

The laxative properties of roughage became progressively more apparent during the second half of the nineteenth century. Allinson (3) in England published a book extolling the advantages of whole meal bread, but as a consequence of his dabbling in commercial enterprises his name was erased from the Medical Register. Around the turn of the century the brothers William and John Kellogg in the United States, no doubt influenced by their Seventh-Day Adventist background, became enthusiastic proponents of foods that were, to the best of their knowledge, healthy. Whole grain cereals were high on their list (4). It was John who founded the breakfast food industry.

Sir Arbuthnot Lane, a prosperous English surgeon and colorful personality, who is remembered eponymously by virtue of surgical instruments bearing his name, began to blame stagnation in the colon as the cause of a range of diseases. He initially dealt with the problem by excising the colon, but fortunately soon came to appreciate that administration of bran was similarly effective in combatting constipation and so abandoned this brutal and unwarranted surgery.

Also worthy of mention are Cowgill and Anderson (5) and Olmsted et al. (6), who, in the 1930s, were extolling the benefits of roughage. Later in that decade Dimock (7) in England recommended the use of fiber for the prevention of constipation and the treatment of hemorrhoids. He was perhaps the first person to object to the term *roughage*, since it wrongly implies that fiber irritates the intestines. In 1953 the term *dietary fiber* was used for the first time when Hipsley in Australia suggested that its consumption is associated with a low incidence of pregnancy toxemia (8).

This steadily emerging appreciation of the role of dietary fiber was to become part of the much greater concept of characteristically Western diseases. In fact, it was this concept, considered in conjunction with the growing interest in fiber that accounted for such an explosion of literature on the subject that the ten or so scientific papers written on fiber before 1968 had passed 500 by 1980.

2. The Concept of Western Disease

At the same time as the vital health role of dietary fiber was being elucidated, the geographical distribution of disease and its relationship to diet was coming under increasing scrutiny. Of major impact was Sir Robert McCarrison, a British army doctor, who in the 1920s and 1930s worked among the hill tribes in northwest India (9). He noticed a remarkable absence among them of the diseases common in Western countries. He attributed this to their excellent diet, which contained much whole grain cereals, particularly wheat. By contrast, the tribes on the plains ate much white rice and were considerably less healthy. McCarrison's main focus was on the diet as a whole. He summed up his observations in the words, "The greatest single factor in the acquisition and maintenance of good health is perfectly constituted food."

Between 1937 and 1981 seven key books and a number of papers were published pointing out that certain diseases common in Western populations were rare in developing ones. The observations were first made with respect to Africa but later included all five continents. Although it was the contrasts in the amounts of carbohydrates that first attracted attention, many other aspects of lifestyle, and diet in particular, were eventually incriminated as causative factors. The seven books dealing with the subject are:

1. Donnison (1937 *[10]*)
2. Price (1939 *[11]*)
3. Gelfand (1944 *[12]*)
4. Trowell (1960 *[13]*)
5. Cleave and Campbell (1966 *[14]*)
6. Burkitt and Trowell (1975 *[15]*)
7. Trowell and Burkitt (1981 *[16]*)

Donnison *(10)* did little more than give an account of certain diseases that he realized were rare in Kenya but knew to be common in Great Britain. Weston Price *(11)*, a Canadian dentist, carried out a profoundly important worldwide study during the 1930s. He detailed the changing disease pattern in some 12 populations who were in the process of adopting a Western diet. Gelfand *(12)* wrote a comprehensive review of medical observations in Africa and mentioned several diseases that were common there but rarely seen in the West. In a 1952 paper I pointed out that several diseases responsible for emergency abdominal surgery in Britain were rare in Africa *(17)*. Likewise, in 1955 Walker *(18)* pointed to several diseases commonly found in the West as being uncommon in the Bantu of South Africa. He speculated that this might be related to their dietary pattern, which was low in fat and high in fiber.

The 1960 book by Trowell *(13)* was a milestone. It lists over 30 diseases, tabulated in appropriate categories of disorders, common in more affluent countries but rare throughout sub-Saharan Africa. It contains an extensive review of African medical literature, and points to the bulky starch-based foods of Africans, which, because of their roughage content, were responsible for their large output of soft stools. These in turn were believed by him to be protective against several Western bowel disorders. This book could and should have been the turning point in the all-important concept of Western diseases, but only a few copies were sold and the book was not referenced in any medical literature outside Africa. The history of medicine is replete with numerous similar examples of important discoveries being ignored for years.

The book by Cleave and Campbell *(14)* published in 1966 was another of true revolutionary thinking. (It was later rewritten by Cleave alone *[19]*.) These two authors had never read Trowell's book. All but one chapter was written by Cleave, then a retired Medical Officer from the Royal Navy. Although never serving overseas other than in naval establishments, he recognized with extraordinary perception and genius not only that many of the commonest chronic disorders in industrialized nations were rare or even absent in preindustrial communities, but, of crucial significance, that these diseases were commonly encountered in black Americans. He confirmed these observations with prodigious correspondence with doctors throughout the Third World and came to the inescapable conclusion that their cause must be sought not in skin color or genetic constitution but in lifestyle. His method was careful collection of data, which was subjected not to statistical analysis but rather to common sense examination with conclusions arrived at by logical deduction.

His main conclusion was that the single factor that most characterized Western diets was that carbohydrate foods are refined. Cleave referred to Western diseases with the all-inclusive term *saccharine disease*, meaning related to sugar. In his view refined carbohydrates, such as sugar, white bread, and cornflakes, are overconcentrated. He referred to sugar as being eight times more dangerous than white flour as it is eight times more concentrated. He assumed that sugar is directly deleterious to health and not merely indirectly via its lack of fiber. Subsequent studies have given scant support to this viewpoint. At the same time he was well aware of the importance of dietary fiber. He specifically referred to lack of fiber as being the mechanism by which refined carbohydrates cause obesity, dental caries, and some colon related diseases. Indeed, as far back as 1941 he published a letter extolling the value of bran for the control of constipation *(20)*. Regrettably, Cleave would not accept that excessive fat could play a role in disease causation.

Cleave's ideas were not accepted by the medical profession and he had, in fact, been viewed by many as a crank. This treatment begs comparison with the similar treatment accorded to Pasteur. However, one person who early on had the foresight to take Cleave's ideas seriously was Sir Richard Doll. In the Foreword to Cleave's book he wrote that

> ... If only a small part of (the predictions made prove to be correct), the authors will have made a bigger contribution to medicine than most university departments or medical research units make in the course of a generation.

Twenty-seven years hence these words now ring true. Alas, Cleave has still not received the recognition he deserves. For instance, according to the Science Citation Index his works were cited only 10 times in 1991.

Peter Cleave, as he was affectionately known to his friends, though his real name was Thomas Latimer Cleave, was a true pioneer with burning zeal and enthusiasm for his cause. It was he more than any other person who revolutionized my thinking and opened my eyes to the breathtaking concept that many, if not most, of the diseases filling hospital beds and doctors' offices in Western Europe and North America must be potentially preventable.

I was introduced to Cleave in 1967 but had not yet become aware of Trowell's 1960 book. I was at that time particularly, and perhaps even uniquely, in a position to test Cleave's hypothesis. I was receiving monthly reports

on patients seen suffering from certain forms of cancer from
over 150 rural hospitals, mostly in Africa, but some in other
parts of the Third World. In addition, a friend, Raymond
Knighton, headed a charitable organization in North
America, Medical Assistance Programs, which had become
one of the largest organizations supplying free drugs to
mission hospitals throughout the world. I realized that
questionnaires sent out together with generous gifts of
drugs and equipment were much more likely to be completed than requests from an unknown person, such as
Cleave, offering nothing in return. I also knew well from
my 20 years in Uganda, and my close association with
mission hospitals, that these doctors were both willing and
able to provide the information I required. Many missionary doctors work for 10, 20, or even 30 years in the same
location. Furthermore, my own sympathy with medical
mission work was widely known. In government-run district hospitals, on the other hand, medical staff are changed
at short intervals and, in consequence, establishing a close
and ongoing relationship is well-nigh impossible.

As a consequence of the questionnaires Knighton sent
round the world for me, I was able to analyze valuable
reports from nearly a thousand rural Third World hospitals. This confirmed, without any question, the rarity of
such diseases as appendicitis, gallstones, hemorrhoids,
varicose veins, coronary heart disease, colon cancer, and
others. Later the rarity of conditions like diverticular disease and hiatus hernia was verified by prospective radio-
logical studies. All these figures could have been criticized
as anecdotal, but when doctors in hundreds of hospitals,
some with over 20 years of experience, all tell the same
story the message is accepted and most critics are silenced.

On the basis of these studies I published, in collaboration with others, a series of papers between 1969 and 1973
(21–26). These papers developed Cleave's concept, both
with regard to the epidemiology of Western diseases as well

as the role of dietary fiber. It is no exaggeration to say that these papers had a tremendous impact and were instrumental in persuading the medical community to seriously consider the concepts of dietary fiber and Western disease.

The question that begs an answer is: Why did these papers have such an impact on the medical and scientific community, whereas Cleave's work did not? There are several reasons for this. A major one is that I was already well known for my work on Burkitt's lymphoma. Another important factor was that Cleave's major work appeared in books, whereas I and my colleagues published papers in medical journals. The apparent importance of these two factors suggests that, had they been in operation at the time Darwin published *The Origin of Species*, that book would have had little impact. Cleave's arguments were also hindered by his less than diplomatic way of dealing with fellow medical investigators: He viewed any criticism as a personal attack and often reacted fiercely. Moreover, he was very resistant to admitting errors. These were tactical mistakes I (and Trowell) avoided. Other factors that certainly helped our case were the new evidence I presented, as pointed out above, and the strong case made that colon cancer is also related to a fiber-depleted diet.

Not surprisingly, our arguments were fiercely resisted by the British milling and baking industry. On one occasion in 1972 my wife and I were invited by the organization representing British millers to a reception at the Mansion House in London. The chairman of a major milling company admitted to my wife that, "At home we follow your husband's advice but do not officially accept it yet in the industry."

Another serendipitous or providential meeting followed my eye-opening meeting with Cleave. In 1970 I arranged one of my fact-finding safaris in Africa to coincide with a medical conference in Kampala, Uganda, celebrating the centenary of the birth of East Africa's most

famous medical missionary, Sir Albert Cook. The Ugandan government provided a ticket for Hugh Trowell, who
had been Sir Albert's personal physician. Trowell had been
my friend and colleague in Uganda for many years. I spoke
on my recent epidemiological studies that were initiated
by my contact with Cleave. Trowell immediately recognized
that I was unknowingly building on the foundations laid
in his 1960 book. As a result of this meeting I became aware
of Hugh's earlier work and asked him, as an internist, to
cooperate with me, a surgeon, in coediting and writing what
was to become *Refined Carbohydrate Foods and Disease
(15)*. Trowell added as a subtitle: *Some Implications of
Dietary Fiber*. After these meetings, Trowell and I were
photographed on the very spot where 13 years previously
we had discussed a patient, which started the study that
led to the discovery of Burkitt's lymphoma.

Where Cleave had focused on sugar, we recognized
that refined sugar could only be produced by removing the
fiber, and turned our attention primarily to the depleted
fiber aspects of Western diets. My close association with
Hugh Trowell continued until his death in 1989 when, in
compliance with his wish, I was privileged to give the
address at his memorial service. We had written many articles together and in two further coedited books he did most
of the work and was senior editor. The most important of
these was *Western Diseases, Their Emergence and Prevention*, in 1981 *(16)*. This book, with over 30 contributors
from all five continents, documented the emergence of
Western diseases in different socioeconomic, racial, and
cultural groups worldwide. It was in fact selected by the
publishers to be enclosed with some others in a time cap-
sule not to be opened for 100 years, when it will be seen
what changes in medical practice have taken place in the
intervening time. The last book dealing with Western diseases and their relationship to dietary fiber was largely

attributable to Hugh Trowell, though Ken Heaton and I played a subsidiary role. Again, it was a multiauthor book. This was published in 1985 under the title: *Dietary Fibre, Fibre-Depleted Foods and Disease (27)*.

The concept of Western diseases, which must be largely preventable once causative factors in the Western lifestyle are identified and then reduced or eliminated, must be one of the most, if not the most, important medical observations made in the last 30 years. The credit must go predominantly to three men, all current or former friends of mine: Peter Cleave, Hugh Trowell, and Alexander Walker of South Africa. Walker has been studying disease patterns in different ethnic groups in South Africa since 1947 and relating these to contrasts in lifestyle and to diet in particular.

The broad river that now embraces the concept of Western diseases, with all the potential that the concept contains, includes all that has been discovered with regard to the role of dietary fiber in the maintenance of health. It started in many small tributaries, each independent of the others but eventually coming together to form a mighty river. The small tributaries represent the many isolated, lonely, ignored, and even ridiculed pioneers, each making his or her particular contribution to the formation of the river.

The importance of these studies has been underlined by Thomas McKeown, formerly Professor of Epidemiology and Vice-Chancellor of Birmingham (UK) University *(28)*, who in 1983 wrote,

> The most important medical advance in the nineteenth century was the discovery that infectious diseases were largely attributable to environmental conditions and could often be prevented by control of the influences that led to them. The most significant advance in the twentieth century is the recognition that the same is true of many noncommunicable diseases.

References

1. Trowell H. Definitions of fibre. *Lancet* 1974; **i**:503.
2. Englyst HN, Cummings JH. Resistant starch, a "new" food component: a classification of starch for nutritional purposes. In: Morton ID, ed. *Cereals in a European Context.* Chichester: Ellis Horwood, 1987; 221–233.
3. Allinson TR. *The Advantages of Wholemeal Bread.* London: Allinson, 1931.
4. Kellogg WH. *The New Dietetics: A Guide to the Scientific Feeding in Health and Disease.* Battle Creek: Modern Medicine, 1923.
5. Cowgill GR, Anderson WE. Laxative effects of wheat bran and "washed bran" in healthy men. *JAMA* 1932; **98**:1866–1875.
6. Olmsted WH, Curtis G, Timm OK. Cause of laxative effect of feeding bran, pentosan and cellulose to man. *Proc Soc Exp Biol Med* 1934; **32**:141,142.
7. Dimock EM. The prevention of constipation. *Br Med J* 1937; **1**:906–909.
8. Hipsley EH. Dietary "fibre" and pregnancy toxaemia. *Br Med J* 1953; **2**:420–422.
9. McCarrison R. *Nutrition and Health.* London: Faber and Faber, 1961.
10. Donnison CP. *Civilization and Disease.* London: Bailliere Tindall Cox, 1937.
11. Price WA. *Nutrition and Physical Degeneration: A Comparison of Primitive and Modern Diets and Their Effects.* New York: Hoeber, 1939.
12. Gelfand M. *The Sick African.* Cape Town: Stewart Printing, 1944.
13. Trowell HC. *Non-Infective Disease in Africa.* London: Edward Arnold, 1960.
14. Cleave TL, Campbell GD. *Diabetes, Coronary Thrombosis, and the Saccharine Disease.* Bristol: Wright, 1966.
15. Burkitt DP, Trowell HC. *Refined Carbohydrate Foods and Disease: Some Implications of Dietary Fibre.* London: Academic Press, 1975.
16. Trowell HC, Burkitt DP. *Western Diseases: Their Emergence and Prevention.* London: Edward Arnold, 1981.
17. Burkitt DP. Acute abdomens. British and Bagunda compared. *E Afr Med J* 1952; **29**:189–194.
18. Walker ARP. Diet and atherosclerosis. *Lancet* 1955; **i**:565,566.
19. Cleave TL. *The Saccharine Disease.* Bristol: Wright, 1974.

20. Cleave TL. Natural bran in the treatment of constipation. *Br Med J* 1941; **1:**461.

21. Burkitt DP. Related disease—related cause? *Lancet* 1969; **ii:**1229–1231.

22. Burkitt DP. Relationship as a clue to causation. *Lancet* 1970; **ii:**1237–1240.

23. Burkitt DP. Epidemiology of cancer of the colon and rectum. *Cancer* 1971; **28:**3–13.

24. Burkitt DP. Varicose veins, deep vein thrombosis, and haemorrhoids: epidemiology and suggested aetiology. *Br Med J* 1972; **2:**556–561.

25. Burkitt DP, Walker ARP, Painter NS. Effect of dietary fibre on stools and transit-times, and its role in the causation of disease. *Lancet* 1972; **ii:**1408–1412.

26. Burkitt DP. Some diseases characteristic of modern Western civilization. *Br Med J* 1973; **1:**274–278.

27. Trowell H, Burkitt D, Heaton K. *Dietary Fibre, Fibre-Depleted Foods and Disease*. London: Academic Press, 1985.

28. McKeown T. A basis for health strategies. A classification of disease. *Br Med J* 1983; **287:**594–596.

Western Diseases and What They Encompass

Denis P. Burkitt

1. The Nature of the Western Disease

As stressed in the previous chapter, the concept of diseases characteristic of more economically developed and affluent countries is the greatest advance toward the prevention of diseases to have emerged in the last several decades. The designation *Western* is not entirely satisfactory, since these disorders are becoming more common in more affluent societies in the east and Middle East. Nor is *affluence* a satisfactory word, since these diseases are often more prevalent in the poorer sections of Western populations. *Industrialized* is also inappropriate, since, for example, they are much more prevalent in largely rural New Zealand than in highly industrialized Czechoslovakia. *Related to Western lifestyle* might be the most appropriate term, but it is too cumbersome, and so *Western* is the term now generally used.

The term *Western* encompasses that large group of disorders that currently have their highest incidence rates in more affluent Western countries yet are still rare, or even unknown, in rural communities throughout the Third World.

The emergence of these diseases in different cultural, socioeconomic, and racial groups throughout the world has

From: *Western Diseases: Their Dietary Prevention and Reversibility*
Edited by: N. J. Temple and D. P. Burkitt Copyright ©1994 Humana Press, Totowa, NJ

been documented by those who have worked among these communities for long periods of time (1).

The diseases were rare or even unknown in Western populations before World War I. They have been shown in numerous situations to increase in prevalence in populations that emigrate from countries with low incidence rates to those with high rates. Examples of this include the descendants of African slaves who now constitute the black population of North America, and the descendants in Hawaii and California of the Japanese immigrants who arrived in the early part of this century.

Of particular importance with regard to efforts to identify causes and in promoting prevention is the observation that these diseases all share the same, or closely similar, epidemiological features. In situations in which one of the diseases is common, the others are also, generally speaking, prevalent. For instance, particular malignant diseases, such as breast and colon cancer, are often both high in the same population, as are various other conditions, such as coronary heart disease (CHD) and diabetes. Some of these diseases appear invariably to emerge long before others, suggesting that the latter are dependent on a longer exposure to the causative factors or require a much greater intensity of these factors (2). For example, diabetes (type II) always seems to appear many years before CHD. Similarly, among gastrointestinal diseases, appendicitis long precedes a rise in incidence of colorectal cancer or the emergence of diverticular disease. Some Western diseases are very late in emerging, sometimes several generations after the others. These include multiple sclerosis, Crohn's disease, and ulcerative colitis. However, the precise etiology of these diseases is uncertain.

As well as sharing epidemiological features these diseases often tend to occur together in the same individuals. Diabetes and CHD are associated, as are hiatus hernia and gallstones. This again underlines the conclusion that from a causative and preventive viewpoint the Western diseases must be considered members of the same family.

Thomas McKeown *(3)* suggested that diseases could profitably be classified according to their causation rather than anatomically according to the structures or the organs involved. He recommended that all diseases could be divided first into two categories: those owing to prenatal causes and those owing to postnatal causes (Fig. 1). The latter, the vast majority, could then be subdivided into Diseases of Poverty, which would include those owing to hazards and deficiencies, and Diseases of Affluence, including those owing to hazards and maladaptation. Hazards include all environmental factors, whether microorganisms, trauma, radiation, or extremes of temperature.

Maladaptation includes all Western diseases, for these, as so ably portrayed by Eaton et al. *(4)*, are now viewed as manifestations of maladaptation to a new environment into which Western humans have plunged precipitously and are not genetically adapted. Modern humans are believed to be adapted to the hunter-gatherer environment in which our ancestors lived for tens of thousands of years in the distant past. Western humans have made more changes to our lifestyle, particularly to our diet, in the last 200 years or eight generations since the Industrial Revolution than our ancestors made in the previous 20,000 years. This concept will be further developed in the chapters dealing with dietary contrasts between populations at high and low risk of developing these diseases (*see* chapters by Walker et al. and Campbell and Chen). An argument in favor of this maladaptation hypothesis is the rarity of all these diseases in wild animals *(5)*.

Although some may reject the argument that these diseases must have a basic shared cause in Western lifestyle, no alternative hypothesis consistent with the epidemiological evidence has been put forward during the two decades or more that this hypothesis has been discussed.

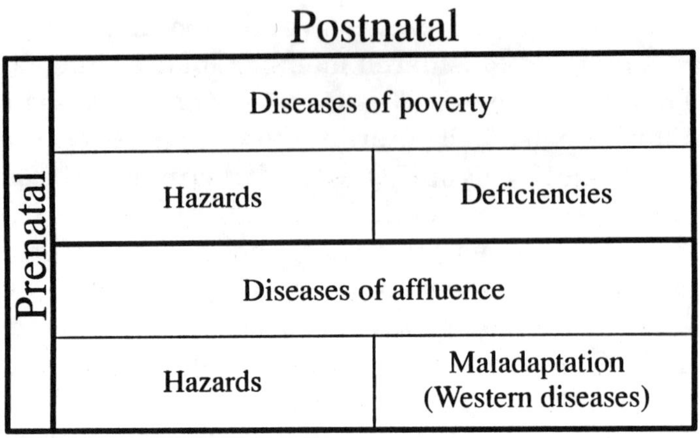

Postnatal

Prenatal	Diseases of poverty	
	Hazards	Deficiencies
	Diseases of affluence	
	Hazards	Maladaptation (Western diseases)

Fig. 1. Classification of diseases according to causation.

2. The Possibility of Prevention

It is profitable at this juncture to remind ourselves why Western diseases must be considered preventable. The epidemiological features of these diseases, which are basically all similar, can only be explained by the assumption that they are all caused by similar or related environmental factors.

2.1. The Contrasts in Prevalence of Western Diseases Between More and Less Economically Developed Countries

These striking contrasts have been observed throughout the world and can no longer be a matter of dispute. The argument that they might be common but remain undetected in less developed populations is no longer tenable. Despite the profuse literature on the subject that has appeared over the last two decades, these diseases have rarely been reported as having a high prevalence in any community that has not radically changed its lifestyle during the past few generations. One of the few such examples is the traditionally high mortality rate in Japan for cerebrovascular disease. This is undoubtedly caused by the high salt intake in that country.

The prevalences of the different Western diseases in varying circumstances in different stages of acculturation in all five continents has been well documented in Burkitt and Trowell's *Refined Carbohydrate Foods and Disease (6)* and in Trowell and Burkitt's *Western Diseases, Their Emergence and Prevention (1).*

An excellent example is afforded by South Africa. The different prevalences of Western diseases in the different ethnic groups in South Africa—the whites, urban and rural Africans, Indians, and colored—has been demonstrated by Walker (7; *see* next chapter).

2.2. Changes in Migrant Populations

If a disease changes in prevalence over a relatively short period of time in evolutionary terms, then the causative factors must be sought in environmental changes since our genes do not change in short time spans.

2.2.1. Black Americans

The most striking example of a population migrating from an environment in which Western diseases are rare to one in which they are common (or, to be more precise, were later to become common) and adopting the new disease pattern is to be seen in black Americans today. Long after they were forcibly removed to the United States they remained in a lifestyle, and a dietary pattern in particular, very different to that of the more affluent white population. Consequently, they became afflicted with Western diseases much later than did the white population. This continued until the second quarter of this century. Now, with basic similarity between the diets of the two groups, disease patterns are also similar. Even with diseases that are the last to appear, such as diverticular disease and hiatus hernia, the prevalences in the black and white populations are the same *(6)*.

2.2.2. Polynesian Islanders Migrating to New Zealand

Many Western diseases rare in the Pacific Islands have increased in emigrants to New Zealand to the extent that within two generations they can even be commoner in their descendants than in the white population *(8)*.

2.2.3. Japanese Emigrants to the United States

The same changes have been observed in the descendants of Japanese emigrants to Hawaii and California. Subsequent generations have disease patterns much closer to those of the host country than to those of Japan *(9,10)*.

2.2.4. Jews Who Emigrated from European, African, or Asian Countries to Israel

The Jews coming from these countries have very different prevalence rates of the Western diseases. After emigration the rates change to reflect the new society. This is particularly the case for CHD, where ethnic groups with very low rates, such as Yemenis, have experienced a massive rise *(11)*.

Other similar examples, in addition to the above four, could be cited. The conclusion, however, seems clear: These diseases are indeed caused by environmental changes, particularly lifestyle, and consequently must be potentially preventable.

2.3. Changes Occurring Within Populations Associated with Economic Development

2.3.1. Western Populations

It is often forgotten that the dramatic rise in Western diseases has only occurred in the last 80 or 100 years. Alas, however, there is a shortage of reliable data. Statistics were not collected for many diseases, particularly nonfatal ones, whereas for others there may have been widespread misclassification. Nevertheless, we can state with confidence that there has been an explosive rise in the incidence of numerous Western diseases over the last century. This is documented in Table 1.

2.3.2. Japan

Since World War II there has been a dramatic increase in the prevalence of many Western diseases in Japan *(23)*. This has been associated with a great increase in consumption of meat, milk, and other dairy products at the expense of rice *(23)*.

2.3.3. Eskimos and Pastoral Tribes in Africa

The Eskimos in their traditional culture have a low prevalence of Western diseases, but they quickly develop when they adopt a Western culture *(24)*. The same applies to certain pastoral tribes in East Africa, such as the Masai, who have a diet rich in animal fat and protein. Both these groups are apparently adapted to an environment into which they have existed for long periods of time.

2.3.4. Other Developing Countries

In various developing countries, India and parts of Africa, for instance, Western diseases such as CHD, diabetes, and appendicitis are becoming much more prevalent, particularly in the upper socioeconomic groups. Similar examples could be cited from numerous other developing countries. For instance, in the eastern province of Saudi Arabia cholecystectomy rates among Saudis increased 13-fold between 1977 and 1986 *(25)*. This indicates a massive increase in the incidence of gallstones and parallels a widespread shift toward a Western diet.

2.3.5. Sudden Changes in Lifestyle

It appears that when a community makes a particularly sudden change in lifestyle toward a Western culture, it becomes even more prone to the development of Western diseases than are communities that have made the

Table 1
Diseases That Have Risen Greatly in Incidence
Since the Last Century[a]

Disease	Place	Period beginning	Reference
Dental caries and periodontal disease	Western countries[b]	Previous centuries	12,13
Lung cancer	United States	1919	14
Coronary heart disease	Britain	Early 20th century	15–17
Diabetes	Western countries	Previous centuries	18
Hiatus hernia	Western countries	Circa 1925	—
Duodenal ulcer	Western countries	Circa 1900	19
Appendicitis	Britain and United States	1890–1900	13,20,21
Diverticular disease	Western countries[b]	Early 20th century	13,22
Renal stones	Europe	1900	13

[a]In each case there has been a large or, more usually, a massive rise in the incidence of the disease during the period beginning as indicated up to the present.
[b]Data refer mainly to Britain and the United States.

change over several generations. The most obvious example is the population of the little Pacific island of Nauru, which now experiences the highest rate of diabetes in the world: About half of those aged over 30 have the condition (26). A rate almost as high is found among the Pima Indians of Arizona (27).

2.4. Lifestyle Changes
Causing a Fall in Western Diseases

It is difficult to identify situations in which populations have voluntarily altered their lifestyles toward that associated with lower prevalences of Western diseases. It was, however, imposed on Britain and other European countries during both World Wars. Whereas a fall in the prevalence of appendicitis was documented in both wars, lower rates of diverticular disease complications, diabetes, CHD, and dental caries were recognized only in World War II *(21,28–32)*. In some countries, particularly the United States and Australia, there has also been a major drop in the incidence of CHD since about 1970. Indisputably, this has been largely because of lifestyle changes.

The best available example of significantly lower rates in the case of many, if not most, Western diseases are the Seventh-Day Adventists, who, besides being largely vegetarian, are also nonsmokers with a low consumption of alcohol. They have been shown to have much lower rates of CHD, diabetes, cerebrovascular disease, and renal stones. Moreover, they suffer less from several cancers common in Western countries (lung and colon in men; lung, breast, and colon in women) *(33,34)*.

In Australia O'Dea has demonstrated that when Aborigines return to a traditional lifestyle, there is clear evidence of a reversal of some Western diseases (*see* chapter by O'Dea).

2.5. Comment

The common factor in all these situations is that populations living in an environment little removed from that in which they have evolved have low prevalences of Western diseases, whereas those who alter their culture and diet become prone to all the diseases of Western culture. Conversely, when Western populations, as a result of such factors as war or religion, revert to a lifestyle nearer to that traditionally followed by humans, then Western diseases become considerably less common.

The above represents a summary of the evidence. It compels the conclusion that the causative factors must be sought in lifestyle rather than in genetic influences. Of course, genetics can play a role in determining which individuals are at high or low susceptibility to the causative environment:

It is this genetic predisposition that helps to explain why numerous diseases run in families. The inevitable conclusion is that a large degree of prevention must be possible if causative environmental factors can be identified and adequate action taken to reduce or eradicate them. It is this practical action, rather than the identification of causes, that is likely to be the most difficult part of the exercise. Nevertheless, it is prevention that must be our primary goal.

3. Diseases Currently Recognized as Western

The diseases listed below are those for which there is strong evidence that they should be considered as Western diseases.

1. Those for which specific causes, mainly dietary, have been postulated, and in many cases demonstrated.
    a. Gastrointestinal
        Constipation
        Hiatus hernia
        Appendicitis
        Diverticular disease
        Colorectal polyps
        Colorectal cancer
        Hemorrhoids
    b. Cardiovascular
        CHD (coronary heart disease)
        Cerebrovascular disease (stroke)
        Essential hypertension
        Deep vein thrombosis
        Pulmonary embolism
        Pelvic phleboliths
        Varicose veins
    c. Metabolic
        Obesity
        Diabetes (type II or noninsulin dependent)
        Cholesterol gallstones

        Renal stones
        Osteoporosis
        Gout
   d. Cancer
        Colorectal (also listed above)
        Breast
        Prostate
        Lung
        Endometrium
        Ovarian
   e. Other
        Dental caries
2. Diseases of obscure etiology
   a. Autoimmune diseases
        Diabetes (type I or insulin dependent)
        Autoimmune thyroiditis
   b. Other
        Crohn's disease
        Celiac disease
        Thyrotoxicosis
        Pernicious anemia
        Ulcerative colitis
        Multiple sclerosis
        Rheumatoid arthritis

The Western type peptic ulcer, with a tendency to bleed or perforate, is rare in Third World communities, but a chronic stenosing ulcer involving the pylorus or first stage of the duodenum is very common in rice eating communities in India, in the Nile-Congo watershed, and in the rain forests of West Africa.

References

1. Trowell HC, Burkitt DP. *Western Diseases: Their Emergence and Prevention*. London: Edward Arnold, 1981.
2. Burkitt DP. Relationships between diseases and their etiological significance. *Am J Clin Nutr* 1977; **30**:262–267.

3. McKeown T. *The Origins of Human Disease.* Abingdon, England: Blackwell, 1988.
4. Eaton BS, Shostak M, Konner M. *The Paleolithic Prescription.* New York: Harper and Row, 1988; 38–68.
5. Leader RW, Hayden DW. *Some Diseases Characteristic of Western Civilization Prevalent in Wild and Domestic Animals.* (In ref. *6*), 311–317.
6. Burkitt DP, Trowell HC. *Refined Carbohydrate Foods and Disease: Some Implications of Dietary Fibre.* London: Academic Press, 1975; 251–277.
7. Walker A. *South African Black, Indian and Coloured Populations.* (In ref. *1*), 285–318.
8. Prior I, Tasman-Jones C. *New Zealand Maori and Pacific Polynesians.* (In ref. *1*), 227–267.
9. Glober G, Stemmermann G. *Hawaii Ethnic Groups.* (In ref. *1*), 319–333.
10. Curb JD, Marcus EB. Body fat and obesity in Japanese Americans. *Am J Clin Nutr* 1991; **53**:1552S–1555S.
11. Modan B. *Israeli Migrants.* (In ref. *1*), 268–284.
12. Adatia A. *Dental Caries and Periodontal Disease.* (In ref. *6*), 251–277.
13. Cleave TL. *The Saccharine Disease.* Bristol: Wright, 1966; 66–69.
14. Ochsner A. My first recognition of the relationship of smoking and lung cancer. *Prev Med* 1973; **2**:611–614.
15. Michaels L. Aetiology of coronary artery disease: an historical approach. *Br Heart J* 1966; **28**:258–264.
16. Yellowlees WW. Sir James Mackenzie and the history of myocardial infarction. *J R Coll Gen Pract* 1982; **32**:109–112.
17. Morris JN. Recent history of coronary disease. *Lancet* 1951; i:1–7 and 69–73.
18. Trowell HC. *Diabetes Mellitus and Obesity.* (In ref. *6*), 227–249.
19. Tovey F. *Duodenal Ulcer and Diet.* (In ref. *6*), 279–309.
20. Yellowlees WW. Modern diseases, seen from a Highland practice. An ecological approach. *Ecol Dis* 1983; **2**:81–91.
21. Burkitt DP. *Appendicitis.* (In ref. *6*), 87–97.
22. Painter NS, Burkitt DP. *Diverticular Disease of the Colon.* (In ref. *6*), 99–116.
23. Oku T. The epidemiological significance of dietary changes in Japan. In: Chen SC, ed. *Proceedings of Kellogg's International Symposium on Dietary Fiber.* Tokyo: Academic Publishers, 1990; 120–135.
24. Schaefer O. *Eskimos (Inuit).* (In ref. *1*), 113–128.

25. Tamimi TM, Wosornu L, Al-Khozaim A, Abdul-Ghani A. Increased cholecystectomy rates in Saudi Arabia. *Lancet* 1990; **336:**1235–1237.
26. Zimmet P, Taft P, Guinea A, Guthrie W, Thoma K. The high prevalence of diabetes mellitus on a Central Pacific Island. *Diabetologia* 1977; **13:** 111–115.
27. Bennett PH, Rushforth NB, Miller M, Le Compte PM. Epidemiologic studies of diabetes in the Pima Indians. *Recent Prog Horm Res* 1976; **32:** 333–376.
28. Marthaler TM. Epidemiological and clinical dental findings in relation to intake of carbohydrates. *Caries Res* 1967; **1:** 222–238.
29. Malmros H. The relation of nutrition to health: a statistical study of the effect of the war-time on arteriosclerosis, cardiosclerosis, tuberculosis and diabetes. *Acta Med Scand Suppl* 1950; **246:**137–150.
30. Strom A, Jensen RA. Mortality from circulatory diseases in Norway 1940–1945. *Lancet* 1951; **i:**126–129.
31. Bransby ER, Knowles EM. A comparison of the effects of enemy occupation and post-war conditions on the incidence of dental caries in children in the Channel Islands in relation to diet and food supplies. *Br Dent J* 1949; **87:**236–243.
32. Trowell H. Diabetes mellitus death-rates in England and Wales 1920-70 and food supplies. *Lancet* 1974; **ii:**998–1002.
33. Phillips RL, Kuzma JW, Beeson WL, Lotz T. Influence of selection versus lifestyle on risk of fatal cancer and cardiovascular disease among Seventh-Day Adventist. *Am J Epidemiol* 1980; **112:**296–314.
34. Berkel J, de Waard F. Mortality pattern and life expectancy of Seventh-Day Adventists in the Netherlands. *Int J Epidemiol* 1983; **12:** 455–459.

PART II

THE CAUSES OF
WESTERN DISEASE

Diet-Related Disease Patterns in South African Interethnic Populations

Epidemiological Perplexities and Future Prospects

Alexander R. P. Walker, Demetre Labadarios, and Ingrid I. Glatthaar

1. Introduction

In South Africa there are four ethnic populations:

1. Blacks (30 million);
2. Coloreds (Euro-African-Malay) (3 million);
3. Indians (1 million); and
4. Whites (5 million).

These populations exhibit considerable differences in patterns of diseases. Indeed, between them they afford probably greater contrasts, as juxtaposed populations, than are encountered elsewhere in the world. Currently, the Indian population exhibits high frequencies of dental caries, obesity, hypertension, and particularly diabetes and coronary heart disease (CHD). In contrast, blacks in *rural* areas have

From: *Western Diseases: Their Dietary Prevention and Reversibility*
Edited by: N. J. Temple and D. P. Burkitt Copyright ©1994 Humana Press, Totowa, NJ

very low frequencies of dental caries, CHD, noninfective bowel diseases (e.g., appendicitis and diverticular diseases), and certain cancers (e.g., colon, breast). As to changes— among *urban* blacks, Indians, and coloreds, rises are occurring, although variably, in all of these disorders and diseases. To exemplify, the teeth of urban black children, previously far superior, are now *inferior* to those of white children. Among urban blacks, frequency of hypertension (WHO criteria) now exceeds that in the white population; this contrasts with the situation in rural blacks among whom, in generations past, blood pressure scarcely rose with age.

Most of the differences depicted, although not all, are related to differences in environmental factors, especially diet. There are great contrasts in the dietary patterns of the different populations and their subpopulations. At the one extreme, the diet of rural blacks (which in pattern resembles that of ancestors of Western populations) is characterized by a relatively low intake of energy, of total protein, especially of animal protein, of total fat, especially of animal fat; their diet until recently was high in fiber-containing foods. In contrast, the diet of whites, and that of the more socioeconomically favored segments of the other populations, is high in energy, total protein, and fat, especially the respective moieties of animal origin; however, dietary fiber intake is low. The first pattern of diet, in contexts of indigence, is well-nigh invariably associated with very low or low frequencies of degenerative disorders and diseases mentioned (although with high frequencies of infections); conversely, the second is associated with very high or high frequencies of the diseases of prosperity.

As to the precise role of diet (whether in promotion of good health or in aggravation of ill-health), it must be kept in mind that patterns of health and disease are determined not only by diet, but by genetic, and by nondietary factors. The latter include degree of physical activity, extent of smoking practice, and of alcohol consumption, stresses (particularly those linked with urbanization and rise in income), and the availability and utilization of preventive medical services. Each of these components has a variable although often powerful influence in the regulation of the frequencies of the diseases under discussion.

2. Disease Patterns in South African Populations

General information on the relative frequencies of diseases in the interethnic populations is given in Tables 1–3. Table 1 concerns diseases of prosperity; Table 2 cancer patterns, and Table 3 noninfective bowel diseases. Patterns of dietary intake are given in Table 4. The low and high frequencies, respectively, of diet related diseases of prosperity in the traditionally living, compared with such in the developed populations, will be readily apparent. Understandably, limitations of space preclude dealing with changes occurring in each disease. Hence, it has been decided to select and to characterize four diseases in detail.

1. *Coronary heart disease*—absent in rural blacks, very uncommon in urban blacks, but particularly common in whites, and still more so in Indians.
2. *Appendicitis*—rare in rural blacks, still uncommon in urban blacks, also in coloreds and Indians, but very common in whites.
3. *Colon cancer*—near absent in rural blacks, still very uncommon in urban blacks, but common in whites; coloreds and Indians have intermediate occurrence.
4. *Dental caries*—rare in rural black children, but common in urban black, colored, and Indian children.

Each disease considered will be discussed in relation to its epidemiology, past and present, causative factors later, perplexing epidemiological situations, and outlook for the future. Finally, there will be a brief general discussion.

Table 1
Frequencies of Some Diseases of Prosperity
in South African Populations

	Rural blacks	Urban blacks	Coloreds	Indians	Whites
Dental caries	+	+++	+++	++++	+++
Femoral fractures	+	+	++	++	+++++
Obesity	+	+++	+++	+++	+++
Hypertension	+	++++	++++	+++	+++
Diabetes	+	+++	++++	+++++	+++
CHD	$-^a$	+	++++	+++++	+++++
Stroke	+	++	+++	+++	++

[a] Implies that occurrence is rare.

Table 2
Cancer Patterns in South African Populations

	Rural blacks	Urban blacks	Coloreds	Indians	Whites
Lung	$-^a$	++	+++	++	++++
Breast	$-^a$	++	+++	+++	+++++
Colon	−	+	++	++	+++++
Stomach	−	+	+++	++	++
Pancreas	−	+	++	+	+++
Liver	$++^b$	++	++	+	+
Esophagus	$++++^b$	+++	++	+	+
Cervix	++	++++	+++	+	+
Prostate	−	+	++	++	++++

[a] Implies that occurrence is rare.
[b] Frequency of occurrence related to some regional but not nationwide populations.

3. Coronary Heart Disease

3.1. Epidemiology

3.1.1. Early and Later History in Western Populations

Centuries ago it was appreciated that coronary atherosclerosis is a disease of antiquity, and was shown by

Table 3
Frequencies of Noninfective Bowel Disease
in South African Populations

	Rural blacks	Urban blacks	Coloreds	Indians	Whites
Hemorrhoids	–[a]	++	++	++	+++
Appendicitis	+	++	++	++	+++++
Ulcerative colitis	–	+	+	+	+++++
Irritable bowel syndrome	–	++	++	+	++++
Diverticular disease	–	+	++	++	+++++
Colon cancer	–	+	++	++	+++++

[a] Implies that occurrence is rare.

Table 4
Dietary Patterns
Respecting Fat, Total Carbohydrate, and Fiber Intakes

	Rural blacks	Urban blacks	Coloreds	Indians	Whites
Energy from fat, %	10–15	20–30	30–35	30–40	35–45
Energy from carbohydrate, %	70–75	65–75	60	60	55
Dietary fiber, g	20–25[a]	10–20	15–20	15–20	15–20

[a] Intake depends on the season; seasonal fruits and spinaches are high in dietary fiber.

thickening and calcification of the coronary arteries in a 50-yr-old person mummified from the XXI Egyptian Dynasty (circa 1000 BC). In India, the physicians Chareka and Sushruta (circa 600–800 BC) knew of the circulation of the blood; furthermore, they were the first to describe anginal pain: "a sense of constriction in the precordium, chest pain and sensations of churning, or bursting or rubbing." The Hippocratic School in Greece probably recognized angina pectoris and apoplexy (they attributed symptoms to advancing age), and possibly cerebral arteriosclerosis.

Historically, up to the turn of the present century, clinicians and pathologists alike were seemingly totally ignorant of myocardial necrosis secondary to obstructive coronary artery disease. It was the classical paper of Herrick (1), published in 1912, that attracted widespread interest, albeit slowly, in part because of the relative rarity of the disease. Ultimately, however, the publication exerted a greater influence in bringing coronary artery obstruction and myocardial infarction to the attention of the medical profession than any previous report. It is insufficiently appreciated that within the lifetime of some living today, indeed, as late as in the 1912 edition of Osler's *Principles and Practice of Medicine* in the chapter on angina pectoris it is stated, "It is a rare disease in hospitals; a case a month is about the average, even in the large metropolitan hospitals" (2). This rarity was also noted in the early experience of Paul Dudley White, President Eisenhower's physician, when working in the hospital wards at the Massachusetts General Hospital, for he wrote of CHD then as being "rare."

From the 1920s onward, there were enormous, indeed, catastrophic rises in the occurrence of CHD. Whereas previously it was responsible for less than 1% of all deaths, in many Western populations its incidence rose to such an extent that until recently it caused up to one-third of all deaths. In 1975 Morris wrote, "An average of one man in five in Great Britain—in the United States probably one man in four—may be expected to develop clinical CHD during middle age" (3).

Within the past two or three decades, however, in some populations there have been major decreases in mortality rates from the disease. The falls appear to have been maximal in the United States and in Australia. In the United States, age-adjusted mortality from CHD has fallen by 40% (4). In Australia it has declined by approx 50%. Falls have been smaller in most other countries; but in some, mainly Eastern European countries, there have been rises.

3.1.2. South African Populations

CHD is absent in the rural black population. Even in black urban dwellers it is relatively rare (5). Mortality in the colored population is relatively high (6), as with the white population (7). That in the Indian population is higher still (8)—an invariable response noted when Indians migrate to more prosperous countries. Mortality rates in whites, coloreds, and Indians have started to fall, even though significant avoiding measures have been lacking (5).

3.2. Causative Factors

Research has revealed hypercholesterolemia, cigaret smoking, hypertension, and physical inactivity to be salient risk factors (9). Regarding diet, in South Africa, as early as 1955, Walker suggested that a dietary pattern low in fat but high in crude fiber was associated with a low occurrence not only of atherosclerosis, but of appendicitis, gallstones, diabetes, and certain forms of cancer (10). This is the first specific mention of a defined pattern of diet being associated with a specific list of diseases.

For the prevention, primary and secondary, of the disease, cardiac societies and heart associations in Western populations have advanced the following recommendations:

1. Reduce weight if excessive to "ideal" weight;
2. Reduce the energy supplied by fat from 40 to 30%; reduce the intake of saturated fats, replacing them in part by polyunsaturated fats; and
3. Double the intake of fiber from fiber-containing foods (coarse cereals, legumes, vegetables, fruit).

3.3. Perplexing Epidemiological Situations

There are numerous perplexing features over CHD's past and present occurrence, over the variable degree of noxiousness of risk factors, and even over the worth of benefits derived from combating the disease.

Salient questions are as follows. In the past, why was CHD so rare in the small segment of the populations who were prosperous? Why did the frequency of the disease rise, and then fall, so steeply? Why do incidence and mortality rates vary so much from population to population, indeed, within particular populations?

As to why the disease rose steeply, certainly some risk factors increased markedly, but they do not explain the magnitude of the rises. Equally, although some risk factors have recently decreased, again such do not wholly account for the extent of the falls. Especially puzzling are the considerable variations in CHD experience in European populations *(11)*. Thus, the mortality rate in Scotland is almost three times higher than that in France. The rate in the United Kingdom is double that in Switzerland. In Greece, as well as in Japan, the rate is remarkably low. As to regional differences, in Australia, current mortality rates from CHD in males and females in Newcastle are 45 and 95% higher, respectively, than those in Perth; it was stated that further explanations should be sought for the large difference noted among the cities.

There are also perplexing situations in the significance of serum cholesterol level. In Scotland and England, adults have the same mean serum cholesterol level, but the two populations have quite different mortality rates. In Scotland, regionally, there are large differences in CHD experience in segments who have the same mean cholesterol level. In the United Kingdom, although mean cholesterol level has not fallen, CHD has decreased by 20%. Not least, a given serum cholesterol level may connote a different risk to different populations. This is well exemplified in India. In one series of CHD patients studied, half of them had cholesterol levels of 4.5 mmol/L or lower *(12)*. Such a level in Western populations would signify individuals or groups to be almost totally free from the disease. For a given level of serum cholesterol, risk of CHD can be elevated threefold by heavy smoking, and raised markedly by hypertension (9). But even taking all these factors into account, there are still numerous puzzles in the frequency of occurrence of the disease in communities and individuals, and in the significance of particular risk factors.

There are, as already indicated, perplexing situations in South African populations. The absence of CHD in black rural dwellers, because of the low prevalence of known risk factors, is understandable. Yet the relative rarity of the disease in black urban dwellers is puzzling, since a proportion of such, at least 10%, have most of the orthodox risk factors, including a high prevalence of diabetes. Also puzzling is the very high mortality rate from CHD in the Indian population, an excess mortality not wholly explicable on the basis of known risk factors.

3.4. Can Coronary Heart Disease Be Avoided, or Its Rise Restrained?

3.4.1. Western Populations

Animal experimental and epidemiological evidence, as well as the results of short-term studies on humans, indicate that the lowering of risk factors is likely to be associated with reduced incidence and mortality rates. It must be noted, however, that the recent major falls in CHD have taken place in populations both in those who have changed little, and in those who have changed much with respect to risk factors. This suggests the operation of additional protective factors whose identities are not known. Further falls in CHD could well take place.

3.4.2. South African Populations

The frequency of the disease among the black population will certainly increase. Rates in the white, colored, and Indian populations are already falling (5).

3.5. Comment

The citing of the largely unexplained variations in the occurrence of CHD in different populations must not be construed as belittling the extent of our knowledge of the disease's development and stigmata, nor of means of combating them. Thus, as to the endeavors of individuals, it has been shown that a reduction in cholesterol level in the hypercholesterolemic, lessens CHD episodes and mortality rate from the disease. Additionally, it has been shown that recommended changes in diet and in other risk factors can cause lesions in the coronary vessels to regress *(13)*. However, the fall in mortality rate from CHD does not betoken a rise in life expectancy, since it has been shown that rises occur in mortality rates from cancer, trauma, and other causes *(14,15)*.

In view of the huge variabilities in occurrence of CHD, two points arise. First, under ideal natural conditions, what freedom from meaningful lesions in, say, the arteries, can persist into middle and old age? Optimum conditions still obtain in developing populations. Thus in 1958 a study on the composition of aortas of young and old urban blacks and whites was carried out in Johannesburg *(16)*. It was noted that one-third of elderly blacks had very few scattered lesions, consistent with their very low occurrence of myocardial infarction. There was scarcely any change in appearance with age; indeed, the aortas of old blacks resembled those of young whites. Doubtless this relative freedom from arterial lesions still holds true for most blacks in rural areas. Although the African experience portrays the extreme, it certainly affirms that degenerative changes are not the inevitable sequelae of aging. The results of a number of interventions are in harmony with the feasibility of a restraint, an arrest, or even a regression of lesions. Thus, the severe privations experienced in postwar Germany (energy from fat was 10%) was associated with a major reduction, perhaps a regression, in severe atherosclerotic lesions and a fall in occurrence of CHD. The second point to be emphasized is the large measure of individual susceptibility to CHD occurrence, which appears little related to the extent of risk factors prevailing. At present only about half of the variation in CHD is explicable from known risk factors.

4. Appendicitis

4.1. Epidemiology

4.1.1. History

There appears to be no record of early physicians—Hippocrates, Galen, down to Moses Maimonides—recognizing appendicitis. Actually, in the fourteenth century the organ was depicted by Leonardo da Vinci in his drawings. Hence, it would seem that early reports toward the end of the nineteenth century regarding an increase in perityphlitis, later called appendicitis, concerned a new phenomenon. The general public was made aware of the disease in 1901—when Sir Frederick Treves operated on King Edward VII for appendicitis. The King's coronation had to be postponed. Remarkably, this great surgeon retired from clinical practice at 50 years of age. Ironically, his daughter died from a failed appendix operation performed by her father. He himself died from peritonitis.

4.1.2. Western Populations

Frequency of the disease rose rapidly from the turn of the century until about 1940–1950 (17). Since then it has fallen to about half of its peak level. At first the disease caused high mortality, up to 22% for acute appendicitis. Subsequently, this declined considerably, to about 0.1% for uncomplicated appendicitis, 0.6% for gangrenous, and 5% for perforated appendicitis. Notwithstanding its decrease, the disease remains the commonest abdominal emergency. Annual incidence in children of 10–19 yr in the United States is 23/10,000; i.e., in this age group roughly 3% of children are likely to be affected *(18)*. For the whole population, approx 6% are likely to suffer from acute appendicitis during their lifetime. Peak incidence is between 15–25 yr, with females slightly predominating.

4.1.3. South African Populations

In Third World populations appendicitis is very uncommon, indeed, in rural populations who pursue their traditional manner of life, the disease is all but unknown. However, among urban and periurban dwellers in Africa, with continuing westernization of diet and other aspects of lifestyle, the disease increases, albeit only slightly. Currently, the incidence rate in black children of 10–19 yr in Soweto, Johannesburg, is 0.6/10,000, i.e., less than 1/20 of the incidence in white children, namely, 9.6/10,000 *(19)*. Those in the colored and Indian children are 1.7 and 1.9/10,000. As to pattern of pathology, in an urban series of 275 black patients, the appendix was normal in 14%, acute in 46%, perforated in 18%, abscessed in 16%, and gangrenous in 4%; mortality was 1.1%. In Nigeria, in 700 cases of acute appendicitis treated at the University of Benin Teaching Hospital, only two deaths were recorded. Mortality, however, is much higher in many regions where there are inadequate medical services.

4.2. Causative Factors

4.2.1. Diet

Before the turn of the century, Graham in the United States and Allinson in the United Kingdom thought that the decreased consumption of bread, and widespread replacement of brown by white bread, tended to increase the occurrence of diseases of the digestive tract. In 1910, Williams suggested that an excessive intake of sugar and meat, i.e., low fiber containing foods, favored the disease. McCarrison, in the 1920s, from his observations made in India, had similar thoughts on the adverse effects stemming from food refinement. In 1920, Rendle Short of Bristol advanced that the increased commonness of appendicitis was caused by the reduced consumption of *cellulose*, which commenced in 1880–1890, particularly that contained in cereals *(20)*. Later, in the 1950s, Walker *(10)* in South Africa, and subsequently Trowell *(21)* and Burkitt *(22)*, from observations made in East Africa and elsewhere, maintained that appendicitis is almost unknown among traditionally living Third World populations, accustomed to a high intake of fiber-containing foods, whether derived from cereals, legumes, tubers, vegetables, or fruits. There are no exceptions in this behavior. Also in the 1950s Cleave *(23)* contended that it was the refinement of cereals plus the increased consumption of sugar that merited chief blame.

It must be stressed, however, that in contexts where dietary fiber intake is or has become low, the intake no longer is a major regulatory factor, since it cannot explain the variable epidemiology. Thus, on the one hand, a low or decreased intake of fiber-containing foods may be consistent with a high appendicitis frequency, as is usual with white populations. At the same time a low intake may also be consistent with a very low frequency of the disease, as still prevails with urban blacks in South Africa, also coloreds and Indians, among whom fiber intake has become far lower than in the past. Clearly, other factors, dietary or nondietary, must be in operation, which promote, or inhibit, the development of the disease.

4.2.2. Other Hypotheses

In the 1930s, Ashoff and others advanced that a normal bacterial inhabitant in the intestinal tract could be changed by some alteration in *milieu interieur*, to become invasive. It was conjectured that this is what happens with appendicitis.

4.2.3. The Hygiene Hypothesis

This was advanced by Barker *(24)* in 1985. He attributed the rise in appendicitis to improvements in water supplies, and in sewage and refuse disposal, during the late nineteenth century. These improvements in hygiene greatly reduced the exposure of infants and children to enteric organisms, which in turn modified response to later virus infections, so that they now triggered appendicitis. The virus did this either by causing vigorous lymphoid hyperplasia, which blocked the appendix, or, conceivably, by devitalizing the appendiceal mucosa so that bacteria could invade. Barker and associates studied the occurrence of appendectomy in three national samples of British children. They found that the risk of having the operation depended on the amenities present, in particular whether or not there was a bathroom. The risk was independent of social class. These and other observations were considered as supporting a relation between acute appendicitis and Western hygiene, which would explain the geographical distribution of the disease and its changing incidence over time. It was maintained that in the developing world, where children grow up in conditions of poor hygiene, there may be outbreaks of appendicitis when housing improves.

4.2.4. Other Aspects of Epidemiology, Risk, and Causation

Apart from diet, vascular disorders and nonspecific viral infections may be important. A cluster of true appendicitis patients has been reported. A stressful life, including emotional problems and depression, may be influential. Some studies have indicated that there is a familial tendency. In females, there is some evidence of a varying frequency of appendicitis in different phases of the menstrual cycle.

4.3. Can Appendicitis Be Avoided, or Its Rise Restrained?

4.3.1. Western Populations

The answer depends in part on whether the hypotheses mentioned can be tested. Thus far, case-control studies in western communities regarding fiber intake and proneness to the disease have not been particularly revealing. In local unpublished studies there has also been a lack of clear-cut differences in the dietary intakes of series of white and black appendicitis patients, compared with an appropriate series of controls.

Regarding *involuntary* dietary and other interventions, there is some evidence that favors the dietary hypothesis. During World War II, military exigencies in many countries imposed dietary restrictions and changes that involved partial reversion to the diet of Western ancestors. Thus, there was a rise in the intake of fiber-containing foods, but falls in intakes of fat and sugar. The occurrence of appendicitis in Switzerland and in the Channel Islands *(25)* was reported to have decreased. In Dutch internment camps in the Far East, appendicitis incidence was reported to have fallen considerably *(26)*. In all of these contexts, the decrease was associated with a considerable rise in the intake of plant foods. It is noteworthy that wartime populations subjected to the most marked of dietary changes, also experienced falls in occurrences of other diet-related diseases, including dental caries, diabetes, and CHD. Frequencies of all of these diseases reverted to pre-war levels once dietary and other restrictions were removed; subsequently, frequencies increased. During the period of hostilities there were no simultaneous changes in hygiene milieu, nor in sanitation, nor in other risk factors, commensurate with the changes that occurred in diet.

In 1939 in an enquiry made by a British Medical Research Council group, it was strongly doubted whether *voluntary* dietary changes (specifically, eating more fiber-containing foods) would be sufficient to reduce appendicitis incidence *(27)*.

4.3.2. Outlook for South African Populations

In Third World populations, with urbanization, fiber intake falls progressively to a level even below that of the white population; this change is likely to promote appendicitis. Moreover, the hygiene context of such populations is continually improving and hence, according to Barker, this change could at first engender an increase, but ultimately contribute to the inhibition of the disease. Appendicitis in African blacks, after allowing for rise in population, is certainly increasing, although far less than might be expected *(28)*. Frequencies are likely to rise in the colored and Indian populations.

4.4. Comment

Prosperity may well confer a high and possibly irreducible level of susceptibility to appendicitis, just as it does to other degenerative diseases associated with changes in lifestyle. Yet it could be argued that the present still high incidence of appendicitis in developed populations is not immutable; for since the frequency of the disease has fallen markedly for reasons not totally clear, still further falls could well occur, despite the low fiber intake of most Western populations.

So what should we do? For the avoidance of appendicitis and other chronic bowel diseases, perhaps the best advice is that emphasized by Sir Richard Doll *(29)* in his Harveian Oration, in 1982, on "Prospects for Prevention." He urged,

> Whether the object is to avoid cancer, coronary heart disease, hypertension, diabetes, diverticulitis, duodenal ulcer, or constipation ... the type of diet least likely to cause the disease is one that provides a large proportion of calories in whole grain cereals, vegetables, and fruit, provides most of its animal protein in fish and poultry; limits the intakes of fats and if oils are to be used, gives preference to liquid vegetable oils; includes very little dairy produce, eggs, and refined sugars; and is sufficiently restricted in amount not to cause obesity.

The diet described approximates to the "prudent" diet advocated in the 1960s *(3)*.

5. Colon Cancer

5.1. Epidemiology

5.1.1. History

Information on colon cancer is meager. Hippocrates mentioned tumors, and in the works of Galen, in the second century AD, several passages concern tumors. However, not until the early twentieth century did really reliable information on cancer epidemiology become available, simply because only a small proportion of cancers were verified. Knowledge of previous cancer rates is limited, but undoubtedly in all developed populations colon cancer occurrence has increased. Thus, in Leeds, United Kingdom, between 1910 and 1931, a rise occurred in the frequency of intestinal polyps, a common precursor of the tumor *(31)*.

5.1.2. Western Populations

Approximately 3% of US-born males and 3.5% of females will develop colon cancer. From 1973 through 1986, annual incidence rate per 100,000 population increased by about 10%. Increases have occurred for all US races, for whites and blacks, and for males and females. Rates for blacks and whites are now similar, whereas the rate for males is sometimes higher than that for females. This cancer is primarily a disease of the elderly. Median age at diagnosis is about 70 years. Risk increases enormously with age.

Prosperous countries that have high rates are New Zealand and the United States. Countries with low rates are Poland and Greece *(32)*. In Japan, the rate is only 20% of that in the United States. There are numerous puzzling features in rates. They can differ considerably between countries, regionally, and even in areas within the same district. Major differences occur *within* national populations, which stem from variations in culture and practice. As examples, compared with national statistics the disease is less common among Seventh-Day Adventists *(33)*, Mormons, and strict vegetarians.

5.1.3. South African Populations

From study of causes of admissions to hospitals, colon cancer is virtually absent in the rural black population. In urban populations, according to the Cancer Registry of 1986 *(34)*, the incidence rate in the white population, 14.5/ 100,000 "world" population, is somewhat lower than that reported for most developed populations. The rate for blacks, 1.4/100,000, is very low, similar to that of black populations elsewhere in Africa. Rates for coloreds and Indians, 4.7 and 4.7/100,000, are intermediate.

5.2. Perplexing Epidemiological Situations

5.2.1. Effects of Prosperity and/or Migration

Locally, the incidence rate for colon cancer in urban blacks is lower than expected. Indeed, from the results of limited incidence studies made previously, and from the records of patients admitted to Baragwanath Hospital, Johannesburg, hardly any increase in occurrence rate of the disease appears to have taken place. This is puzzling in view of known changes in lifestyle that have occurred *(28)*. It is noteworthy that in Birmingham, United Kingdom, admissions to hospital for colon cancer, of immigrant black populations from the West African and the Caribbean countries, were far fewer, proportionally, than those in the white population *(35)*.

In contrast is the occurrence of colon cancer among Japanese immigrants to Hawaii, and the United States. In Japan, colon cancer has a very low incidence rate, 3/ 100,000 "world" population, almost as low as that of African populations *(32)*. In Japan, fat intake is low; it still contributes little more than half of that usual in Western diets. Fiber intake, once high, is now decreasing. Colon cancer incidence rate is rising slowly. The point of narrating the foregoing is that Japanese immigrants to Hawaii and California soon experience tremendous increases in the occurrences of polyps and colon cancer *(36)*. In both regions, first-generation immigrants attain almost *double* the frequency of sigmoid colon cancer, compared with the rates in their white neighbors. Reasons are not clear. It was stated that findings suggest that, "patterns of risk in relation to migration are complex and defy simple dietary or other interpretations" *(36)*.

5.3. Risk Factors

5.3.1. Dietary Factors

The components most implicated from experimental animal and human studies are fat and fiber-containing foods. In a study made in Wisconsin, vegetable consumption was consistently protective against colon cancer *(37)*. Odds ratios (ORs) for the most significant dietary components (based on high vs low consumption) for proximal colon cancer were: salad, 0.29; miscellaneous vegetables, 0.58: cruciferous vegetables, 0.59; processed lunchmeat, 2.04; panfried foods, 1.79; eggs, 1.75. For distal colon cancer, ratios were: salad, 0.43; cruciferous vegetables, 0.44; cheese, 0.62; processed lunchmeat, 0.79; panfried foods, 1.55. Thus, results support the view that the "prudent diet" (low fat, high vegetables) may reduce risk of colon cancer.

In a case-control study made in Utah, crude fiber consistently decreased risk associated with colon cancer *(38)*. In males, OR was 0.4, and in females, 0.5. The highest quartiles of intake of fruits and vegetables were also associated with a decreased risk of colon cancer in males, with OR, 0.3 and 0.6, respectively; and in females the figures were 0.6 and 0.3, respectively, when compared with the lowest quartile of intake. A high intake of grains was not protective.

In Argentina, a case-control study on colon cancer showed, from conditional regression models, that dietary fiber was highly protective (OR = 0.07/19.0 g/day). The intake of vegetables, fish and poultry was associated, in the highest compared with the lowest quartile, with decreasing risk (ORs of 0.075, 0.39, and 0.39, for the three foodstuffs) *(39,40)*.

In a colon cancer case-control study on American Japanese men, a negative association was found concerning dietary fiber and colon cancer risk among men with low fat intake (<61 g/day) *(41)*. In this subgroup, the men consuming <7.5 g/day of fiber had an adjusted relative risk (RR) for colon cancer of 2.28, compared to those consuming >14.8 g/day of fiber. Also noted among the total group of subjects was a significant negative association between vitamin C intake and the risk of colon cancer. Men in the lowest quintile of vitamin C intake (<37 mg/day) had an adjusted colon cancer RR of 1.87, compared to men in the highest quintile (>160 mg per day).

As to non-Caucasian populations, in Singapore, a hospital-based case-control study among Chinese was carried out on diet and colorectal cancer *(42)*. For colon cancer, significant observations were a protective effect of a high intake of cruciferous vegetables (i.e., green vegetables), OR, 0.50 ($p < 0.01$), and conversely, a predisposing effect of a high meat, low vegetables consumption ratio, OR, 1.77 ($p < 0.05$).

As to the prevention of colon cancer, in this connection there was a recent and much publicized study made in the United States *(43)*. Over a 4-year period, in a chemoprevention trial on large bowel neoplasia, 58 patients with familial adenomatous polyposis were treated with 4 g of ascorbic acid (vitamin C)/day plus 400 mg of α-tocopherol (vitamin E)/day, alone or with a bran supplement (22.5 g/day of fiber). The results from this randomized, double-blind, placebo-controlled study provided evidence for the inhibition of benign large bowel neoplasia by grain fiber supplements, at least in patients with rectal polyps. Findings are consistent with the hypothesis that dietary grain fiber and total dietary fat act as competing variables in the genesis of large bowel cancer.

In the past, cereals comprised particles of large size from stone ground milling compared with the present with steel roller mill products. It is intriguing that large compared with small particle size can make a major difference in bowel behavior, by increasing fecal excretion as well as by shortening transit time. This has been demonstrated in relation to bread and bran particles, and, remarkably, even in the case of indigestible plastic particles (44). This latter evidence indicates a mechanical stimulation of mucosal receptors. Ramifications of the phenomenon regarding effects on bile acid metabolism, also on excretion of mutagens, await investigation.

5.3.2. Other Risk Factors

Evidence is accumulating that parity may regulate the subsequent development of colorectal cancer. In a study made in Alberta, protective associations between previous pregnancies and colorectal cancer were found in women over age 50 at diagnosis; the OR was 0.5. Thus, parity may have a protective effect, similar to the effects reported in the cases of breast, endometrial, and ovarian tumors. Regarding estrogen use, in one study in which such were partaken of, OR was reported to be 0.6.

As to alcohol, although a role has been postulated, many doubt it. Respecting smoking, although one study noted a somewhat lower incidence among smokers, it transpired that when diagnosed, smokers had the disease in a more advanced state.

5.4. Possible Mechanisms of Protection

A low or a reduced consumption of fat is associated with increased excretion of free primary bile acids and also short-chain fatty acids. These metabolites are also affected by level of fiber intake.

A high consumption of fiber-containing foods, characteristically, is associated with the following sequelae.

1. With faster transit time, there is a shorter period of exposure of the gut to potentially pre- or cocarcinogenic substances.
2. There is a low or a decrease in fecal pH value, a parameter known to be high in colon cancer patients, but low in less prone populations (45).
3. There is an associated higher concentration of short-chain fatty acids, which are believed to enhance colonic health.
4. There is a high excretion of free primary bile acids (near absent from consumers of low fiber diets, and also evoked by low fat intake) associated with reduced microbiological activity in the bowel, and with lower levels of tumor-promoting secondary bile acids.
5. There is a decrease in the production and/or excretion of mutagens in the stools.
6. There is a tendency for a reduction in the number and size of polyps (46), the usual sites of carcinogenesis in white populations.

Regarding the use of exogenous hormones, such may decrease the risk of large bowel cancer in women through the medium of changes in the bile acid pool.

5.5. Can Colon Cancer Be Avoided, or Its Rise Restrained?

5.5.1. Western Populations

It is of interest that in pharmacological intervention, aspirin use, 16 or more times a month, for at least a year, has been found to reduce relative risk of colon cancer to 0.6 (47). The mechanism of protection is unclear. In an editorial comment it was stated,

A pill, even one that required regular administration, would probably be more acceptable to the public than an overhaul of lifestyle or diet. Despite the promise of these agents, however, current information is clearly inadequate to support their widespread use for cancer prevention *(48)*.

The only major risk factor that lends itself to alteration is diet. The consensus from reviews is that the dietary factor is of more moment in the etiology of colon cancer than is the case with other diet-related cancers *(49)*. Persons at risk, genetically, should be foremost in making changes.

5.5.2. South African Populations

The foregoing considerations apply equally to inter-ethnic populations.

As mentioned previously regarding CHD, it is important to bear in mind that there is a large measure of individual susceptibility to diseases, little or not at all related to the presence of risk factors. This is the case with numerous cancers. It has been stated that,

> carcinogenic influences in colorectal cancer may depend on differences in individual susceptibility *(50)*. In a proportion of cases, a person at greater risk is attributed variously to genetic proneness, to individual susceptibility, or, as Burkitt has put it, to "the luck of the draw" *(51)*.

To give an example of variability of response, in unpublished studies on local white adults eating the same pattern of diet, a tenth had fecal pH values as high as values reported for colon cancer patients, and a tenth had low values like those of rural blacks, who are virtually free from the disease. Virtually all of the subjects studied were on a low fiber diet. The foregoing implies that one segment of a community may have to make relatively small changes in diet to put themselves into a category far less susceptible to colon cancer. In contrast, another segment of the population may have to make heroic changes in diet (halving fat intake and doubling or trebling fiber intake) to significantly lessen their proneness to colonic carcinogenesis. As emphasized, one group at particular risk who will certainly have to make the abovementioned changes are those with familial predisposition (present in 25% of cases), with polyposis, also patients with inflammatory bowel diseases.

5.6. Comment

For the avoidance of colon cancer the cardinal protective measure is the adoption of the "prudent" diet, principally, reducing fat intake and increasing the intake of fiber foods, especially vegetables and fruit. These changes, unfortunately, are difficult to achieve. In national communities, despite authoritative advocacy of the "prudent" diet, dating from almost three decades ago, minimal changes in these respects are occurring. A major fall in fat intake (from supplying 40% down to 30% of energy) can be achieved by doubling or more the consumption of bread. This change is highly unlikely. It can also be achieved by halving the intake of all visible fat; this cannot be accomplished without a fall in the palatability of the diet. As mentioned before, to accomplish the rise of fiber intake recommended, there should be *five* helpings of vegetables or of fruits each day. Currently, in the United States only *half* of families are consuming two helpings of vegetables or fruit each day *(52)*.

As mentioned and as stressed by Sir Richard Doll *(29)*, the measures advocated for the avoidance of colon cancer, or for other diet-related cancers, are the same as those urged for the avoidance of CHD and other important degenerative diseases.

6. Dental Caries

6.1. Epidemiology

6.1.1. History

Interestingly, in Biblical times King Solomon complimented his attractive girlfriend by saying, in captivation, "Your teeth are white ... perfectly matched, without one missing." Yet the elderly were bidden, "Let your lips be tightly closed while eating when your teeth are gone." Seemingly, teeth were excellent in the young, but caries in the elderly were common. Some reports have indicated that ancient populations had negligible caries, but others have claimed that significant lesions were widespread, although of course nowhere near the high frequency that until recently characterized Western populations (53).

In European countries, changes toward the modern picture of caries occurrence began in the early 1600s, and by the middle of the twentieth century the prevalence had doubled compared with what it was at the turn of the century. Through the intervening years caries frequency rose progressively, until by 1960–1975, prevalences in children had reached their maxima.

6.1.2. Western Populations

During the period 1965–1980 mean caries scores of populations were very high. Thus, in Australia, mean DMFT (decayed-missing-filled-total) score for pupils aged 15 yr was 12.5; other values reported were—New Zealand, 15.8, and for slightly older children in Denmark, 16.6. In passing it is insufficiently appreciated that even within living memory, some populations of Western children had excellent teeth. For example, as late as 1938, rural pupils of 13–15 yr in the Island of Lewis, off the west coast of Scotland, had a very low mean DMFT score of 2.2; almost one-third of the pupils were caries-free (54).

As to falls, it must be remembered that during World War II, in certain countries, the dental caries situation in children improved considerably. This was reported in the United Kingdom (55), the Scandinavian countries, and also in Japan. However, after the war, with the return to prewar conditions, caries scores reverted to their previous levels, and then rose further.

In the last 10–15 years, in numerous Western populations, major, although variable, falls in caries prevalence have occurred. As examples, in Sydney, Australia, from 1963 to 1982, caries in 12-yr-olds fell from a mean DMFT of 8.5 to 1.5, a most remarkable reduction of 84%; prevalences of decayed and missing teeth fell by 95 and 94%; respectively (56). In New Zealand, there has been a substantial decline, especially in the last decade, both in areas with and without any fluoridation, and without any evidence of change in the pattern of sugar consumption. In that country, in 1978, 28% of 12-yr-olds were caries-free in nonfluoridated communities, and 42% in fluoridated communities; in contrast, in 1950 the caries-free figure was far lower, 13%. Currently, children in fluoridated areas have 18% fewer cavities than those drinking untreated water.

6.1.3. South African Populations

A few years ago, as indicated, in Western populations mean DMFT scores of school pupils of 16–18 yr ranged from 10 to 17. In enormous contrast, among corresponding children in Third World populations, e.g., in blacks, the mean DMFT scores were 1–2 (57). This former disparate situation is now changing dramatically, partly because of the decline in caries in children in numerous Western populations; and partly, conversely, because of the rises occurring in the Third World urban dwellers. To exemplify, in Western Australia the mean DMFT of school pupils of 15 yr has now fallen to 4.95, whereas that of urban blacks of the same age in Johannesburg, South Africa, has now risen to 5.1. More remarkable are the changes which have occurred in 12-yr-olds. As already cited, in a group of children studied in Sydney, the mean DMFT was 1.5—an enormous contrast to the figure of 8.5 that prevailed 19 years previously *(56)*. In 12-yr-old black children studied in Nairobi, Kenya, the mean DMFT remains very low, 1.4, yet the increased mean figure for a series of black 12-yr-olds studied in Cape Town was 6.5 *(58)*. Thus, urban black children in Africa no longer hold superiority in caries experience compared with the situation now obtaining in middle- and upper-class white children. Indeed, in the near future it is likely that the dichotomy described will widen, to the disadvantage of the African. Caries scores in colored and in Indian children are higher than those of black children.

6.2. Causative Factors

Neither the rise in caries occurrence nor the falls that have taken place, nor the current variable epidemiology of caries, are explicable in simple terms. It is unfortunate that, of risk factors, almost all attention has been focused on the roles of sugar consumption and snack frequency.

Historically, up to the sixteenth and seventeenth centuries, sugar continued to be a luxury, and was consumed sparingly except among the very small segment of the affluent. By 1750, however, sugar started to become an everyday foodstuff on the tables of the rural poor, and in time became extremely popular, filling an enormous role in the diets of populations. By 1900, sugar supplied about 15–18% of the energy requirements in the general population, equivalent to an average intake of about 100–120 g daily *(59)*. In Third World populations, especially among rural dwellers, little sugar was consumed until a generation ago. Currently, in the more favorably placed populations, sugar contributes about 10–15 and 15–18% of energy in rural and town dwellers, respectively. As mentioned, until recently, numerous authorities blamed sugar as the chief factor, indeed, the almost exclusive factor, in causing dental caries. Many have insisted, as expressed in the recent COMA Report *(60)* in the United Kingdom, that a fall in sugar intake would be associated with a fall in caries scores. There is little doubt, as will be indicated later, that sugar's contributory responsibility is being exaggerated, because, *inter alia*, the recent spectacular falls in caries occurrence have taken place with minimal changes in sugar intake or in snack frequency *(61)*.

Regarding other changes in caries frequency, as already mentioned, during World War II the lesser availability of sugar and other foodstuffs caused diets to revert in measure to being less rich, less refined diets. In many countries, caries incidence in children improved considerably, as was reported in the United Kingdom *(55)*, in Scandinavian countries, as well as in Japan. But, as often pointed out, wartime diets were characterized not only by a fall in sugar intake, but also by major changes in consumption of other foodstuffs, especially fiber-containing foods.

No studies have shown convincingly that a voluntary decrease in sugar intake *alone* in the everyday diet of children, say of 20–30%, has been followed by a meaningful decrease in caries prevalence. In recognition of the incompleteness of knowledge, Rugg-Gunn and Murray *(62)* have stressed, "There is a need for studies which assess sugar-eating habits over a defined period of time, and which related these to caries which develop over the same period."

6.3. Perplexing Situations

In the United Kingdom, in a recent commentary from Westminister, percentages of 13-yr-olds "with active decay in permanent teeth" were stated to be: Northern Ireland 64%, Scotland 51%, Wales 40%, and England 31% *(63)*. In Australia, the mean DMFT of 12-yr-olds was reported to be 5.0 in Victoria, yet 2.5 in New South Wales. In rich vs poor communities, differences in children's mean DMFT may be as much as threefold, with the disadvantage to the poor, as reported from some centers in the United Kingdom. Both in developed and developing populations, scores are lower in rural than in urban populations. Not least of the puzzling features is the very wide range of caries in all populations, which prevails from individual to individual, in both the young and old.

There are numerous other inexplicable findings, mostly relating to sugar intake.

1. In general, mean DMFT scores of segments in the upper vs the lower quartiles of total sugar intake differ very little.

2. In South African interethnic groups, segments in the lower quartile, consuming low means of 30–50 g sugar *per diem*, had mean DMFT scores that ranged very widely, from 1.6 to 8.7 *(61)*.

3. Mean DMFT scores of 13-yr-olds in two regions in Sweden differed, 5.9 and 11.4; ironically, in the former region the children were exposed to a higher sugar intake and snack frequency.

4. In Johannesburg, Jewish compared to non-Jewish adolescents had significantly lower caries scores, although they had on average a higher sugar intake and snack frequency.

5. In a prospective study on middle-school children, the segment who developed no caries ate on average 113 g total sugars and 53 g confectionery each day; these data differ trivially from the data of the segment who developed most caries, who, on average consumed 122 g total sugars and 62 g confectionery each day *(64)*.

6. Correlations between the prevalence of caries with total sugar intake, also with snack frequency, have been marked in some studies, but correlations have been slight or even absent in others *(61)*. In recent longitudinal studies carried out in teenagers, relatively little (<6) of the total variance in caries could be explained by the identifiable factors *(65)*.

6.4. Can Dental Caries Be Avoided, or Its Rise Restrained?

6.4.1. Western Populations

Many authorities have cautioned that such are our limitations of knowledge on the cause of recent falls, that we are not in a position to predict whether further decreases will occur, whether the position will become steady, or whether perhaps even an increase in caries may take place. In Australia the opinion has been expressed that caries severity was likely to decrease over the period 1980–1990, but with evidence of a plateauing for younger adolescents by 1995.

6.4.2. South African Populations

In developing populations, in so far as dietary changes are concerned, especially with urban dwellers, further rises in caries occurrence are judged to be inevitable. In so far as they are promotive, both sugar consumption and snack frequency will rise. As to ameliorating factors, an increase in the use of fluoridated toothpaste is likely, as is improved oral hygiene.

6.5. Comment

Because of the many puzzling epidemiological situations described, the extent of the future changes in caries occurrence likely to take place in both developed and developing populations is uncertain.

7. Changes in the Prevalence of Other Disorders and Diseases

7.1. Obesity

In the past, among rural blacks, weight increased little during adulthood. Nowadays, obesity is common in both rural and urban dwellers, especially in adult females, reaching as high as 60% (BMI > 30) in the cities (6). The disorder is common in the colored and Indian populations, reaching 28% in the latter (8). The proportion in white adult females has been reported as 18%.

7.2. Hypertension

As with obesity, previously, blood pressure increased little in the adulthood of rural blacks. Nowadays, hypertension is common in rural dwellers, 10–15%, and especially so in urban dwellers, 28% (WHO criteria) (66). This is higher than the percentage among the white population, 20–25%. The proportion in the Indian population is 18% (8).

7.3. Diabetes

Previously diabetes was rare in black rural dwellers. It is now rising in incidence, more especially among urban dwellers (although still consistent with a low occurrence of CHD). Its frequency is high in colored adults (6), and particularly so in the Indian population, 16% (8). As with CHD, a very high frequency of the disease prevails in all Indian immigrant populations.

7.4. Osteoporosis Sequelae

Among rural black females hip fractures are rare. Even in town dwellers, the frequency remains very low, a tenth or so of that in white adult females (67). This superiority prevails in the black population despite a habitually low calcium intake, and losses of calcium from often high parity and long lactations. In this population, their lower intake of protein and higher level of everyday physical activity contribute measures of protection. An ethnic element must also play a part, as also prevails in the black population in the United States, although the precise mechanism of protection is not understood. Hip fracture frequencies in colored and Indian women are not known.

8. General Health Outlook
for South African Interethnic Populations

In the white population, mortality rates for many degenerative diseases either are steady or are falling (principally those of CHD and stroke). As with like populations overseas, many whites are persuaded of the benefits likely to accrue from habitual consumption of a "prudent" diet. Yet enquiries indicate, again, as with overseas populations, that less than a tenth are putting their beliefs into everyday practice. Smoking frequency is falling in males, but far less so in females. Alcohol consumption is steady, possibly increasing. Level of physical activity, although rising in the younger segments, is changing little. In brief, whatever decreases have taken place, reversals could be considerably accelerated by the application of known and fully proven public health measures.

In the urban black, colored, and Indian populations, rates for almost all infections (especially childhood infections), except tuberculosis, are falling. In these populations in transition, unquestionably, the more palatable diet pattern of the white population is favored. To illustrate, a recent unpublished survey on the dietary choices of young black men revealed that the huge majority would love to have eggs, bacon, and sausages, for everyday breakfast; they wish for more butter, and for far more sugar and carbonated drinks. The desire for more fruit and vegetables, although certainly present, is secondary to these choices. Scarcely any wished to eat more legumes—beans (previously widely consumed), peas, and lentils. So intense were the selections expressed that any educational attempts directed by the media toward consumption of a "prudent" diet, are deemed to be almost nonstarters. As to other health determinants, in these populations in transition, smoking practice is increasing, although frequency remains relatively low in both black and Indian females. As prevails with all Third World populations, alcohol consumption is rising. Level of physical activity, although higher than that in the white population, is far lower than that of these populations in the past. In brief, although there have been gains regarding the lessening of diseases of low socioeconomic state, there have been rising mortalities from diseases of prosperity.

In the discussion of patterns of health and ill-health of South African interethnic populations, there is thankfulness in some respects, but dismay in others. Notwithstanding the latter, it is imperative to take heart over the extremely important gains that have taken place in primary vital statistics. The infant mortality rates of black populations in big cities are almost as low as the mean IMR of blacks in the United States. Those of the colored and Indian populations have become greatly reduced. In all populations there have been major rises in expectation of life at birth. The cardinal regret is that although tremendous efforts and determination would be required, there could certainly be meaningful improvements in health, especially in the lengthening of disability-free years. As acknowledged in a recent major review on cancer prevention and control, "The knowledge is there—the intention is not" *(68)*.

9. Summary

A description is given of changing patterns of health/ ill-health in various South African interethnic populations—blacks, whites, coloreds, and Indians—as affected by changing patterns of influencing factors. In respect to four diseases, namely, CHD, appendicitis, colon cancer, and dental caries, descriptions are given in some detail of their epidemiology, risk factors, puzzling occurrences, and outlooks for the future. Discussion indicates that reluctance to make alterations in diet and in nondietary risk factors will hinder efforts, according to the population, to accelerate falls, or to restrain rises, in these and other diseases of prosperity.

References

1. Herrick JB. Clinical features of sudden obstruction of the coronary arteries. *JAMA* 1912; **59**:2015–2020.
2. McCrae T. *Osler's Principles and Practice of Medicine*. London: Appleton, 1912; 836.
3. Morris JM. Primary prevention of heart attack. *Bull NY Acad Med* 1975; **51**:62–74.
4. Sytkowski PA, Kannel WB, D'Agostino RBD. Changes in risk factors and the decline in mortality from cardiovascular disease. *N Engl J Med* 1990; **322**:1635–1641.
5. Walker ARP, Adam A, Küstner HGV. Changes in total death rate and ischaemic heart disease death rate from 1978 to 1989 in South African interethnic populations. *S Afr Med J* 1993; **83**:602–605.
6. Steyn K, Rossouw JE, Joubert G. The coexistence of major coronary heart disease risk factors in the coloured population of the Cape Peninsula (CRISIC study). *S Afr Med J* 1990; **78**:61–63.
7. Rossouw JE, Jooste PL, Steenkamp HJ, Thompson ML, Jordaan PCJ, Swanepoel ASP. Socio-economic status, risk factors and coronary heart disease: The CORIS baseline study. *S Afr Med J* 1990; **78**:82–85.
8. Seedat YK, Mayet FGH, Khan S, Somers SR, Joubert G. Risk factors for coronary heart disease in the Indians of Durban. *S Afr Med J* 1990; **78**:447–454.
9. Tunstall-Pedoe H, Smith WCS. Cholesterol as a risk factor for coronary heart disease. *Br Med Bull* 1990; **46**:1975–1987.
10. Walker ARP. Diet and atherosclerosis. *Lancet* 1955; **i**:565,566.
11. Pisa Z, Uemura K. International differences in developing improvements in cardiovascular health. *Ann Med* 1989; **21**:193–197.
12. Mukerjee AB. Heart diseases in India. *J Indian Med Assoc* 1975; **65**:156–158.
13. Ornish D, Brown SE, Scherwitz LW, Billings JH, Armstrong WT, Portis TA, McLanahan SM, Kirkeeide RL, Brand RJ, Gould KL. Lifestyle changes and heart disease. *Lancet* 1990; **336**:741,742.

14. Muldoon MF, Manuck SB, Matthews KA. Mortality experience in cholesterol-reduction trials. *N Engl J Med* 1991; **324:**922,923.
15. Hulley SB, Walsh JMB, Newman TB. Health policy on blood cholesterol: time to change directions. *Circulation* 1992; **86:**1026–1029.
16. Anderson M, Walker ARP, Lutz W, Higginson J. Chemical and pathological studies on aortic atherosclerosis. *AMA Arch Path* 1959; **68:**380–391.
17. Pokorny WJ. *Appendicitis Principles and Practice*. Oski FA, De Angelis CD, Feigin RD, Warshaw JB, eds. Philadelphia: JB Lippincott Co., 1990; 1737–1739.
18. Addiss DG, Shaffer N, Fowler BS, Tauxe RV. The epidemiology of appendicitis and appendectomy in the United States. *Am J Epidemiol* 1990; **132:**910–925.
19. Walker ARP, Walker BF. Appendectomy in South African interethnic school pupils. *Am J Gastroenterol* 1987; **82:**219–222.
20. Short AR. *The Causation of Appendicitis*. Bristol: Wright, 1946.
21. Trowell HC. *Non-Infective Diseases in Africa*. London: Edward Arnold, 1960.
22. Burkitt DP. The aetiology of appendicitis. *Br J Surg* 1971; **58:**695–699.
23. Cleave TL. *The Saccharine Disease*. Bristol: John Wright, 1974.
24. Barker DJP, Osmond C, Golding J, Watsworth, MEJ. Acute appendicitis and bathrooms in three samples of British children. *Br Med J* 1988; **296:**956–958.
25. Banks AL, Magee HE. Effects of enemy occupation on the state of health and nutrition in the Channel Islands. *Mon Bull Min Health Pub Health Lab Serv* 1945; **4:**184–187.
26. Black J. Vegetable consumption and acute appendicitis. *Br Med J* 1986; **293:**52.
27. Young M, Russell WT. *Appendicitis. A Statistical Study*. Med Res Council Spec Rep Ser No 133, London: HMSO, 1939.
28. Segal I, Walker ARP. Low-fat intake with falling fiber intake commensurate with rarity of noninfective bowel diseases in blacks in Soweto, Johannesburg, South Africa. *Nutr Cancer* 1986; **8:**185–191.
29. Doll R. *Prospects for Prevention. Harveian Oration of 1982*. London: Royal College of Physicians, 1982.
30. Jolliffe N, Rinzler SH, Archer M, Maslansky E, Rudensey F, Simon M, Faulkner A. Effect of a prudent reducing diet on the serum cholesterol of overweight middle aged men. *Am J Clin Nutr* 1962; **10:**200–211.

31. Stewart MJ. Precancerous lesions of the alimentary tract. *Lancet* 1931; **ii:**669–675.
32. Muir C, Waterhouse J, Mack T, Powell J, Whelan S. *Cancer Incidence in Five Continents, Vol V.* (IARC Scientific Publication No 88) Lyons: International Agency for Research on Cancer, 1987.
33. Phillips RL. Role of life-style and dietary habits in risk of cancer among Seventh-Day Adventists. *Cancer Res* 1975; **35:**3513–3522.
34. *Cancer Registry of South Africa, 1986.* Johannesburg: South African Institute for Medical Research, 1988.
35. Potter JF, Dawkins DM, Pandha HS, Beevers DG. Cancer in Blacks, Whites and Asians in a British Hospital. *J Roy Coll Phys (London)* 1984; **18:**231–235.
36. Shimizu H, Mack TM, Ross RK, Henderson BE. Cancer of the gastro-intestinal tract among Japanese and white immigrants in Los Angeles County. *J Natl Cancer Inst* 1987; **78:**223–228.
37. Young TB, Wolf D A. Case-control study of proximal and distal colon cancer and diet in Wisconsin. *Int J Cancer* 1988; **42:**167–175.
38. Slattery ML, Sorenson AW, Mahoney AW, French TK, Kritchevsky D, Street JC. Diet and colon cancer: assessment of risk by fiber type and food source. *J Nat Cancer Inst* 1988; **80:**1474–1480.
39. Iscovich JM, L'Abbe KA, Castelleto R, Calzona A, Bernedo A., Chopita NA, Jmelnitzsky AC, Kaldor J. Colon cancer in Argentina. I. Risk from intake of dietary items. *Int J Cancer* 1992; **51:**851–857.
40. Iscovich JM, L'Abbe KA, Castelleto R, Calzona A, Bernedo A., Chopita NA, Jmelnitzsky AC, Kaldor J, Howe GR. Colon cancer in Argentina. II. Risk from fibre, fat and nutrients. *Int J Cancer* 1992; **51:**858–861.
41. Heilbrun LK, Nomura A, Hankin JH, Stemmermann GN. Diet and colorectal cancer with special reference to fiber intake. *Int J Cancer* 1989; **44:**1–6.
42. Lee HP, Gourley L, Duffy SW, Esteve J, Lee J, Day NE. Colorectal cancer and diet in an Asian population—a case-control study among Singapore Chinese. *Int J Cancer* 1989; **43:**1007–1016.
43. DeCosse JJ, Miller HH, Lesser ML. Effect of wheat fiber and vitamins C and E on rectal polyps in patients with familial adenomatous polyposis. *J Natl Cancer Inst* 1989; **81:**1290–1297.

44. Tomlin I. A survey of normal bowel habit. *Br J Clin Pract* 1975; **29:**289–291.
45. Thornton JR. High colonic pH promotes colorectal cancer. *Lancet* 1981; **i:**1081,1082.
46. Segal I, Cooke SA, Hamilton DG, Ou Tim L. Polyps and colorectal cancer in South African blacks. *Gut* 1981; **22:** 653–657.
47. Thun MJ, Namboodiri MM, Heath CW. Aspirin use and reduced risk of fatal colon cancer. *N Engl J Med* 1991; **325:**1593–1596.
48. Baron JA, Greenberg ER. Could aspirin really prevent colon cancer? *N Engl J Med* 1991; **323:**1644–1646.
49. Willett WC, Stampfer MJ, Colditz GA, Rosner BA, Speizer FE. Relation of meat, fat, and fiber intake to the risk of colon cancer in a prospective study among women. *N Engl J Med* 1990; **323:**1664–1672.
50. Higginson J. In: Maltoni C, ed. *Advances in Tumour Prevention. Detection and Characterization. Part II.* Amsterdam: Excerpta Medica, 1974; 3.
51. Burkitt DP. In: Reilly RW, Kirsner JB, eds. *Fiber Deficiency and Colonic Disorders.* New York: Plenum, 1975, p. 156.
52. Block G, Lanza E. Dietary fiber sources in the United States by demographic group. *J Natl Cancer Inst* 1987; **79:**83–91.
53. Mandel D. Caries through the ages: a worm's eye view. *J Dent Res* 1983; **62:**926–929.
54. King JD. *Dental Disease in the Island of Lewis.* Med Res Coun Spec Rep Ser No 241. London: HMSO, 1940.
55. Bransby ER, Knowles EM. A comparison of the effects of enemy occupation and post-war conditions on the incidence of dental caries in children in the Channel Islands in relation to diet and food supplies. *Br Dental J* 1949; **87:**236–243.
56. Burton VJ, Rob MI, Craig GG, Lawson JS. Changes in the caries experience of 12-year-old Sydney schoolchildren between 1963 and 1982. *Med J Aust* 1984; **140:** 405–407.
57. Walker ARP, Dison E, Walker BF, Segal AF. Contrasting patterns of caries profile and dental treatment in pupils of 16–18 years in South African ethnic groups. *Comm Dent Oral Epidemiol* 1982; **10:**69–73.
58. Steyn NP, Albertse EC. Sucrose consumption and dental caries in twelve-year-old children residing in Cape Town. *J Dent Assoc S Afr* 1987; **42:**43–49.
59. Mintz SW. *Sweetness and Power: The Place of Sugar in Modern History.* Viking: Elizabeth Sifton Books, 1985; 67.

60. COMA Report. *Dietary Sugars and Human Disease: Conclusions*. Department of Health and Social Subjects. No 37. London: HMSO, 1990.
61. Walker ARP, Cleaton-Jones PE. Sugar intake and dental caries: where do we stand? *J Dent Child* 1989; **Jan/Feb:**30–35.
62. Rugg-Gunn AJ, Murray JJ. The role of sugar in the aetiology of dental caries. I. Sugar and the antiquity of caries. *J Dent* 1983; **11:**190–199.
63. Dietch R. Commentary from Westminster: Another ferocious but unavailing stand against fluoridation. *Lancet* 1985; **i:**231.
64. Rugg-Gunn AJ, Hackett AF, Appleton DR, Jenkins GN, Eastoe JE. Relationship between dietary habits and caries increment assessed over two years in 405 English adolescent school children. *Arch Oral Biol* 1984; **29:**983–992.
65. Dummer PMH, Oliver SJ, Hicks R, Kingdon A, Addy M, Shaw WC. Factors influencing the initiation of carious lesions in specific tooth surfaces over a 4-year period in children between the ages of 11–12 years and 15–16 years. *J Dent* 1990; **18:** 190–197.
66. Seedat YK, Seedat MA, Hackland DBT. Biosocial factors and hypertension in urban and rural Zulus. *S Afr Med J* 1982; **61:**999–1002.
67. Solomon L. Bone density in ageing Caucasian and African populations. *Lancet* 1979; **ii:**1326–1330.
68. Muir CS, Sasco AJ. Prospects for cancer control in the 1990's. *Ann Rev Public Health* 1990; **11:**143–163.

Diet and Chronic Degenerative Diseases

A Summary of Results
from an Ecologic Study in Rural China

T. Colin Campbell and Junshi Chen

1. Introduction

In 1983, a cross-sectional survey of diet, lifestyle, and disease mortality characteristics was carried out in 130 villages located in 65 counties of rural China (1). A wide variety of characteristics were recorded, ultimately yielding 367 items of information on each of a total of 6500 adults aged 35–64 years (50 subjects/village, half of each sex).

This study was designed, in part, to better understand the causes of disease within a broad context of diets, lifestyles, and disease. Investigation of "causes of disease within a broad context" means having the opportunity of examining, through various perspectives, a common cause of related diseases. To accomplish this purpose, a database ideally should include a cluster of diseases of common cause, a cluster of exposure characteristics of these common causes, and a cluster of endogenous characteristics (biomarkers) of these common causes. Although this project has expanded the investigational framework of dis-

From: *Western Diseases: Their Dietary Prevention and Reversibility*
Edited by: N. J. Temple and D. P. Burkitt Copyright ©1994 Humana Press, Totowa, NJ

ease causality only minimally when compared with what Nature has wrought, it nonetheless has extended the contextual framework of the more traditional research, which focuses on isolated causes and isolated diseases and isolated "mechanisms of action."

In this chapter, the relationship of diet as a cause of the so-called chronic degenerative diseases* will be considered. These are the diseases that chiefly afflict individuals and societies who appear to have exceeded the bounds of optimal nutrition. The causation of these diseases has attracted increasing attention during the past 10–20 years in Western societies, obviously in response to the awareness that the majority of citizens prematurely and unnecessarily succumb to these diseases at great societal costs. This same diet–disease relationship also has begun to attract attention among less developed societies who are moving toward greater industrialization and wealth. They are beginning to recognize the extraordinary costs to their future development, both in total costs and in the economic disequilibrium thereby imposed, if the emergence of these diseases is allowed to flourish.

To say, however, that we will investigate the role of diet in the causation of these diseases is bound to provoke considerable skepticism among research purists, because a "diet" is enormously complex and variable and "these

---

*These diseases include most if not all of the cancers, most of the cardiovascular diseases, diabetes, and certain other ailments associated with the more industrialized societies; they also have been referred to as "diseases of affluence" (2), "diseases of misdevelopment" (3), or in the older literature, as "diseases of civilization" (4). We propose the term "diseases of extravagance," because the total cost of these diseases to society is a summation of the costs of producing the causes plus the costs of treating the consequences. In this chapter, we will refer to these diseases as "degenerative diseases," because of their association with biological degeneration, or aging.

diseases," even if limited to those commonly found among the richer and more extravagant societies, also are known to have specific and unique dietary and lifestyle causes. But it is our contention that, although specific causes of the "diseases of extravagance" must be discovered and their mechanisms must be elucidated, these same diseases also have some very profound common causes that, when properly understood, can be used to control disease development, even to control the activities of specific causes.* Therefore these common causes, usually highly complex in their composition, may be the more significant, both for individuals and for societies, and research procedures must be found for these complex causes to be investigated within a broad context. Common causes cannot be fully understood by limiting investigations to the generalization of details of specific cause–effect relationships, a process that invites too much error, as has happened in this field too often in the past (5). Synthesis of detailed information into a larger picture is required. We do not mean to denigrate studies on details, only to re-emphasize, when making choices of which details to investigate and which interpretations to apply, that this be done within context. It is doubtful whether there is any other area of science so in need, yet so bereft of this understanding, than the discipline of nutrition. The endless digestive and metabolic interactions that occur between food constituents, the profound time-dependent adaptations that integrate and alter nutrient activities, and the psychosocial determinism of nutrient choice, on the one hand, and of nutritional determinism of genetic expression, on the other, are but some

---

*For example, nutrient intakes, at appropriate levels, are able to modify carcinogen metabolism and control the development of clonally expanded lesions. Thus, nutrition, as the common cause, is able to control the activity of a chemical carcinogen as the specific cause.

of the very profound phenomena seldom considered in this field of study. These phenomena must be understood and integrated into our studies in this field if the infinitely complex interface between diet and disease is ever to be understood.

Thus, one of the more important objectives for this study was to investigate both the broad and specific abilities of diet to reduce degenerative diseases, particularly the abilities of a diet enriched in foods of plant origin. In this regard, several unique and unusually valuable features characterized this study. First, wide variations in disease mortality rates (both cancers and noncancers) were observed; variation in mean county rates were several dozen-, if not several hundred-fold, in contrast to the two- to threefold variation commonly observed among geographic regions of highly industrialized societies (6). Second, the subjects under investigation exhibited highly stable residence patterns, with 90–94% still residing at the time of the survey in the county of their birth (1). Third, the food consumed at each study site is, except for certain grain rations, produced locally and is well known to reflect stable dietary composition patterns because of the localized geographic and climatic conditions of agriculture. Fourth, this study included the important study design feature of having duplicate survey villages in each county, thus enabling comparison of intracounty homogeneity with intercounty heterogeneity—an indicator of data quality. And finally, nutritional exposures in rural China sharply contrast with those for highly industrialized societies, a feature that permits investigation of diet–disease associations within nutritional exposure ranges rarely, if ever, available in studies of Western subjects.

However, this survey also employed a cross-sectional study design that often has been considered to yield the weakest kind of evidence on causality, particularly when considering the dietary causes of degenerative diseases. Reasons for this concern primarily include possible confounding of the factors under investigation by unmeasured factors (7,8), the use of different experimental procedures for the populations being compared (8), and the possibility of differing genetic backgrounds for the study populations.* In contrast, we suggest that this study in rural China has substantially overcome these pitfalls, especially when considering diet–disease hypotheses within a broader context. A large number and variety of nutritional factors (nutrient intakes, biochemical markers of nutritional status, frequencies of food consumption, and anthropometric indicators) were measured, data were obtained by the same procedures for all survey sites, and individuals of reasonably common genetic background were the subjects under study (more than 80% of the Chinese population belong to the Han ethnic group [15]).

Much of the motivation for this study arose from the report that cancer rates (and presumably their causes) were highly localized geographically (16), thus offering opportunities to formulate new hypotheses on causes of cancer. There was further interest, however, in examining the relationship between diet and disease within a context much broader in scope than that which compares the relationships between individual diseases and individual

---

*There already is substantial evidence from so-called migrant studies showing that whatever genetic differences in disease susceptibility may exist for individuals, the ultimate risk of disease is determined by environmental factors, not by ethnic or genetic backgrounds (9–12); thus, we do not regard this concern as very important. We agree with the views of Higginson and Muir (13) and Knudsen, 1977, as stated by Bertram (14), "… that not more than 2% of all human cancer is attributable to purely genetic or congenital factors" (p. 3013).

nutrients. The opportunity to undertake this project arose at the time when the 1982 US National Academy of Sciences (NAS) report on diet, nutrition, and cancer *(17)* was being published. This report, which was coauthored by the first author of this paper, reviewed and summarized the available diet–cancer literature according to individual nutrient groups, but nonetheless cautioned against drawing inferences that these cancers might be prevented by individual nutrients. Unfortunately, subsequent developments have proven that researchers, particularly those outside of the field of nutrition, are much more interested in studying the effects of individual nutrients and other food constituents than investigating the more holistic effects of the entire diet* *(see* the following chapter for further discussion). Therefore, this study in China has become, since its inception and at least in part, an attempt to investigate the broader disease prevention role of the entire diet that was emphasized by the NAS report but that mostly has been ignored in subsequent research.

This chapter presents a progress report on the diet–disease associations** thus far interpreted and published from this study in rural China. These findings will be considered especially in reference to the widely publicized dietary recommendations for the prevention of the degenerative diseases *(17,28–30)* now being advocated in many industrialized societies.

*The NAS report explicitly stated in the Executive Summary that their recommendations did not apply to individual nutrient supplements *(17)*. However, this same NAS report may have sown the seeds of its own misinterpretation of its admonition against single nutrients by its specific recommendation on dietary fat.

**The data from this study also have been used for other purposes, including evaluation of novel diet–disease hypotheses *(18–22)*, the development of Chinese policies on the use of tobacco products *(23)*, and investigations of the role of socioeconomic conditions in various aspects of national development *(24–27)*.

2. Study Design Characteristics

In the late 1970s the Chinese Academy of Medical Sciences, acting on the suggestion of the late Premier Chou En Lai, carried out a retrospective nationwide survey of causes of death for the years 1973–1975. This survey covered a total population of more than 800 million people (about 96% of the nationwide total) residing in approx 2400 counties. Causes of death were classified into approx 50 different disease and aggregate disease categories.

The first publication of these data appeared in 1981 under the authorship of Junyao Li and colleagues at the Chinese Academy of Medical Sciences, with the release of the *Atlas of Cancer Mortality (16)*. This document included impressive color-coded maps showing unusual patterns of sex-specific mortality rates by county for about a dozen different types of cancer. Of particular interest was the extraordinarily wide range of cancer mortality rates that demonstrated a high degree of geographic concentration, with ranges from a few dozen- to a few hundredfold. The geographic localization of these diseases was truly remarkable when compared with the mere two- to threefold range for the same cancers observed in highly industrialized countries such as the United States *(6)*. It was this observation, then, that initially gave rise to the inception of a cross-sectional ecologic survey to identify the principal dietary and lifestyle correlates of these unique disease patterns. An additional attraction was the fact that if the survey were limited to the more rural and semirural counties, then there would be considerable stability of residence and food consumption patterns based on locally produced foods.

A survey of 65 counties, two villages per county, thus was organized and carried out during the fall of 1983 under the auspices of the Chinese Academy of Preventive Medicine (with additional data on nitrosamine exposure collected during the fall of 1984). Impressive residence stability was observed, since it was found that 90–94% of the survey population were born in their counties of residence *(1)*.

The details of the study design, the biochemical methods and procedures used for the analyses of the blood, urine, and food samples, and the data tabulation methods are given in the monograph published by Chen et al. *(1)*. Briefly, the 65 counties were selected to represent the full range of mortality rates for seven different cancers. Within each county, two villages were chosen randomly and within each village, 25 adults 35–64 yr of age residing within individual households, also randomly chosen, were invited to give blood and to complete a questionnaire concerning their individual lifestyle habits. In one of the two villages of each county urine samples also were obtained. And finally, a 3-day dietary survey (weighed intakes of food) was conducted within each household and, in each village, the principal plant foods being consumed were collected for later analyses.

From 1984 to about 1988, the blood, urine, and food samples were analyzed for various constituents*—particularly those concerned with nutritional status. And from about 1987–1990, data were computer coded,** were checked for possible reanalyses, and finally were published in the monograph by Chen et al. *(1)*. Also included in the monograph are the 1973–1975 retrospective mortality data representing not only the various cancers previously published *(16)* but also some newly compiled rates for about three dozen noncancer diseases. A total of 367 items of information are displayed in the monograph in various

---

*These analyses were coordinated by Linda Youngman at Cornell University and Feng Zulin at the Chinese Academy of Preventive Medicine.

**The computer programming of data was coordinated by Julian Boreham and Richard Peto at the University of Oxford.

forms, including a tabulation of original data, a summary of simple statistics, univariate correlation coefficients relating each variable with every other variable, histograms of data distribution, scatterplots comparing sexes and villages in the same counties, and the geographic distribution of the data range by quartile. An extensive variety of dietary, lifestyle, and mortality characteristics are available for interpretation, including information on mortality rates, food intakes, tobacco and alcoholic beverage use, selected viral exposures, reproductive characteristics, anthropometric indicators, toxic chemical exposures, drinking water sources, geographic characteristics, and demographic parameters. These variables are summarized in Table 1. In general, the data quality was considered to be quite good as indicated by the observation that 85% of the blood variables measured in both villages of each county were intercorrelated at $p < 0.05$ and 78% at $p < 0.01$. Thus, the relative homogeneity within counties when compared with the heterogeneity across all counties was highly favorable for detecting and testing the relationship between various factors.

Since 1989, we have been involved in publishing interpretations of specific associations of selected variables. The publications and other project-related reports thus far produced have included consideration of the effects of tobacco use (31), chronic viral (21,22,32) and bacterial (19,33) infestation, reproductive/hormonal relationships to disease (34–36), and dietary/nutritional activities (references summarized later). National policies have been developed from the data on tobacco use (23) and dietary practices (25,37), with the latter receiving major circulation in the government newspaper, *The China Daily* (26) as well as numerous major publications in the United States and elsewhere.

In summary, two features of this study distinguish it from most other cross-sectional ecologic studies. First, there is a large array of life and death characteristics and,

Table 1
Abbreviated List of Variables Collected in China:
Cross-Sectional Study on Diet, Lifestyle, and Disease

Causes of death (age-standardized mortality rates, 1973–1975)
    Total deaths
    Neoplastic diseases
    Cardiovascular diseases
    Respiratory diseases
    Reproductive and metabolic diseases
    Infectious diseases
Nutrient status indicators
    Lipid (blood lipids, fat intakes)
    Protein (blood indicators, intakes)
    Carbohydrate (blood glucose, varied fiber intakes)
    Vitamins (intakes and blood indicators)
    Minerals (intakes, blood indicators, urinary metabolites)
Food intakes
    Three-day intakes of foods at time of survey
    Frequency of consumption of main staples
    Government rations
Viruses (blood factors)
Hormones (blood levels of reproductive hormones)
Anthropometric, reproductive, and fecundity characteristics
Food contaminants
Smoking characteristics
Drinking water sources
Geographic features

---

second, the range of values for many of these characteristics show that this is a population with substantial variation of dietary habits when compared with the far more studied populations of industrialized societies.

Cross-sectional ecologic studies are widely characterized as possessing several experimental weaknesses, including:

1. The ecologic fallacy, i.e., imputing to the individual the characteristics of the population;
2. The inability to assign causality;

3. The inability to reflect long-term experience from a single point in time; and
4. Spurious relationships caused by unmeasured and uncontrolled factors.

Accordingly, relationships between causes and their effects only can be hypothesized with such data; they cannot be proven because of the countless unmeasured and uncontrolled factors that also may account for the observed associations. To establish causal relationships, study designs that control all factors except the one under investigation are required. As a result of these inherent problems, cross-sectional studies generally are considered to be limited in their experimental value either to the generation of new hypotheses or to the provision of new evidence on old hypotheses.

This study in rural China, however, is rather different in that it offers an opportunity not only to generate new hypotheses (and there are many) but also, because of its large number of variables, to add important evidence to confirm a hypothesis, a seemingly heretical idea to some, considering that these data were obtained from a cross-sectional study.

The main hypothesis under investigation in this study concerns the comprehensive effects of diet upon the development of chronic degenerative diseases commonly found in the more industrialized societies. In essence, this hypothesis assumes, in addition to specific causes, a common nutritional etiology for these diseases. It is possible in this study to simultaneously investigate multiple nutritional features (nutrient intakes, biochemical "indicators," anthropometric parameters) and a variety of chronic degenerative diseases hypothesized to share this common nutritional etiology, a feature that provides a decided advantage for the investigation of a hypothesis as broadly based as this one. For example, if we hypothesize that a diet enriched in a variety of plant foods minimizes development of the chronic degenerative diseases commonly observed in populations who use nutritionally rich diets*—and this is the main hypothesis under investigation *(38)* *(see also* chapter by Campbell)—then we are examining a biologically complex relationship which is characterized by the causation of a wide variety of diseases by a broad array of nutrients and other food constituents. It is important to note that this hypothesis is not focused on the independent effects of individual nutrients, such as fat, but is addressed to the comprehensive effects of the entire diet on the development of all degenerative diseases. Thus it is essential that we have access to a comprehensive array of dietary and disease variables, preferably all collected in one study population. Such is the case with this study in China. Investigation of this hypothesis is not based on an unreasoned statistical model with all of the specified and obviously highly correlated dietary constituents, but rather is a careful and critical examination of a series of associations (and adjusted associations) of the various constituents with the prevalence of these diseases. Biological interpretation in the context of the current knowledge base is the key.

Other cross-sectional studies are considered to produce only weak evidence on causality because of their focus on single or few causes of disease among a background of many unmeasured variables. In contrast, we propose that this study is capable of obtaining much more reliable evidence on causality if the hypothesized cause is comprised

---

*We intentionally use the phrase "nutritional richness" perjoratively, to mean that "richness" is equivalent to "extravagance," meaning that such societies enriching their nutritional status are willing to pay more money for expensive causes of diseases that, in turn, cause the expenditure of more funds to cure the disease—a double cost.

of many characteristics and if these characteristics are available in the study. This, essentially, is the situation concerning the hypothesis in this report, namely, that a diet enriched in foods of plant origin reduces the risk for degenerative diseases, a hypothesis that must be examined from many perspectives.

3. Summary of Findings

To illustrate the differences in dietary and nutritional characteristics between rural China and the United States, a few comparable dietary, biochemical, and anthropometric characteristics, as previously published by Chen et al. *(1)*, are shown in Table 2. These characteristics, which include only a few of those summarized by Chen et al. *(1)*, demonstrate a sharp contrast in dietary composition between the United States and rural China (urban dwellers and herdspeople of northwestern China were not included in this comparison). Average dietary fat content in rural China is far below that of the United States, with a range of 6–24% of calories. This is achieved primarily through a marked reduction in the intake of animal foods. For example, the proportion of total protein in China as animal protein is 10.8%, whereas in the United States it is 69% *(1,41)* (data not shown). Note also the much lower intake in China of retinol, exclusively found in animal foods, along with the higher intakes of the plant food constituents such as vitamin C, dietary fiber, and starch and starch-like material. Average intakes of dietary fiber and fat in China also substantially exceed, in opposite directions, the recommended intakes advocated by the Western dietary guidelines for fat, at 30% *(29,30)*, and dietary fiber, at 25–35 g/d *(42)*. Energy intake in China is much higher than in the United States, being more than 30% higher after adjustment for body weight *(38)*. In spite of this greater energy consump-

Table 2
Comparison of Selected Dietary, Biochemical,
and Anthropometric Characteristics of Rural China
with the United States[a]

Variable	China	United States
Anthropometry		
Body wt, kg	55 (M)	77.2 (M)
	49 (F)	69.5 (F)
Body ht, cm	164 (M)	173 (M)
	154 (F)	163 (F)
Plasma		
Cholesterol, mmol/L[b]	2.58–4.91 (3.75)	4.01–7.08 (5.48)
Iron, mg/dL	.073–.247 (M)	.060–.150 (M)
	.064–.146 (F)	.050–.130 (F)
Hemoglobin, g/dL	11–16 (M)	14–17 (M)
	10–15 (F)	12–15 (F)
Intakes		
Total protein, g/d	64.1	91.0
Dietary fiber, g/d	33.3	11.1
Fat, percent kcal	14.5	38.8
Carbohydrates, g/d[c]	473	245
Energy, kcal/d	2641	2360
Iron, mg/d	34.4	18.4
Retinol, RE/d	27.8	990
Vitamin C, mg/d	140	73

[a]Data indicate "normal" ranges or means, as specified in Table 5.5 of Chen et al (1). Separate anthropometric and plasma values are given for males (M) and females (F), whereas means of intakes are given for "reference man." In China, data are for a "reference man," which is an adult male of 65 kg body wt and undertaking light physical work; in the United States, data are for individual adult males (39) whose average body wt is 77.2 kg (40); both intakes are consumption at the household level and are not disappearance data.

[b]Upward adjustment of 15% owing to methodological underestimate (41).

[c]Starch and starch-like material fraction.

tion, obesity is much less common in China. For example, average body mass indices (wt/ht^2) for adult males are 20.5 and 25.8 for China and the United States (38), respectively.

Plasma cholesterol levels in rural China, as measured at the time of the survey, are substantially below those of the United States, although these values may have been underestimated by about 10–15% because of the sample assay method used *(1)*. If 15% is added to the observed mean level of 3.28 mmol/L (127 mg/dL) to give 3.77 mmol/L (146 mg/dL), this would compare with values of 4.01–4.13 mmol/L (155–160 mg/dL) obtained for Guangzhou rural workers by Tao et al. *(43)*. (This underestimate, if true, is not important for consideration of the associations of plasma cholesterol with other experimental variables because between-village, within-county cholesterol means are highly correlated [$r = .77$, $p < .001$], thus indicating consistency of error.)

Many previous studies have indicated a positive relationship between blood cholesterol levels and prevalence of coronary heart disease among populations with traditionally high cholesterol levels *(44,45)*. However, there have been persistent reports that cholesterol levels below about 180–190 mg/dL (4.65–4.91 mmol/L) are associated with increasing numbers of deaths from other causes *(46–48)*, particularly of colon cancer, although some of these observations have been ascribed to a depressing effect of clinically undetected disease on cholesterol levels before they were measured *(49–51)*. Thus, this study in rural China *(1)*, comprised of mortality rates for a wide variety of diseases along with cholesterol levels well below 200 mg/dL (5.17 mmol/L), offered an opportunity to examine the nature of disease associations with very low blood cholesterol levels. For example, of the various diseases in China, we observed that there is geographic clustering of two groups, one associated with lesser economic development, the second with the emergence of more wealth (Table 3) *(24)*. The chief correlate of these so-called diseases of extravagance (or affluence) is plasma cholesterol ($p < .01$). In turn, plasma cholesterol is associated with the intake of meat and total fat and inversely associated with the intakes of legumes and certain fiber fractions (data not shown),

Table 3
Self-Clustered Disease Groups[a]

Diseases of poverty	Diseases of affluence
Pneumonia *(16)*	Stomach cancer *(5)*
Intestinal obstructions *(12)*	Liver cancer *(10)*
Peptic ulcer *(13)*	Colon cancer *(9)*
Other digestive disorders *(17)*	Lung cancer *(16)*
Nephritis *(12)*	Breast cancer *(1)*
Pulmonary tuberculosis *(10)*	Leukemia *(15)*
Infectious diseases	Diabetes *(2)*
(other than tuberculosis) *(17)*	Coronary heart disease *(1)*
Parasitic diseases	Brain cancer (ages 0–14 yr) *(13)*
(other than schistosomiasis) *(10)*	
Eclampsia *(13)*	
Rheumatic heart disease *(13)*	
Metabolic and endocrine disease	
(other than diabetes) *(10)*	
Diseases of pregnancy and birth	
(other than eclampsia) *(15)*	

[a]Each disease category, when significantly correlated ($p < .05$) with any other category is positive for disease categories in its own group and negative for disease categories in the other group (there are no exceptions). Numbers in parentheses indicate the number of correlations that are statistically significant at $p < .05$ (from a total of 20 comparisons). Originally published in *Ecology of Food and Nutrition (24)*.

being analogous therefore to similar correlations obtained for Western populations at much higher blood cholesterol levels. This finding is considered rather remarkable in view of the low level of this plasma constituent and the already hypocholesterolemic intake of its chief determinants. Thus, in China, only small intakes of meat and very modest elevations of dietary fat—within the 6–24% fat-as-calories range— are associated significantly with increases in plasma cholesterol, which associates significantly, in turn, with the emergence of diseases that are typical of Western countries.

In-depth investigations of the chief correlates of liver
(21) and breast cancers *(36)* also are revealing. In the case
of liver cancer, very common in many parts of China, the
main correlates are plasma cholesterol ($p < .01$) and chronic
infection with hepatitis B virus ($p < .001$). Aflatoxin, which
is widely regarded as a cause of human liver cancer *(52)*,
is not related to the mortality rates for this disease ($r = -.17$).
This unusual finding suggests that the combination of
chronic HBV viral infectivity along with the consumption
of only small amounts of the dietary determinants of
plasma cholesterol are capable of causing this disease,
essentially without modification by aflatoxin as a carcinogen initiating agent. These results are in accord with experimental animal studies in our laboratory *(53–55)*. We
believe that these rather surprising results of a null effect
of aflatoxin should be taken seriously because this human
study on aflatoxin relationships was more comprehensive
and statistically sensitive than all other studies combined,
and was in accord with our experimental animal studies
(53–55), thus providing a biologically plausible relationship.

In a more recent investigation of a smaller cohort of
1556 men selected from the larger cohort *(56)*, it was found
that among the 238 (15%) who were chronic, viremic car-
riers of the HBV antigen (HBsAg+), plasma cholesterol
fractions were lower, 4.2% for total cholesterol ($2p < .05$)
and 7.0% apolipoprotein B (apo B) cholesterol ($2p < .001$).
In other words, this finding on individuals, at first glance,
appears to suggest an *inverse* relationship between plasma
cholesterol and liver cancer rates (i.e., higher cholesterol
levels among viremic persons), thus appearing to be at odds
with our earlier finding *(21)* of a *direct* relationship between
plasma cholesterol and liver cancer rates for our cross-sectional analysis of county level populations. Furthermore,
this inverse association of Chen et al. *(56)* is similar to the
inverse associations observed by other investigators for

plasma cholesterol and several other cancers—in *prospective studies of individuals*, including lung *(49,57–61)*, breast *(62)*, colon *(49,50,57,59,61,63,64)*, and leukemia *(57,59,61)*, and, in another study in China, of liver *(56)*. Thus, these contrasting observations between Chen et al. *(56)* and Campbell et al. *(21)* appear to provide different conclusions for studies of viremic *individuals* at high risk for liver cancer *(56)* (depressed plasma cholesterol) than for studies of *populations* at high risk of this disease *(21)*. A possible clue explaining this anomaly was revealed when viremic status and plasma cholesterol comparisons among individuals were stratified for groups of villages ranked according to their mean viremia prevalence. The previously observed inverse relationship between plasma cholesterol and HBsAg+ carrier status (and, therefore, liver cancer risk) *was reversed* to produce a highly significant positive relationship between plasma cholesterol and HBsAg+ carrier prevalence. The quartile of villages with the highest HBsAg+ carrier prevalence exhibit plasma cholesterol levels about 20% higher than the quartile of villages with the lowest plasma cholesterol levels; however, within each of these village quartiles, plasma cholesterol levels are about 4% lower for viremic individuals than for nonviremic individuals. Put another way, villages with the richer nutritional status (20% higher plasma cholesterol) exhibit a higher prevalence of viremia and liver cancer, but *within* these villages the viremic individuals most prone to this disease exhibit slightly (4%) but significantly lower ($2p <$ .05) plasma cholesterol levels.

These seemingly paradoxical findings are important to reconcile because they often have been observed for several other cancers as well *(63,65,66)*, as reviewed by McMichael *(67)*. The most frequently cited explanation for these inverse relationships between plasma cholesterol and cancer risk, as observed in case control studies of individuals, is that the subclinical presence of the disease depresses plasma cholesterol, the more so, the nearer the time when the disease begins to makes its presence known clinically *(49,66,68)*. Similarly, in these studies in China, it would seem reasonable to suggest that chronic infection with HBV, a severe irritant to normal liver function, could readily impair cholesterol metabolism (either decreased synthesis and/or increased breakdown) in a manner somewhat analogous to neoplasia, albeit by somewhat different mechanisms. Moreover, it is unlikely that HBV infection would exhibit its cholesterol lowering effect only during the late stages of infection, thus this "inverse" association between active viremia and plasma cholesterol could exist for many years prior to clinical manifestation of disease. As with neoplasia, it could be that, as the body begins to experience the onset of lesion development, even in its very early stages, it defends or copes by several mechanisms, including the metabolic restriction of nourishment, as indicated by slightly depressed plasma cholesterol. In spite of these small decreases in plasma cholesterol induced by the onset of lesion development, however, the mean levels of cholesterol among populations of individuals at greatest risk for disease still are higher than those not at risk. It might be thought of as a nutritional tug-of-war occurring in the tissues of individuals who begin to form degenerative lesions, exogenous nutritional excess (through dietary means) favoring lesion development and endogenous nutritional deprivation (through body defense mechanisms) favoring lesion inhibition. There is every reason to suspect that this process would start very early, as soon as biochemical aberrations are detected, thereby inducing well-known biochemical feedback mechanisms to control further development. In these very early stages, perhaps many years before clinical manifestation, these efforts to deprive unwanted lesions of their nourishment likely would remain undetected, statistically speaking, because of the highly localized and diluted nature of the process within whole body metabolism. With advancing lesion development, nutritional deprivation becomes more vigorous and pronounced to produce a statistically detected reduction of circulating cholesterol among individuals nearing clinical manifestation of the disease. Naturally, this process is enriched and more readily detected among individual cases, as in case control or cohort studies, but remains undetected among high risk populations both because the relative effects of nutritional excess are much more pronounced and because the impending cases within the population are relatively few in number.

In respect to other cancers, plasma cholesterol is associated directly with all cancer mortality rates measured in this study. Most notably, these associations are statistically significant for eight different cancers, including the singularly interesting colon cancer ($p < .01$ for males and $p < .001$ for females) *(46,47,69)*. In a brief overview of the relationship of plasma cholesterol with aggregate cancer rates, expressed as cumulative mortality from cancer by age 65 in the absence of other causes of death, Peto et al. *(70)* observed that, "there was no evidence that those (counties) with particularly low plasma cholesterol concentrations had high cancer rates" (p. 1249). Indeed, when compared with cholesterol concentrations in Britain, these data suggest that at these much lower levels of plasma cholesterol in China, if anything, "... the opposite trend (i.e., high cholesterol, high cancer) was indicated."

More complete analyses of the associations of the various blood lipid (erythrocyte fatty acids and plasma triglycerides) and cholesterol constituents (plasma total, HDL, non-HDL, apo B, and apo A1) with coronary heart disease/myocardial infarction (CHD/MI) mortality rates are currently underway, or are under review for publication (Stamler, Chen, and Campbell, unpublished). Very briefly, preliminary conclusions show that the mean national mortality rate for CHD/MI in China is far lower than that for the United States. Annual age standardized rates, truncated for 35–64 yr of age, in China, are $11.5/10^5$ and $9.5/10^5$, but in the United States are $198/10^5$ and $56/10^5$ for males and females, respectively. In certain areas of China, this disease is exceedingly rare, being less than 2 cases/10^5/yr for people of middle age (35–64 yr of age) for four of the 65 counties, as recorded during the years 1973–1975 *(1)*. Whereas these very low mortality rates limited analytic sensitivity to detect potential determinants of disease, the range of mortality rates for the entire set of survey counties is sufficiently broad to partially offset this limitation.

Within China, CHD/MI is not significantly associated with plasma total cholesterol but is directly associated with apo B ($p < .01$), a cholesterol fraction somewhat equivalent to LDL cholesterol in its biological properties *(71)*. In turn, apo B is directly associated with animal protein intake ($p < .05$) and frequency of meat consumption ($p < .01$) but inversely with plant protein intake ($p < .01$) and selected complex carbohydrate intakes. In addition, there is a north–south geographic gradient in disease mortality rates, with more disease in the north. This tends to produce a puzzling positive association with the intakes of certain foods generally consumed in the north, such as wheat and carrots (the latter probably being an unstable correlation because of its relatively low intakes).

In another paper concerning the associations of protein intake with plasma lipids and other characteristics (Zhao, Campbell, Parpia, and Chen, unpublished), it was found that total protein intake in China is 64.1 g/d *(1)*, which amounts to 9.7% of total energy intake, with 10.8% of this protein being from foods of animal origin (including fish). Comparable data for the United States show 106 g/d, which amounts to 16% of total energy intake *(72)*, with 69% of this protein being from foods of animal origin *(73)*. Thus, when expressed as a percent of calorie intake, animal protein intake in the United States is more than 10 times that of China. Total cholesterol was shown to be inversely correlated with the intake of legumes ($p < .01$), an important source of plant protein, and positively correlated with the intake of meat ($p < .05$), a major source of animal protein. Non-HDL (LDL) cholesterol exhibits the strongest associations, being associated positively with animal protein intake ($p < .0001$) and inversely with plant protein intake ($p < .01$). That these relationships are this significant in view of the low intake of animal protein and low range of plasma cholesterol is a provocative finding. Similar associations are seen with dietary fat intake, although these are weaker, thus being similar to the human metabolic studies of Sirtori et al. *(74)*, who showed a stronger hyper-cholesterolemic effect for animal protein than for dietary fat. A rather surprising finding was observed for the relationship between dietary protein and adult body size. Plant protein intake showed a significant association with attained adult body size, whereas animal protein intake failed to exhibit this association. Thus, genetic potential for body size may be attained with a diet mostly comprised of plant protein, thus bringing into question the widely held view that animal protein best serves this purpose.

Investigation of the chief correlates of breast cancer has been approached from two perspectives, the first concerned with fat intake *(36)* and the second with circulating hormone concentrations *(34,35)*. The mean mortality rate for breast cancer in China, for women 0–64 yr of age, is only one-sixth of the British rate *(34)*, although this difference in mean national rates may narrow somewhat with older aged women, as is illustrated by comparing postmenopausal women between China and the United States *(36)*. In contrast, these differences may be even larger when comparing certain areas of China, since the mean rate for the highest quintile counties is 5.3 times greater than the mean rate for the lowest quintile counties *(36)*.

It may be more informative first to consider the role of hormones since they seem to reflect both dietary exposure, on the one hand, and disease outcome, on the other, acting therefore as a biochemical intermediary of dietary effects on disease risk. Pike et al. *(75)*, for example, found that sixfold lower breast cancer rates in Japan, when compared to the United States (although the gap now is closing *[76]*), were explained mostly by differences in age at menarche (Japanese girls were later) and postmenopausal weight, and suggested that this may have been a result of lower levels of circulating hormones in Japanese women during the premenopausal period. A number of studies *(77–80)*, reviewed by Key et al. *(34)*, have shown lower plasma estrone and estradiol concentrations among Oriental women when compared with Western women. On the dietary connection, Rose et al. *(81)* have found that a reduction of fat from 35 to 21% of energy for three months in premenopausal women with cystic breast disease and mastalgia causes a decrease in serum total estrogens, estrone, and estradiol concentrations. Experimental animal studies *(82,83)*, along with the possibility of similar interpretations of human studies *(83)*, have shown that an increase in dietary fiber intake alters circulating estrogen activity in a manner consistent with decreased mammary tumor formation. In summary, low fat, high fiber diets are associated with lower levels of female hormones and decreased risk for breast cancer, although the constituent(s) of the diet to which this should be attributed is not yet established.

In an elegant series of studies in the laboratory rat, in work spanning many years, Hawrylewicz and coworkers have shown that either a decrease in the intake of animal protein, or its replacement with plant protein during the growth period of the animal's life, inhibits the development of chemically induced mammary tumors *(84–86)*, elevates the estral surges of progesterone, estradiol, and prolactin *(87)*, increases mammary duct proliferation and morphologic development *(88)*, and accelerates body growth rate and shortens the time for sexual maturation *(84,88)*. These findings are impressive, first, because the protein requirements of laboratory rats *(89)* and humans are very similar *(90)*—thus the dietary protein levels used to produce these effects (8–24% of calories) are comparable to human intakes—and, second, because these observations are consistent with those for humans.

Thus it was of interest to begin exploration of the China data along these same lines, as has been done by Key and colleagues *(34,35)* and Marshall et al. *(36)*. As already noted, fat and animal protein intakes in rural China are exceptionally low, whereas dietary fiber intake is unusually high *(1)*, each of which ought to favor a lower risk of breast cancer. Marshall et al. *(36)*, after adjusting the Chinese data for age at menarche and body weight, found a statistically significant association of fat intake with breast cancer risk, a remarkable finding, given the very low range of fat intake in China (6–24% of calories). Moreover, this analysis showed, by different methods, that fat intake clearly is a stronger predictor of risk than total calorie intake. Although these results are not explained by calorie intake, they may be explained by other correlates of fat intake, such as the intake of animal protein ($r = 0.70$, $p < .0001$), plant protein ($r = -0.36$, $p < .01$), and/or dietary fiber ($r = -.37$, $p < .01$).* Unfortunately, the association of these constituents with breast cancer have not been analyzed as of this writing.

---

*Analysis designed to compare the relative contributions to risk of fat and protein intakes did not consider protein sources separately—an important oversight (on the part of the first author of this chapter), given the opposite correlations of dietary fat intake with plant and animal protein intakes.

Key et al. *(34)* compared hormone concentrations in age pooled samples from 3250 Chinese women residing in all 65 counties and from 300 British women of the same age, using the same analytic method and the same analyst. Mean estradiol concentrations among British women were higher by 36% (p = .04), 90% (p < .001), and 171% (p < .001) for the 35–44, 45–54, and 55–64 yr age groups. Mean testosterone concentrations, about 5% those of men, also were higher among the British women, perhaps related to their greater body weight *(35,91–93)*. Sex hormone binding globulin (SHBG) concentrations were similar in the 35–44 and 45–54 yr age groups, but were 15% lower (p = .002) among British women in the 55–64 yr age group, a finding consistent with earlier reports *(80,94,95)*. There also is some evidence that vegetarian women, perhaps analogous to Chinese women, have higher SHBG concentrations than nonvegetarian women *(96)*, although this was not reported in another study *(97)*. The explanation for such a finding is that higher SHBG levels bind more estrogen, thus leaving less for biological activity. Prolactin concentrations were not different. Chinese women also exhibited later age at menarche, earlier age at first birth, higher parity, and earlier age at menopause, when compared with British women *(1)*, and each of these factors have been associated with lower breast cancer incidence, as Key et al. *(34)* pointed out. Among the various hormones measured in this study, testosterone was found to be the best predictor of breast cancer risk *(35)*, particularly when using the measurements of the premenopausal women (p = .008 for 35–44 yr age group and p = .053 for the 45–54 yr age group), a finding in agreement with the observations of other studies *(98–102)*. Rather surprisingly, plasma estradiol concentrations were not related with breast cancer risk, perhaps because of the variation associated with the random menstrual cycles sampled in these women.

Key et al. *(34)* made the interesting observation that,

> The differences between the Chinese and British
> women in mean (estradiol) concentrations are con-
> sistent with an estrogen hypothesis for breast can-
> cer, and are large enough when taken in conjunction
> with the other risk factor differences to explain the
> *whole of the difference* (our emphasis) in breast can-
> cer rates (as originally put forth by Pike and col-
> leagues *[75,103]*).

We regard this conclusion, particularly since it refers
to a population on a very low fat, very high fiber diet, as
being especially salient to the debate on the dietary determinants of breast cancer among Western women (*see* next
chapter for further discussion).

Perhaps surprising to many observers, the Chinese
were found to consume one-third more energy/U body wt
than did counterpart Americans, yet showed 20% less mean
body mass, thus demonstrating a striking difference in
energy utilization and body fat deposition. That is, obesity is much less common in China than in the United
States, despite a substantial elevation of energy intake.
We propose two contributions to this effect: greater physical activity and consumption of a low fat, low animal protein diet, which should favor higher energy expenditure
(104). Quite remarkable in this regard is the high energy
intake for Chinese adult males who, when standardized
for a reference man, are, by definition, undertaking very
light physical activity (i.e., office type work).

Intakes of 14 complex carbohydrate and fiber fractions
were measured in this study, with the intent of determining whether particular fiber fractions were associated with
particular diseases, especially with cancer of the large
bowel. Average dietary fiber intake in China is about three
times higher than in the United States *(1)*, with one county
mean being as high as 77 g/d. As of this writing, only a

preliminary report of these data has been published *(105)*, with a more complete manuscript to be completed in the near future. However, based on an overview of the univariate correlations *(1)*, colon and rectal cancer mortality rates exhibited consistently inverse correlations with all fiber/complex carbohydrate fractions except for pectin, which was essentially zero. These relationships, however, appear to be rather weak because only rhamnose-containing complex carbohydrate intakes reached statistical significance for cancer of the colon ($r = .33, p < .05$). It would appear, then, that within the range of 7–77 g/d, where mean intakes of 29 of the 65 counties were above the upper US recommendation limit of 30–35 g/d *(42)*, there is evidence of a weak inverse relationship between cancer of the large bowel and the intakes of multiple complex carbohydrate and dietary fiber fractions.

There has been some concern that high intakes of dietary fiber, i.e., above 30–35 g/d, might compromise mineral status because of gastrointestinal chelation *(106–108)*. The only data thus far interpreted in this regard are those for iron status. Iron deficiency anemia generally has been regarded as one of the top three or four nutritional problems in developing countries *(109,110)*, perhaps owing in part to the elevated intakes of plant foods containing various fibrous materials, in part to intestinal parasitism, and in part to an inadequate intake of the more highly absorbable heme iron only available in animal foods *(111)*. Thus, on the basis of these considerations, it has been widely believed that individuals who consume a very low fat diet naturally high in fibrous chelating constituents and low in heme iron might be more disposed to iron deficiency. Based on these considerations, assessment of iron status in China was therefore of considerable interest, particularly in view of the very low, almost negligible, intake of heme iron, and the very high intake of iron chelation constituents of certain foods and the iron-binding polyphe-

nols of tea (108), a widely used beverage in China. To put
these observations on iron nutrition in perspective, it may
be helpful to note that in the most recent edition of the
RDA publication for iron in the United States (90), a level
of 15 mg/d was considered to be adequate to "provide a
sufficient margin of safety" to "cover the needs of essentially all the adult women ... except for those with the most
extreme menstrual losses" (p. 200).

Iron intake in China is surprisingly high, averaging
about 34 mg/d, especially when compared with the United
States intake of about 18–19 mg/d (1). Also, at least 95%
of the mean iron intake is in the nonheme form. Analysis
of iron status was undertaken by using six measures of
iron status:hemoglobin levels for 6500 individuals, iron
intakes calculated from the household survey of food
intake, iron intakes derived from direct analysis of food
samples collected at each survey site, and means of plasma
pools of ferritin, iron binding capacity, and iron. Data quality was judged to be very good because the biochemical
indicators of iron status (hemoglobin, plasma ferritin, and
plasma iron binding capacity) demonstrated highly significant self-correlations ($p < .001$) within county and between
sexes. An analysis of these data was undertaken by Beard
et al. (112,113) who concluded that iron status is not compromised by a mainly vegetarian Chinese diet and that
there is no suggestion that iron deficiency is a causal factor of anemia in this population. The fact that these results
were obtained for individuals who consumed large amounts
of tea polyphenols, along with intakes of dietary fiber in
excess of the upper recommended limits of intake in the
United States, makes all the more impressive the suggestion that iron status is not compromised by high fiber diets.
In fact, in China, the dietary fiber content of food is highly
correlated with its iron content ($r = .70$, $p < .001$). These
data therefore suggest that the perceived compromise of
iron status putatively caused by iron chelating materials

in foods and beverages, and the preponderance of the less absorbable nonheme iron in the diet did not materialize. Apparently, these hypothetical deficits are more than offset by the very high intakes of iron that accompany the consumption of relatively unprocessed plant foods as well as the consumption of other factors favoring iron digestion, absorption, and metabolism.

These data also have been examined for relationships between the intakes of various antioxidants typically found in foods of plant origin and degenerative disease mortality rates (33,114–116). Kneller et al. (33), using a stepwise linear regression model, found that for stomach cancer, the leading cause of cancer mortality in China, the most significant and consistent risk factors are lower intakes of green vegetables ($p < .003$ for males and $p < .006$ for females), higher intakes of eggs ($p < .0008$ for males and $p < .05$ for females), and higher intakes of salted vegetables ($p < .09$ for males and $p < .02$ for females). This is a particularly interesting finding because the protective effect of green vegetables on stomach cancer risk, also observed in several other studies (117–120), is undoubtedly owing to the natural antioxidants and other constituents present in these foods, which counterbalance the effects of the nitrosamines and other carcinogenic compounds (121,122) known to exist in salted vegetables. In other words, fresh native nutrients from these foods are protective, whereas artificial preservation procedures of the same foods, at least in this survey, appear to have more than offset these natural protective effects. The authors (33) suggested that improved transportation and refrigeration opportunities would "... permit substitution of fresh for salt-preserved foods" (p. 115) (further analysis of nitrosamine associations with this and other cancers, using more sophisticated and more recent data, are now under way).

Stepwise regression analysis of blood micronutrient associations with stomach cancer mortality rates produced results consistent with the above. Plasma selenium ($p <$ .002 for males and $p <$ .0001 for females) and β-carotene ($p <$ .08 for males, $p <$ .01 for females), both reflective of antioxidant status, are the most significant protective factors, whereas plasma albumin ($p <$ .07 for males and $p <$ .0002 for females) exhibit the most significant positive association. On the basis of Pearson correlation coefficients, plasma ascorbate also is inversely associated (inverse for both sexes and $p <$ .05 for males) with this cancer, whereas copper (positive for both sexes and $p <$ .05 for males) and ferritin (positive for both sexes and $p <$ .05 for females) are positively associated. These biochemical markers indicate that those factors that are associated with antioxidation (β-carotene, ascorbic acid and selenium) are inversely associated with stomach cancer while those which were associated with pro-oxidation or nutritionally richer diets (copper,* ferritin, and albumin**) are positively associated with this cancer. The only apparent exception is plasma α-tocopherol, which is positively associated with stomach cancer mortality rates for both sexes ($p <$ .10 for males). However, the plasma concentration of this lipid-soluble constituent is dependent on plasma lipid concentrations (associations with total cholesterol, $p <$ .01, non-HDL cholesterol, $p <$ .01, triglyceride, $p <$ .05, and apo A1, $p <$ .001, are all positive) and this must be taken into account when assessing an independent effect of this fat soluble vitamin. In a separate analysis by Chen et al. *(116)*, computation of partial coefficients showed that the positive associations of α-tocopherol with liver and lung cancer mortality rates become insignificant when this adjustment is made.

*Circulating copper, largely bound to albumin, is a pro-oxidant, causing lipid peroxidation, among other oxidations *(123,124)*.
**See* footnote on page 90.

In a separate study of a subset of 1882 individual plasma samples of men residing in 46 counties, Forman et al. *(19)* found that chronic infection with *Helicobacter pylori*, as indicated by *H. pylori* IgG antibodies, was associated with stomach cancer mortality ($2p < .02$), the only cancer showing a significant association. It has been reported that this organism can survive under the acidic conditions of the stomach *(125)*. Both a greater use of salted vegetables associated with a lack of refrigeration and a greater prevalence of *H. pylori* infection would be more common under poorer socioeconomic conditions, thus explaining why this cancer is more common in these areas. In contrast, it would appear to be fortunate that the intake of foods of plant origin, even though they may not be fresh, is greater in these areas, thus attenuating an already high risk environment. This illustrates the importance of taking into consideration as many factors as possible before concluding that consumption of foods of plant origin, as conventionally used in poor societies, causes an increased incidence of this cancer, as might be concluded from international correlation studies. The greater prevalence of this cancer in poor countries, therefore, is likely a result of the manner in which plant foods are preserved and prepared—as a consequence of limited resources and poor public health conditions, and not of the native constituents present in these plant foods.

Other investigations of these data for relationships of tissue antioxidant status and cancer mortality rates *(114–116)* showed that plasma ascorbate exhibits the strongest, most consistent, and most significant inverse associations with the various cancers, with the strength of these associations being substantially greater for males than for females. When plasma ascorbate was compared with mortality rates for all cancers, Pearson correlations were $r = -.43$ ($p < .001$) for males and $r = -.14$ (not significant) for females *(115)*.

Among the various cancers, plasma ascorbate exhibits the strongest inverse associations for esophageal cancer ($p <$ .001) *(114,115)*. With respect to the intakes of specific foods the association for this cancer is for fruit ($p < .01$ for both sexes) *(114)*. The cumulative stomach cancer mortality rate was 5.4-fold greater for males and 8.5-fold greater for females when the highest quartile of fruit intake was compared with the lowest quartile.

Plasma β-carotene, which is the second most consistent inverse association for the various cancers, exhibits an inverse association for stomach cancer independent of retinol *(116)*. In fact, the relationships of plasma retinol were singularly unimpressive, perhaps partly because of its several significantly positive associations with plasma lipid concentrations and partly because of its narrow, homeostatically controlled ranges of concentrations.

4. Commentary

This study possesses several unique features not available in most other studies on diet and disease, especially not in other cross-sectional studies. Other cross-sectional studies generally are considered to offer, at best, opportunities for the generation of new hypotheses, certainly not opportunities to provide reliable evidence for the "testing" of hypotheses. However, we propose that in addition to the organizational features discussed earlier, this study, in fact, offered some unusual advantages for generating hypotheses not normally available in the cross-sectional studies.

First, this study was unusually comprehensive, meaning that it included a wide variety of dietary, nutritional, clinical, lifestyle, and disease characteristics. This feature not only expanded opportunities to generate new hypotheses on rather specific diet and disease associations but also provided a fertile ground to investigate rather broad hypotheses. The hypothesis under investigation states that a diet comprised of a wide variety of good quality (fresh) foods of plant origin, perhaps including a diet entirely comprised of such foods with little or no added fat, minimizes the occurrence of degenerative diseases. This hypothesis can be, and should be, investigated by examining the associations of many different constituents of these foods on the development of the many different diseases of tissue degeneration. Thus, to accomplish these objectives, this survey documented as many characteristics as possible, simply to increase the opportunities for investigation of as many relevant associations as possible. The larger is the number of associations supporting the hypothesis, the greater the reliability of concluding that the hypothesis is true; thus, this study, in its comprehensiveness, offers a *test* of the hypothesis.

Second, the study included a population that experiences dietary and disease characteristics that are unique when compared with the far more commonly studied Western populations. For example, plasma cholesterol concentrations are much lower in rural China, dietary fiber intakes are much higher, animal protein intake is far lower, and cereal grain consumption is much higher, just to name a few of the more interesting differences. These ranges of exposure, thus being substantially different, offered opportunities to investigate associations beyond the "dose–response" curves traditionally observed in Western populations. This should be an especially valuable asset when it is recognized that these ranges are, more or less, those to which recent dietary recommendations in Western countries are directed. Thus they must be investigated. Moreover, the consequences of these dietary recommendations should be investigated in a population that is accustomed to the hypothetically desired ranges of nutrient intakes and associated biochemical markers, because nutritional and dietary adaptation oftentimes may take many

months, even years, for optimal nutrient activities to be
fully expressed. In other words, it is not enough to limit
investigation of the consequences of dietary recommendations to the wrong range of exposures or to rely on experimental subjects who have had only a relatively short time
to become accustomed to their new diet.

These Oriental vs Occidental differences also offered
opportunities for discovery of intriguing associations that
cause questioning of existing assumptions. The fact that
calorie intake is so much higher in China, yet body weight
is so much lower, strikes a response among those who
assume that an increase in calorie consumption will cause
a gain in body weight and, perhaps, its disease sequelae.
Similarly, the fact that higher energy consumption is associated with lower risk for cancer, particularly those cancers that are common to Western societies and that are
thought to be particularly responsive to calorie intake,
strikes at another hard and fast assumption in Western
societies. It is likely that this association between energy
intake and energy disposition cannot be attributed to the
extra physical activity of the Chinese. Can dietary composition, especially at the Chinese end of the dietary spectrum, also affect energy disposition and its sequelae of
adverse effects, perhaps through increased energy expenditure? Some observations (126) on the use of low protein,
low fat diets, for example, would support this contention.
Another example that caused us to question existing
assumptions was that concerned with the recommended
intakes of the B vitamin, riboflavin. The discovery that,
by Western standards, recommended intakes of this micronutrient may not apply in China, indeed may not even
apply in the United States, was prompted by an analysis
of some surprising results from China (18), a subject not
discussed in this paper. The observation that plasma levels of β-carotene in China are somewhat below Western
subjects, even though intakes are higher, brings into ques-

tion other factors that should be considered when assessing the activities of this micronutrient. And the *positive* correlations of plasma α-tocopherol with the mortality rates for some of the degenerative diseases was, perhaps surprising, until it was recognized that this fat soluble vitamin was associated with circulating lipids.

These Occidental vs Oriental differences also offer some broader-ranging insights into what each society may learn from the other. The West learns from China what health value there is in consuming a low fat, mostly plant-based diet (*see* the brief summary that follows). China learns what is likely to happen when and if they emulate too closely what the West has done with its diet during the past century or so—producing rapidly rising health care costs while overconsuming the world's natural resources. Both societies learn, perhaps once again, the value of using fresh foods, therefore the value of developing healthful systems to improve food storage and distribution. The finding that salted vegetables are associated with a greater risk for stomach cancer, but that fresh vegetables are associated with less risk for this cancer is a case in point (*33*). Thus, these differences have begun to offer some fascinating insights into some rather rigid assumptions that may need challenge.

The third uniquely valuable feature of this study was its statistical sensitivity when compared to previous correlation studies obtained by including a relatively large number of highly diverse survey sites, perhaps even achieving enough sensitivity to obtain important evidence to seriously question previous hypotheses. For example, aflatoxin, previously regarded by many as a human carcinogen with "sufficient" evidence (*52*), was found not to be associated with primary liver cancer, a surprising observation to many, particularly to those whose career interests rely on the status quo. Yet the number of sites used in this analysis was far greater than all previous studies on aflatoxin and liver cancer combined. The absence of a significant association, although not proving the negative, provides sufficient evidence to question the significance of its alleged human carcinogenicity, especially when this human evidence coincides with experimental animal evidence. This observation also could be placed within the context of other possible causes measured in this study, thus prompting the hypothesis that this disease largely may be the result of a viral-nutrition link and not a viral-carcinogen link or even a carcinogen-based disease. And there were other fundamental lessons to be suggested by this example, such as raising the question of the extent to which small amounts of putative chemical carcinogens really matter, especially when operating within a population of heterogeneous nutritional experiences that can be used as a *primary* means of controlling disease. And there are other examples where the relative statistical sensitivity of this survey may have provided important new information. The discovery of several diet–disease associations near one end of the dietary spectrum, where they otherwise might have been lost, is a case in point.

A fourth feature of this study is the most obvious, namely the opportunity to generate new hypotheses, most of which undoubtedly remain to be formulated, interpreted, and published. For example, we find particularly interesting the significant association between the prevalence of Herpes Simplex II antibodies and CHD mortality rates, particularly since there is supporting biochemical evidence for this association (J. Shih, personal communication); the association of plasma copper, a strong pro-oxidant, with cancer rates; the puzzling association of circulating albumin levels with total aggregate cancer rates (highly significant, $p < .001$, $r = 0.50$), the highly significant association between rectal cancer mortality and schistosomiasis prevalence ($r = .88$, $p < .001$), and the highly significant ($p < .01$) but undoubtedly surprising association between tissue levels of ω-3 fatty acids with aggregate childhood cancer (the opposite is observed for the ω-6 fatty acids). These are but a few of the possible hypotheses that need further investigation.

The chief purpose of this paper is to assess the information thus far interpreted from this survey so as to determine the many ways that a diet rich in plant foods relates to degenerative disease outcomes. The diet of this population is very rich in foods of plant origin, since about 90% of the total protein, for example, is provided by plants. It is a population that experiences very little CHD and certain major cancers (breast and large bowel) thought to be related to nutritionally rich* diets typically used in wealthy countries. It is a population of individuals who exhibit, by Western standards, very low levels of plasma cholesterol, a reasonably consistent biomarker of nutritional richness or extravagance. And finally, it is a diet that, on average, contains only 14.5% of its energy as fat spread over a very low range of 6–24% (county averages). Thus, this diet, very low in fat and very high in dietary fiber (mean of 33.3 g/d), is considerably beyond what is called for in the dietary recommendations put forth in Western countries in recent years to avoid degenerative diseases. These ranges approach the limits of dietary and disease possibilities, perhaps being beyond the thresholds where no further diet–disease associations are possible. From the many perspectives thus far explored for this hypothesis, no obvious thresholds were observed. That is, the richer the diet is in native nutrients provided by plant matter, the greater are the reductions of degenerative diseases and the more beneficial are their biochemical indicators.

Two of the best known antioxidant micronutrients, the water soluble vitamin C and the fat soluble β-carotene, when elevated in plasma, are associated with lower rates

*See footnote on page 78.

for several cancers. Also, selenium, whose concentration
in foods is mostly determined by soil conditions and that
probably expresses its antioxidant effects indirectly, is
inversely associated with some cancers. In general, antioxidant nutrients, mostly or entirely supplied by foods of
plant origin, prevent development of degenerative diseases
that are promoted, in turn, by pro-oxidant conditions or
constituents found in diets enriched in foods of animal origin. Breast cancer is positively related with dietary fat
intake, after controlling for body size and age at menarche.
This cancer also was found to be most reliably predicted
by female testosterone (even though concentrations are
very low), which is highly significantly associated with
plasma cholesterol (at the higher, more statistically sensitive male concentrations). In turn, plasma cholesterol is
significantly positively associated with the intakes of meat,
animal protein ($p = .06$), and fat, and inversely associated
with the intakes of plant protein, various fiber constituents, and legumes. Animal protein intake, even though very
low in China, is specifically associated with higher levels
of several of the plasma cholesterol moieties, whereas plant
protein intake is associated with lower levels of these cholesterol constituents. CHD, although not correlated with
total cholesterol, was correlated with apo B cholesterol.
When the aggregates of disease mortality rates were examined, the diseases typically found among Western countries, "diseases of extravagance," tend to occur in the same
geographic areas, namely the areas near urban centers
where industrialization and acquisition of capital have led
to higher rates of literacy and, apparently, to the age-old
belief that with more personal resources, it is time to enrich
the diet with more fat and animal foods—just like Western societies did a century or so ago.

In slightly more sophisticated analyses designed to
isolate and compare the contributions of different dietary
components to disease risk, again there is evidence that

nutritional richness, i.e., elevated plasma cholesterol levels, among other markers, favor disease development. Liver cancer, long known to be much more common among the world's poorer societies, which mostly consume foods of plant origin and which have low plasma cholesterol levels, was actually shown to be highly significantly *positively* associated with the plasma cholesterol level, and in various multiple regression model specifications, to exhibit associations that were even slightly more significant than those for chronic infection with HBV. In this analysis, aflatoxin, commonly consumed at higher levels in poorer societies, was not associated with this disease. Although this nonassociation may reflect inadequate assay methodology, or an inability to correlate past exposures with present exposures, this finding agrees with our experimental animal data showing that the development of this cancer strongly depends on the same nutritional link that also enhances the development of other cancers. Thus we propose that liver cancer primarily results from the consumption of nutritionally rich diets by HBV chronically infected people who may be more sensitive than noninfected people to the hazards of nutritional richness. Intervention with aflatoxin, or other mutagenic agents, is not necessary for initiation because constant exposure to HBV activity and a large number and variety of other initiating agents, are capable of doing the same thing. In the final analysis, nutrition becomes the controlling factor for disease formation. On stomach cancer, an analogous phenomenon occurs. Poor countries experience much more of this disease, not because they consume plant-rich diets and have relatively low plasma cholesterol levels, but because they rely on the consumption of foods preserved with agents that initiate and/or promote the formation of this disease. If, in contrast, the *native* nutrient components of fresh foods are examined, then the consumption of plant foods is associated with lower rates of this cancer. Also, infection with a relatively little known bacterium, *Heliocobacter pylori*, appears to enhance the development of this disease. Thus, the over-arching chief message from these data is to consume fresh plant foods, not those that require large amounts of chemical salts and other preservatives or, perhaps, encourage infection with this and other disease enhancing organisms. Thus less industrialized societies experience more stomach cancer not because they eat plant rich diets, but because they do not have adequate resources to use *fresh, high quality* plant foods. Indeed, it is fortunate that the fortuitous consumption of these foods probably limits an otherwise explosive epidemic of disease occurrence.

These are but some of the observations that comprehensively indicate the beneficial nutritional effects of foods of plant origin. And there are many more yet to be interpreted both using the 1983 survey data set and especially using a new enlarged data set from the 1989 survey. The truly remarkable finding from these 1983 survey data is the repeated demonstration that statistically significant associations favoring the plant foods hypothesis were found, even though consumption of foods of plant origin already is exceptionally high, whereas consumption of foods of animal origin is exceptionally low. It is as if even a *small* intake of "animal foods," which begin to alter the intake of countless nutrients and other constituents, is capable of significantly elevating plasma cholesterol and similar biomarkers, thence to elevate the development of degenerative diseases.

In an analysis of this rather remarkable relationship, however, it is important to re-emphasize the confounding and countervailing effects sometimes accompanying the intakes of "plant foods," those that occur because these foods have not been preserved in a way to retain the nutritional value of their native constituents. The biological associations that illustrate this relationship are multifaceted, oftentimes extraordinarily dynamic and complex, and certainly are far more than the few so far interpreted from this data set. Nonetheless, of those which are presented here, the number of statistically significant associations existing near the end of the dietary spectrum of plant foods richness is relatively large. The number and diversity of these observations are sufficient, in our view, to warrant the overall conclusion that the evidence for the plant foods hypothesis is substantial, perhaps even overwhelming.

Two caveats must be noted, however. First, if it is true that a diet entirely comprised of foods of plant origin is the optimum for prevention of degenerative diseases—at least for all who experience even the slightest predisposition for these diseases, and at least three-fourths of all Westerners must do so since they die of these diseases—then are there any particularly significant adverse effects to be expected? We have not yet considered this question in any particular depth, but would note that, if iron nutrition and prevalence of anemia, one of the more sensitive indicators of inadequate nutrition, is any indication, then there may be few if any adverse conditions to be expected, if—and this is important—the quality of these foods is sufficiently good so as not to introduce infectious and chemical toxicities and if there is sufficient food variety to insure adequate nutrient intakes. Nutritional adequacy achieved by food diversity is obviously important so as not to incur elevated risks from the infectious and classical nutrient deficiency diseases (beriberi, pellagra, scurvy, and so on).

The second caveat concerns the possible bias introduced in analyzing these data. We have found a substantial number of highly significant associations supporting the use of a diet rich in foods of plant origin, perhaps even supporting the use of a diet entirely comprised of these foods, but we acknowledge that among the 5000+ significant correlations in this data set, there are occasional ones that seem to run in the opposite direction and therefore remain unexplained. Perhaps these are there by chance, perhaps they need more in-depth statistical analysis to determine if they are surrogates for other factors, and perhaps they are real and need rationalization. Future investigation of these data, along with the much larger set from the more recent 1989 survey, should help to clarify these few discrepancies.

Acknowledgments

This research was supported in part by NIH Grant 5RO1 CA33638, the Chinese Academy of Preventive Medicine, the United Kingdom Imperial Cancer Research Fund, the United States Food and Drug Administration, the American Institute for Cancer Research, and several American industry groups.

References

1. Chen J, Campbell TC, Li J, Peto R. *Diet, Life-Style and Mortality in China. A Study of the Characteristics of 65 Chinese Counties*. Oxford, UK; Ithaca, NY; Beijing, PRC: Oxford University Press; Cornell University Press; People's Medical Publishing House, 1990.
2. Trowell HC, Burkitt DP. *Western Diseases: Their Emergence and Prevention*. London: Edward Arnold, 1981.
3. Dumont R. *Mes combats*. Paris: Plon, 1989.
4. Walshe WH. *The Nature and Treatment of Cancer*. London: Taylor and Walton, 1846.
5. O'Connor TP, Campbell TC. Scientific evidence and explicit health claims in food advertisements. *J Nutr Ed* 1988; **20:** 87–92.
6. American Cancer Society. *Cancer Facts & Figures—1989*. Atlanta, GA: American Cancer Society, Inc., 1989.
7. Kinlen L. Fat and cancer. *Br Med J* 1983; **286:**1081,1082.
8. Willett W. *Nutritional Epidemiology*. New York: Oxford University Press, 1990.
9. Williams WR. *The Natural History of Cancer, with Special References to Its Causation and Prevention*. London: William Heinemann, 1908.

10. Staszewski J, Haenszel W. Cancer mortality among the Polish-born in the United States. *J Natl Cancer Inst* 1965; **35**:291–297.
11. Haenszel W, Kurihara M, Seig M, Lee RKC. Stomach cancer among Japanese in Hawaii. *J Natl Cancer Inst* 1972; **49**: 968–988.
12. Buell P. Changing incidence of breast cancer in Japanese-American women. *J Natl Cancer Inst* 1973; **51**:1479–1483.
13. Higginson J, Muir CS. Environmental carcinogenesis: misconceptions and limitations to cancer control. *J Natl Cancer Inst* 1979; **63**:1291–1298.
14. Bertram J, Kolonel LN, Meyskens FL, Jr. Rationale and strategies for chemoprevention of cancer in humans. *Cancer Res* 1987; **47**:3012–3031.
15. Population Census Offices. *Atlas of China.* Beijing: State Council of the People's Republic of China and Institute of Geography of the Chinese Academy of Science, 1987.
16. Li J-Y, Liu B-Q, Li G-Y, Chen Z-J, Sun X-D, Rong S-D. Atlas of cancer mortality in the People's Republic of China. An aid for cancer control and research. *Int J Epidemiol* 1981; **10**:127–133.
17. Committee on Diet Nutrition and Cancer. *Diet, Nutrition and Cancer.* Washington, DC: National Academy Press, 1982.
18. Campbell TC, Brun T, Chen J, Feng Z, Parpia B. Questioning riboflavin recommendations on the basis of a survey in China. *Am J Clin Nutr* 1990; **51**:436–445.
19. Forman D, Sitas F, Newell DG, Stacey AR, Boreham J, Peto R, Campbell TC, Li J, Chen J. Geographic association of *Helicobacter pylori* antibody prevalence and gastric cancer mortality in rural China. *Int J Cancer* 1990; **46**:608–611.
20. Wenxun F, Parker R, Parpia B, Qu Y, Cassano P, Crawford M, Leyton J, Tian J, Li J, Chen J, Campbell TC. Erythrocyte fatty acids, plasma lipids, and cardiovascular disease in rural China. *Am J Clin Nutr* 1990; **52**:1027–1036.
21. Campbell TC, Chen J, Liu C, Li J, Parpia B. Non-association of aflatoxin with primary liver cancer in a cross-sectional ecologic survey in the People's Republic of China. *Cancer Res* 1990; **50**:6882–6893.
22. Berge P, Parpia B, Chen J, Peto R, Campbell C, Armstrong D. *Candida albicans, Nutritional Factors and Nasopharyngeal Cancer in The People's Republic of China.* Symposium Advances in Clinical Nutrition, Thirty-First Annual Meetings of the American College of Nutrition. Albuquerque, New Mexico: 1990.

23. Mackay J. *China's Landmark Tobacco Law, a Memorandum*. Asian Consultancy on Tobacco Control, Kowloon, Hong Kong, July 31, 1991.

24. Campbell TC, Chen J, Brun T, Parpia B, Qu Y, Chen C, Gerssler C. China: from diseases of poverty to diseases of affluence. Policy implications of the epidemiological transition. *Ecol Food Nutr* 1992; **27:**133–144.

25. Chen J, Campbell TC. More meat does not mean better health. *World Health Forum* 1991; **12:**262–264.

26. Chen X, Chen J. Changes in Chinese diet a sign of the times. *China Daily*. Beijing, PRC. June 30, 1992; 4.

27. Knight J, Song L. The length of life and the standard of living: economic influences on premature death in China. *J Devel Stud* 1993; (in press).

28. American Heart Association. Committee report, rationale for the diet-heart statement of the American Heart Association. *Circulation* 1982; **65:**839A–854A.

29. United States Department of Health and Human Services. *The Surgeon General's Report on Nutrition and Health*. Washington, DC: Superintendent of Documents, US Government Printing Office, 1988.

30. National Research Council, Committee on Diet and Health. *Diet and Health: Implications for Reducing Chronic Disease Risk*. Washington, DC: National Academy Press, 1989.

31. Peto R. *Death from Cigarettes*. Memorandum submitted to Chinese Ministry of Health, University of Oxford, Oxford, UK, 1986.

32. Hsing AW, Guo W, Chen J, Li J, Stone BJ, Blot WJ, Fraumeni JF, Jr. Correlates of liver cancer in China. *Int J Epidemiol* 1990; **20:**54–59.

33. Kneller RW, Guo W-D, Hsing AW, Chen J, Blot WJ, Li J, Forman D, Fraumeni JF, Jr. Risk factors for stomach cancer in sixty-five Chinese counties. *Cancer Epi Biomarkers Prev* 1992; **1:**113–118.

34. Key TJA, Chen J, Wang DY, Pike MC, Boreham J. Sex hormones in women in rural China and in Britain. *Br J Cancer* 1990; **62:**631–636.

35. Wang DY, Key TJA, Pike MC, Boreham J, Chen J. Serum hormone levels in British and rural Chinese females. *Br Cancer Res Treatment* 1991; **18:**S41–S45.

36. Marshall JR, Qu Y, Chen J, Parpia B, Campbell TC. Additional ecologic evidence: lipids and breast cancer mortality

among women age 55 and over in China. *Eur J Cancer* 1991; **28A:**1720–1727.

37. Chen X, Campbell TC. *Sixty-Five County Reports on Public Health Prevention Strategies.* Institute of Nutrition and Food Hygiene, Chinese Academy of Preventive Medicine, 1989.

38. Campbell TC. The role of nutrition in the aetiology of cancer and other degenerative diseases. In: Chen J, Campbell TC, Li J, Peto R, eds. *Diet, Life-Style and Mortality in China.* Oxford, UK; Ithaca, NY; Beijing, PRC: Oxford University Press; Cornell University Press; People's Medical Publishing House, 1990; 54–66.

39. Department of Health and Human Services. *Dietary Intake Source Data: United States 1976–1980.* Hyattsville, MD: National Health Survey, 1983.

40. National Center for Health Statistics. *Weight by Height and Age for Adults 18–75 Years.* Publication no. PHS 79-1656. Hyattsville, MD: US Department of Health, Education and Welfare, 1979.

41. Zhao X, Campbell TC, Parpia B, Chen J. Evaluation of dietary protein status in China. *J Nutr* 1992 (submitted).

42. Butrum RR, Clifford CK, Lanza E. NCI dietary guidelines: a rationale. *Am J Clin Nutr* 1988; **48:**888–895.

43. Tao S, Huang Z, Wu X, Zhao B, Xiao Z, Hao J, Li Y, Cen R, Rao X. CHD and its risk factors in the People's Republic of China. *Int J Epidemiol* 1989; **18:**S159–S163.

44. Cambien F, Ducimetiere P, Richard J. Total serum cholesterol and cancer mortality in a middle-aged male population. *Am J Epidemiol* 1980; **112:**388–394.

45. Holme I. An analysis of randomized trials evaluating the effect of cholesterol reduction on total mortality and coronary heart disease incidence. *Circulation* 1990; **82:**1916–1924.

46. Oliver MF. Serum cholesterol—the knave of hearts and the joker. *Lancet* 1981; **ii:**1090–1095.

47. Goldbourt U, Holtzman E, Nuefeld HN. Total and high-density lipoprotein cholesterol in the serum and risk of mortality: evidence of a threshold effect. *Br Med J* 1985; **290:**1239–1243.

48. Taylor WC, Pass TM, Shepard DS, Komaroff AL. Cholesterol reduction and life expectancy: a model incorporating multiple risk factors. *Ann Intern Med* 1987; **106:**605–614.

49. International Collaborative Group. Circulating cholesterol level and risk of death from cancer in men aged 40 to 69 years.

Experience of an international collaborative group. *JAMA* 1982; **248**:2853–2859.

50. Knekt P, Reunanen A, Aromaa A, Heliovarra M, Hakulinen T, Hakama M. Serum cholesterol and risk of cancer in a cohort of 39,000 men and women. *J Clin Epidemiol* 1988; **41**:519–530.

51. Tornberg SA, Lars-Erik H, Carstensen JM, Eklund GA. Cancer incidence and cancer mortality in relation to serum cholesterol. *J Natl Cancer Inst* 1989; **81**:1917–1921.

52. International Agency for Research on Cancer. *Aflatoxins*. Lyon, France: International Agency for Research on Cancer, 1987.

53. Appleton BS, Campbell TC. Effect of high and low dietary protein on the dosing and postdosing periods of aflatoxin B1-induced hepatic preneoplastic lesion development in the rat. *Cancer Res* 1983; **43**:2150–2154.

54. Dunaif GE, Campbell TC. Relative contribution of dietary protein level and Aflatoxin B1 dose in generation of presumptive preneoplastic foci in rat liver. *J Natl Cancer Inst* 1987; **78**:365–369.

55. Youngman LD, Campbell TC. Inhibition of aflatoxin B1-induced gamma-glutamyl transpeptidase positive (GGT+) hepatic preneoplastic foci and tumors by low protein diets: evidence that altered GGT+ foci indicate neoplastic potential. *Carcinogenesis* 1992; **13**:1607–1613.

56. Chen Z, Keech A, Collins R, Slavin B, Chen J, Campbell TC, Peto R. Prolonged infection with hepatitis B virus: a factor contributing to the association between low blood cholesterol and liver cancer. *Br Med J* 1993; **306**:890–894.

57. Sherwin RW, Wentworth DN, Cutler JA, Hulley SB, Kuller LH, Stamler J. Serum cholesterol levels and cancer mortality in 361 662 men screened for the multiple risk factor intervention trial. *JAMA* 1987; **257**:943–948.

58. Schatzkin A, Hoover RN, Taylor RR, Ziegler R, Carter L, Larson DB. Serum cholesterol and cancer in the NHANES I epidemiologic follow-up study. *Lancet* 1987; **ii**:298–301.

59. Isles CG, Hole FJ, Gillis CR, Hawthorne VW, Lever AF. Plasma cholesterol, coronary heart disease, and cancer in the Renfrew and Paisley survey. *Br Med J* 1989; **298**:920–924.

60. Smith GD, Shipley M, Marmot MG, Rose G. Plasma cholesterol concentration and mortality—The Whitehall Study. *JAMA* 1992; **267**:70–76.

61. Neaton JD, Blackburn H, Jacobs D, Kuller L, Lee D.-J., Sherwin R, Shih J, Stamler J, Wentworth D. Serum choles-

terol level and mortality: findings for men screened in the Multiple Risk Factor Intervention Trial. *Arch Intern Med* 1992; **152:**1490–1500.

62. Potischman N, McCulloch C, Byers T, Houghton L, Nemoto T, Graham S, Campbell TC. Associations between breast cancer, triglycerides and cholesterol. *Nutr Cancer* 1991; **15:**205–216.

63. Rose G, Blackburn H, Keys A. Colon cancer and blood cholesterol. *Lancet* 1974; **i:**181–183.

64. Kagan A, McGee DL, Yano K, Rhoads GG, Nomura A. Serum cholesterol and mortality in a Japanese-American population. *Am J Epidemiol* 1981; **114:**11–20.

65. Kark JD, Smith AH, Hames CG. The relationship of serum cholesterol to the incidence of cancer in Evans County, Georgia. *J Chronic Dis* 1980; **33:**311–322.

66. Stemmermann GN, Nomura AMY, Heilbrun LK, Pollack ES, Kagan A. Serum cholesterol and colon cancer incidence in Hawaiian Japanese men. *J Natl Cancer Inst* 1981; **67:**1179–1182.

67. McMichael AJ, Jensen OM, Parkin DM, Zaridze DG. Dietary and endogenous cholesterol and human cancer. *Epidemiol Revs* 1984; **6:**192–216.

68. Rose G, Shipley MJ. Plasma lipids and mortality, a source of error. *Lancet* 1980; **i:**523–526.

69. Taylor YSM, Scrimshaw NS, Young VR. The relationship between serum urea levels and dietary nitrogen utilization in young men. *Br J Nutr* 1974; **32:**407–411.

70. Peto R, Boreham J, Chen J, Li J, Campbell TC, Brun T. Plasma cholesterol, coronary heart disease, and cancer. *Br Med J* 1989; **298:**1249.

71. Mahley RW, Innerarity TL, Rall SCJ. Plasma lipoproteins: apolipoprotein structure and function. *J Lipid Res* 1984; **25:**1277–1294.

72. US Department of Health and Human Services and US Department of Agriculture. *Nutrition Monitoring in the United States: A Progress Report from the Joint Nutrition Monitoring Evaluation Committee.* DHHS publication no. (PHS) 86-1255. Hyattsville, MD: National Center for Health Statistics, 1986.

73. Food and Agriculture Organization. *Food Balance Sheets, 1961–65 Average 1967 to 1977.* Rome, Italy: FAO, 1980.

74. Sirtori CR, Noseda G, Descovich GC. Studies on the use of a soybean protein diet for the management of human hyperlipoproteinemias. In: Gibney MJ, Kritchevsky D, eds. *Animal and Vegetable Proteins in Lipid Metabolism and Atherosclerosis.* New York: Liss, 1983; 135–148.

75. Pike MC, Krailo MD, Henderson BE, Casagrande JT, Hoel DG. 'Hormonal' risk factors, 'breast tissue age' and the age-incidence of breast cancer. *Nature* 1983; **303:**767–770.

76. MacMahon B. Incidence trends in North America, Japan, and Hawaii. In: Magnus K, ed. *Trends in Cancer Incidence*. New York: McGraw-Hill, 1982; 249-62.

77. Hayward JL, Greenwood FC, Glober G, Stemmeman G, Bulbrook RD, Wang DY, Kumaokas S. Endocrine status in normal British, Japanese and Hawaiian-Japanese women. *Eur J Cancer* 1978; **14:**1221–1228.

78. Gray GE, Pike MC, Hirayama T, Tellez J, Gerkins V, Brown JB, Casagrande JT, Henderson BE. Diet and hormone profiles in teenage girls in four countries at different risk for breast cancer. *Prev Med* 1982; **11:**108–113.

79. Goldin BR, Adlercreutz H, Gorbach SL, Woods MN, Dwyer JT, Conlon T, Bohn E, Gershoff SN. The relationships between estrogen levels and diets of Caucasian American and Oriental immigrant women. *Am J Clin Nutr* 1986; **44:**945–953.

80. Bernstein L, Yuan JM, Ross RK, Pike MC, Hanisch R, Lobo R, Stanczyk F, Gao YT, Henderson BE. Serum hormone levels in pre-menopausal Chinese women in Shanghai and white women in Los Angeles: results from two breast cancer case-control studies. *Cancer Causes Control* 1990; **1:**51–58.

81. Rose DP, Boyar AP, Cohen L, Strong LE. Effect of a low-fat diet on hormone levels in women with cystic breast disease. I. Serum steroids and gonadotropins. *J Natl Cancer Inst* 1987; **78:**623–626.

82. Schultz TD, Howie BJ. In vitro binding of steroid hormones by natural and purified fibers. *Nutr Cancer* 1986; **8:**141–147.

83. Rose DP. Dietary fiber and breast cancer. *Nutr Cancer* 1990; **13:**1–8.

84. Hawrylewicz EJ, Huang HH, Kissane JQ, Drab EA. Enhancement of the 7,12-dimethylbenz(a)anthracene (DMBA) mammary tumorigenesis by high dietary protein in rats. *Nutr Reps Int* 1982; **26:**793–806.

85. Hawrylewicz EJ, Huang HH, Liu J. Dietary protein enhancement of *N*-nitroso-methylurea-induced mammary carcinogenesis, and their effect on hormone regulation in rats. *Cancer Res* 1986; **46:**4395–4399.

86. Hawrylewicz EJ, Huang HH, Blair WH. Dietary soybean isolate and methionine supplementation affect mammary tumor progression in rats. *J Nutr* 1991; **121:**1693–1698.

87. Huang HH, Hawrylewicz EJ, Kissane JQ, Drab EA. Effect of protein diet on release of prolactin and ovarian steroids in female rats. *Nutr Rpts Int* 1982; **26:**807–820.

88. Sanz MCA, Liu J-M, Huang HH, Hawrylewicz EJ. Effect of dietary protein on morphoplogic development of rat mammary gland. *J Natl Cancer Inst* 1986; **77:**477–487.

89. Subcommittee on Laboratory Animal Nutrition. *Nutrient Requirements of Laboratory Animals.* Second revised edition, number 10. Washington, DC: National Academy Press, 1972.

90. National Research Council. *Recommended Dietary Allowances.* Tenth edition. Washington, DC: National Academy Press, 1989.

91. Kopelman PG, Pilkington TRE, White N, Jeffcoate SL. Abnormal sex steroid secretion and binding in massively obese women. *Clin Endocrinol* 1980; **12:** 363–370.

92. Bates GW, Whitworth NS. Effects of obesity on sex steroid metabolism. *J Chron Dis* 1982; **35:**893–896.

93. Wild RA, Umstot ES, Andersen RN, Ranney GB, Givens JR. Androgen parameters and their correlation with body weight in one hundred thirty-eight women thought to have hyperandrogenism. *Am J Obstet Gynecol* 1983; **146:** 602–606.

94. Moore JW, Clark GMG, Takatani O, Wakabayashi Y, Hayward JL, Bulbrook RD. Distribution of 17 β-estradiol in the sera of normal British and Japanese women. *J Natl Cancer Inst* 1983; **71:** 749–754.

95. Moore JW, Key TJA, Bulbrook RD, Clark GMG, Allen DS, Wang DY, Pike MC. Sex hormone binding globulin and risk factors for breast cancer in a population of normal women who had never used exogenous sex hormones. *Br J Cancer* 1987; **56:**661–666.

96. Armstrong BK, Brown JB, Clarke HT, Crooke DK, Hähnel R, Masarei JR, Ratajczak T. Diet and reproductive hormones: a study of vegetarian and nonvegetarian postmenopausal women. *J Natl Cancer Inst* 1981; **67:**761–767.

97. Fentiman IS, Caleffi M, Wang DY, Hampson SJ, Hoare SA, Clark GM, Moore JW, Bruning P, Bonfrer JM. The binding of blood-borne estrogens in normal vegetarian and omnivorous women and the risk of breast cancer. *Nutr Cancer* 1988; **11:**101–106.

98. McFadyen IJ, Forrest APM, Prescott RJ, Golder MP, Fahmy DR, Griffith SK. Circulating hormone concentrations in women with breast cancer. *Lancet* 1976; **i(May 22, 1976):** 1100–1102.

99. Hill P, Garbaczewski L, Kasumi F. Plasma testosterone and breast cancer. *Eur J Cancer Clin Oncol* 1985; **21**:1265,1266.

100. Secreto G, Recchione C, Cavalleri A, Miraglia M, Dati V. Circulating levels of testosterone, 17β-oestradiol, luteinizing hormone, and prolactin in postmenopausal breast cancer patients. *Br J Cancer* 1983; **47**:269–275.

101. Secreto G, Recchione C, Fariselli G, Di Pietro S. High testosterone and low progesterone circulating levels in premenopausal patients with hyperplasia and cancer of the breast. *Cancer Res* 1984; **44**:841–844.

102. Secreto G, Toniolo P, Pisani P, Recchione C, Cavalleri A, Fariselli G, Totis A, DiPietro S, Berrino F. Androgens and breast cancer in premenopausal women. *Cancer Res* 1989; **49**:471–476.

103. Pike MC. Reducing breast cancer in women through lifestyle-mediated changes in hormone levels. *Cancer Detect Prev* 1990; **14**:595–607.

104. Rothwell NJ, Stock MJ. Regulation of energy balance. *Ann Rev Nutr* 1981; **1**:235–256.

105. Campbell TC, Wang G, Chen J, Robertson J, Chao Z, Parpia B. Dietary fiber intake and colon cancer mortality in The People's Republic of China. In: Kritchevsky D, Bonfield C, Anderson JW, eds. *Dietary Fiber*. New York: Plenum Publishing Corporation, 1990; 473–480.

106. Monsen ER, Hallberg L, Layrisse M, Hegsted DM, Cook JD, Mertz W, Finch CA. Estimation of available dietary iron. *Am J Clin Nutr* 1978; **31**:134–141.

107. von Dokkum W, Wesstra A, Schippers F. Physiological effects of fiber-rich types of bread. *Br J Nutr* 1982; **47**:451–460.

108. Gillooly M, Bothwell TH, Torrance JD, MacPhail AP, Derman DP, Bezwoda WR, Mills W, Charlton RW. The effects of organic acids, phytates, and polyphenols on the absorption of iron from vegetables. *Br J Nutr* 1983; **49**:331–342.

109. WHO (World Health Organization). *Nutritional Anaemias*. Report of a WHO Scientific Group. Geneva: World Health Organization, 1968.

110. FAO (Food and Agriculture Organization). *Requirements of Vitamin A, Iron, Folate and B12*. Report of a Joint FAO/WHO Expert Consultation. Rome: Food and Agriculture Organization, 1988.

111. Bezwoda WR, Bothwell TH, Charlton RW, Torrance JD, MacPhail AP, Derman DP, Mayet F. The relative dietary importance of haem and non-haem iron. *S Afr Med J* 1983; **64**:552–556.

112. Beard JL, Campbell TC, Chen J. Iron nutriture in the Cornell-China diet cancer survey. *Am J Clin Nutr* 1988; **47(Abst. 56):**771.

113. Beard JL, Wang G, Chen J, Campbell TC, Smith SM, Tobin B. Iron nutriture in the Cornell-China diet cancer study. *J Nutr* 1992; under review.

114. Guo W, Li J, Blot WJ, Hsing AW, Chen J, Fraumeni JF, Jr. Correlations of dietary intake and blood nutrient levels with esophageal cancer mortality in China. *Nutr Cancer* 1990; **13:**121–127.

115. Campbell TC, Chen J, Parpia B, M. L. *Diet and Cancer Mortality Rates in a Survey of 65 Counties in the People's Republic of China: Vitamin Status Indicators.* Pennington Conference on Cancer and Micronutrients. Baton Rouge, LA: 1992.

116. Chen J, Geissler C, Parpia B, Li J, Campbell TC. Antioxidant status and cancer mortality in China. *Int J Epidemiol* 1992; **21:**625–635.

117. Nomura A. Stomach. In: Schottenfield D, Fraumeni JF, eds. *Cancer Epidemiology and Prevention.* Philadelphia: W.B. Saunders Co., 1982; 624–637.

118. You WC, Blot WJ, Chang YS, Ershow AG, Yang ZT, An Q, Henderson B, Xu GW, Fraumeni JF, Jr, Wang TG. Diet and high risk of stomach cancer in Shandong, China. *Cancer Res* 1988; **48:**3518–3523.

119. Hu J, Zhang S, Jia E, Wang Q, Liu S, Liu Y, Wu Y, Cheng Y. Diet and cancer of the stomach: a case-control study in China. *Int J Cancer* 1988; **41:**331–335.

120. Buiatti E, Palli D, Decarli A, Amadori D, Avellini C, Bianchi C, Bonaguri C, Cipriani F, Cocco P, Giacosa A, Marubini E, Minacci C, Puntoni R, Russo A, Vundigni C, Fraumeni JF, Jr, Blot WJ. A case-control study of gastric cancer and diet in Italy: II. Association with nutrients. *Int J Cancer* 1990; **44:**611–616.

121. Liu SH, Ohshima H, Bartsch H. Recent studies on *N*-nitroso compounds as possible etiological factors in oesophageal cancer. In: O'Neill IK, Von Borstel RC, Miller CT, Long J, Bartsch H, eds. N-*Nitroso Compounds: Occurrence, Biological Effects and Relevance to Human Cancer.* Lyon, France: IARC Scientific Publication No. 57, 1984; 947–953.

122. Poirier S, Humbert A, de-The G, Ohshima H, Bourgade MC, Bartsch H. Occurrence of volatile nitrosamines in food samples collected in three high-risk areas of nasopharyngeal carcinoma. In: Bartsch H, O'Neill IK, Schulte-Hermann R,

eds. *Relevance of* N-*Nitroso Compounds to Human Cancer: Exposures and Mechanisms*. Lyon, France: IARC Scientific Publication No. 84, 1987; 415–419.

123. Dillard CJ, Tappel AL. Lipid peroxidation and copper toxicity in rats. *Drug Chem Toxicol* 1984; **7:**477–487.

124. Simpson JA, Cheeseman KH, Smith SE, Dean RT. Free-radical generation by copper ions and hydrogen peroxide. Stimulation by hepes buffer. *Biochem J* 1988; **254:**519–523.

125. Orman JE, Talley NJ. *Heliocobacter pylori*: controversies and an approach to management. *Mayo Clinic Proc* 1990; **65:** 414–426.

126. Rothwell NJ, Stock MJ, Tyzbir RS. Mechanisms of thermogenesis induced by low protein diets. *Metabolism* 1983; **32:**257–261.

The Dietary Causes of Degenerative Diseases

Nutrients vs Foods

T. Colin Campbell

1. Introduction

For a very long time, food has been recognized as being important in the development of disease, at least back to the time of Hippocrates 2500 years ago when he said, "Whoever gives these things (food) no consideration, and is ignorant of them, how can he understand the diseases of man?" Much later, in 1849, John Hughes Bennett, Senior Professor of Clinical Medicine, whose textbook on medicine appears to have been the standard of its day in Britain, pursued the idea that dietary fat was important in the causation of cancer when he first stated that, "The circumstances which diminish obesity, and a tendency to the formation of fat, would seem *a priori* to be opposed to the cancerous tendency" (p. 250) *(1)*, then later, in 1865, urged that, "In carcinoma ... a diminution of this element (fat) in the food should be aimed at" *(2)*. In more modern times, the studies of Tannenbaum and his colleagues *(2–4)* during the 1940s and 1950s on the role of nutrition in the development of tumors in experimental animals and the research

From: *Western Diseases: Their Dietary Prevention and Reversibility*
Edited by: N. J. Temple and D. P. Burkitt Copyright ©1994 Humana Press, Totowa, NJ

of Ancel Keys and colleagues on diet and cardiovascular disease *(5,6)* often have been assigned landmark status.

More relevant to the present chapter, however, are the recommendations published in recent years to reduce by dietary means the prevalence of cardiovascular and neoplastic diseases, commonly called chronic degenerative diseases *(7)*.* Much focus, in particular, has been given to the need to reduce dietary fat to 30% or less of total calorie intake. In fact, it would appear that this figure of 30% dietary fat calories has become a benchmark to gage the acceptability of healthy diets.

The development and promulgation of national dietary recommendations to avoid degenerative diseases, particularly with reference to nutrient intake recommendations, has not been without considerable difficulty, both within and without the scientific research community. Unquestionably, the development of these recommendations has had a long and tumultuous history, sometimes focused on the concept that the nutrients present in food have little to do with disease causation (with the more significant factors being genetics, viruses, and chemical carcinogens or other chemical agents), sometimes on the concept that disease is acceptably controlled through cure, sometimes on the concept that the nutritional benefits of food are described best in terms of a few "magic bullets" (nutrient

---

*These diseases include most if not all of the cancers, most of the cardiovascular diseases, diabetes, and certain other ailments associated with the more industrialized societies; they also have been referred to as "diseases of affluence" *(8)*, "diseases of misdevelopment" *(9)*, or in the older literature, as diseases of "civilization" *(10)*. We propose the term "diseases of extravagance" because the total cost of these diseases to society is a summation of the costs of producing the causes plus the costs of treating the consequences. In this chapter, we mostly refer to these diseases as "degenerative diseases" because of their association with biological degeneration, or aging.

supplements and designer foods), and sometimes on the concept that it is the total diet that produces the most desirable health. These are widely divergent views that, on closer inspection, contrast the broad disease modifying effects of foods vs the narrower effects of nutrients, or the narrow effects of nutrients vs the also narrow effects of nonnutrients. Oftentimes these views are mutually exclusive, sometimes are meant to be exclusive—and diversionary, usually resulting from the tendency of research investigations to be highly focused—and reductionist. This latter feature concerning reductionist research strategies, the main thesis in this paper, may be the biggest casualty of this debate on dietary recommendations. The flood of research details issued to the public often has led to misleading generalities obtained from specific, even specious data, leaving in its wake an appreciation for the all-important effects of nutritional synergy.

The concept that disease prevention is best controlled through a synergistic total diet effect has been a particularly difficult one to grasp, undoubtedly related to the difficulty of experimentally testing and evaluating such a complex dietary mixture. A "total diet" is infinitely variable in its composition of foods and quantities of food eaten, thus investigation and comprehension of which particular mixtures are most appropriate for which disease, for whom, and under what conditions, is generally regarded as an extremely difficult research task. Understanding the effects of the whole diet is not necessarily made easier by limiting research investigations to single nutrients studied in isolation; consideration also must be given to the concept that the "whole is greater than the sum of its parts." Dietary synergy cannot be ignored.

Grasping this concept of the disease modifying effects of a "total diet" also is not made easier when so many research studies use populations or individuals who are adapted (maladapted?) to the diets now hypothesized to be problematic. With such populations, not only is little or no consideration given to the effects of dietary adaptation *(11)*, a process that, when absent, distorts interpretation of short-term results, but also dietary exposure ranges are likely to be too narrow and beyond effect modification. Nutritional adaptation involves the body's continual struggle to minimize, at least in the short run, the adverse effects of inappropriate nutrient exposures (too high or too low); it is not necessarily an ideal steady state for long-term health. If adaptation, which may take months or even years to complete, is not considered (as is usually the case), then the interpretation of results from short-term studies may be seriously compromised. For example, it is not sufficient to make definitive conclusions on the adverse effects of a nutritionally rich Western diet by studying only those individuals maladapted to such a diet. It is also important to investigate populations or individuals who are fully adapted to the types of diets thought to be beneficial, in order to gain a more complete understanding of the hypothesized long-term dietary effects.

In summary, without consideration being given either to the effects of nutrient–nutrient (or other constituent) interactions, or to the effects of long-term adaptations, both of which are important when extrapolating or generalizing experimental results to free living populations, the publication of results from highly reductionist studies is likely to misinform and to mislead.

2. Guidelines Focused on Dietary Fat

As already mentioned, considerable activity has been devoted during the past two to three decades to the development of dietary recommendations to reduce the prevalence of various degenerative diseases commonly found in Western industrialized countries *(7,12–14)*. Many national governments and many agencies within countries have published very similar dietary recommendations (15). Although advocating increased consumption of foods of plant origin (vegetables, fruits, grains), these guidelines have focused particularly on the need to reduce dietary fat to a level of about 30% of calories.*

Although a variety of evidence has implicated an effect of excess fat intake upon the development of many of these diseases—particularly certain cancers and coronary heart disease, the primary evidence originally used to support these dietary recommendations was obtained from international correlation studies, which included an unusual breadth of fat intake, from around 9 to 42% (14,16). In these reports, correlations between fat intake and prevalence of some of these degenerative diseases were very impressive, being as high as 0.8–0.9 (14,17–20). For those populations consuming the lowest fat diets, some of these diseases were almost nonexistent. In addition to these international correlation studies, laboratory animal studies that have a history going back for several decades (21–23), also have shown that increased consumption of dietary fat, when tested over a wide range of intake and under controlled conditions, independently promotes the development of experimental tumors (13). A recent meta-analysis of 100 separate studies of 7838 rats and mice, for example, found a highly significant ($p < 0.0001$) effect of fat on experimental mammary tumors alone (24). Thus the combined evidence from these two kinds of studies, along with substantial evidence demonstrating biological plausibility of an effect of excess fat intake (13,25–27), furnished the primary evidence for this special recommendation.

The special emphasis given to dietary fat is signified both by its being singled out from other nutrients for an independent recommendation and by the numerical speci-

---

*Unless otherwise specified, fat intake will be considered in terms of dietary fat as a percentage of total calories.

fication of a desirable quantity of intake. However, the origin of this particular level of intake was not the result of analyzing the benefits and risks of a full range of possible intakes. Nor was it based on evidence suggesting that there is a threshold intake of fat below which no further prevention of disease might be achieved. Nor was the evidence particularly persuasive to show that excess intakes of this nutrient *independently* accounted for the majority of degenerative diseases associated with the consumption of high fat diets. The NAS report that published the first dietary recommendations on cancer prevention *(13)*, concluded that, "The scientific data do not provide a strong basis for establishing fat intake at precisely 30% of total calories. Indeed, the data could be used to justify an even greater reduction" (p. 15). They considered a 30% level of intake to be a "moderate and practical target." A more recent NAS Committee on Diet and Health *(14)*, when reviewing the effect of dietary fat intake on all degenerative diseases, concluded that, "... there is evidence that further reduction in fat intake (below 30%) may confer even greater health benefits" (p. 14). They went on to say that, "... the recommended levels (of 30%) are more likely to be adopted by the public because they can be achieved without drastic changes in usual dietary patterns and without undue risk of nutrient deficiencies."*

The special recommendation on dietary fat, therefore, originally was considered only to be a "practical" target *(13)*, essentially meaning a modest reduction of fat intake to an integer remaining within the same decade of contemporary intake (30 vs 38%), and a level of intake fairly easily achieved by using less added fats and oils and more

*The recognition of 30% dietary fat as a quantity having special meaning undoubtedly occurs, even though the more recent recommendations *(7,14)* have suggested that this amount should be considered an upper bound intake; a definite number is much better remembered than its qualifiers.

foods of the low fat variety. But a change of this magnitude, as customarily occurs, may be quite trivial. If, for example, the dietary change is limited to the use of less added fat and more low fat dairy foods and leaner cuts of meat, as is the common practice, without actually replacing these foods with those of plant origin, then the aggregate intake of foods of animal origin will remain about the same, and may even be increased. Consequently, intakes of countless other food constituents known to be involved in the development of these diseases also will remain about the same, leaving in doubt whether the overall changes will consistently* or significantly alter risk for most of these diseases.

In contrast, if dietary fat were to be reduced much below 30%, perhaps all the way to 10%, then it would be necessary to substantially reduce, perhaps even eliminate, low fat animal foods, most of which still retain relatively high amounts of fat at about 25–40% *(28)*. A major exchange of foods of animal origin for foods of plant origin therefore would result in major changes in the intakes of nutrients and other constituents that are distinguished primarily by their relative concentrations in these foods. This is an important consideration because constituents that are particularly effective in enhancing or inhibiting the development of degenerative diseases are concentrated in foods of animal or plant origins, respectively (the subject of a separate paper). For example, there now are 500–2000 natural chemical entities** that have been shown to

---

*"Consistency" means that, for all degenerative disease risks enhanced by high fat diets, a change to a low fat diet will reduce disease risk to a greater or lesser extent (not that each disease risk will be reduced to the same extent).

**According to personal communications of Herbert Pierson, James Duke, and Christopher Beecher, the number of such chemicals depends on the inclusion of synthetic chemicals and known antioxidants, as well as on the screening criteria used to assess biological activity.

exhibit activity hypothesized to "chemoprevent" cancer and, undoubtedly, there are tens of thousands more yet to be discovered. The discovery of these chemicals, virtually all from foods of plant origin, illustrates not only the importance of dietary factors in preventing degenerative diseases but, more particularly, the *ability of diets enriched in plant foods* to prevent these diseases. In effect, the original *(13)* and subsequent decisions *(7,14)* to make the dietary fat recommendation "practical" (and special) very likely has become diversionary, in that attention has been drawn away from the possibility of achieving far more consistent prevention of these diseases by simultaneously making all the necessary changes in food intake.

The false impression that a 30% fat diet may be sufficient to produce the desired benefits of disease prevention also is encouraged by the assertion that the public is not likely to accept diets containing less than 30% fat *(29)*. Such an assertion, especially when made by public health institutions, tends to be a self-fulfilling prophecy, causing consumers to assume some measure of scientific certainty with the 30% figure. This assertion, however, is a paradox. It suggests to many consumers, on the one hand, that reducing fat intake to 30% *may be adequate*, yet on the other hand, by beginning with the assumption that consumers are not likely to accept further change, it therefore infers that this change *may not be adequate*. Confusion, both of consumers and researchers alike, is bound to occur. Faced with this uncertainty, consumers therefore must choose among several hypothetical options for reducing disease risk by dietary means. To facilitate further understanding of these options, one possible set of options is presented in Table 1. Option A presumes that reducing fat intake to 30% of calories is adequate, whereas option B presumes that it is not. If not adequate, then a further choice must be made either to reduce fat still further (suboption B1) and/or to supplement the diet with indi-

Table 1
Dietary Options to Reduce Degenerative Disease Risk

Strategy focused on changing the intakes of individual nutrients, leaving food choices intact
  Option A—Reducing dietary fat to 30% is adequate.
  Option B—Reducing dietary fat to 30% is not adequate.
    Suboption B1—Reducing fat to 20% fat may be adequate.
    Suboption B2—Addition of nutrient supplements may be adequate.

Strategy focused on changing the intakes of foods
  Option C—Decrease intake of foods of animal origin and increase intake of foods of plant origin; use less added fat.

---

vidual nutrients or other constituents thought to prevent these diseases (suboption B2). Thus, there are three options (A, B1, and B2) that are highly focused on the disease-altering effects of *single* constituents and that allow food selection to remain more or less the same. It is important to note, however, that none of these options theoretically requires modification of the two primary food groups (foods of either plant or animal origin) that determine the intakes of virtually all nutrients and other constituents known to modify disease risk. By focusing on the potential activities of single nutrients and other constituents, these three options are highly reductionist in concept, meaning that "a system can be fully understood in terms of its isolated parts" (p. 1006) *(30)*. Also, note here the appearance of the notion that if a simple and modest reduction of fat intake does not achieve the desired results, perhaps the addition of single nutrients will, an idea to which the major dietary recommendation reports have given support *(7,13,14)* but without sufficient evidence.

The option to use nutrient supplements or so-called "designer foods" *(14)* to prevent disease (suboption B2) is underscored by the extraordinary efforts being undertaken

to promote the use of these supplements to prevent cancer
(31–34). The enthusiasm for this approach also is illustrated
by the highly visible and well-funded research programs
being undertaken under the auspices of the US National
Institutes of Health and other research funding agencies
to investigate, in double-blinded randomized clinical trials in humans, the ability of *individual* nutrients and other
chemical constituents to prevent cancer *(29,32,33)* and
other clinical markers of degenerative disease occurrence
(reviewed by Chisholm *[35]*). These trials originally were
justified not only on the assumption that consumers probably would not accept *proscription* of dietary advice (to
reduce fat intake below 30%) but might be willing to accept
prescription of individual "chemopreventive" nutrients
(neutriceuticals) and other chemicals *(29)*.* The impact of
this assumption on research investigation cannot be over-
stated; it underlies both the substantial funding support
for the randomized clinical trials and the extensive efforts
being undertaken to develop marketing schemes to produce and advertise specific chemical preparations and special foods designed to chemoprevent cancer *(36)*, among
other degenerative diseases *(37)*.

In contrast to the ease with which prescription
research on nutrient supplements is undertaken, a pro-
posal on proscription to investigate the effects of a so-called
low fat diet (containing somewhat less than 30% of fat) on
the incidence of breast cancer has been much more pro-

*In a later report (1990) Boone *(33)* from the National Institutes of Health also used these same words but defined "proscription" as elimination or avoidance of carcinogens in the environment.
This is not what was intended in the original definition *(29)* by this
same agency, which defined this term as the difficulty of "attempting major modification in dietary habits and lifestyle" (p. 120). Both
reports, but particularly that of Boone *(33)*, seem to reflect a highly
reductionist interpretation of the diet–disease relationship that
lends itself to a highly reductionist pharmacologic solution.

tracted and contentious and, until recently *(38)*, this pro-
posal remained unfunded *(37,39,40)*. Even now, this newly
funded study still remains highly focused on the effects of
a reduction mostly of dietary fat, and primarily its effects
only upon two cancers, breast and colon.

In contrast to these highly reductionist options, there
is a further option (option C in Table 1), which is concep-
tually more holistic. Rather than beginning with the
assumption that excess dietary fat is the chief cause of
these diseases, this option begins with the hypothesis that
an improper balance of "animal" to "plant" foods is the chief
cause. This hypothesis relies on the overwhelmingly com-
pelling observations that suggest that the nutritional characteristics of foods of animal origin enhance degenerative
diseases, whereas the nutritional characteristics of foods
of plant origin inhibit the development of these diseases.
Although this idea is not new, it is only partially considered by the various dietary recommendations *(7,12–14)*.
Indeed, there have been intentional efforts to downplay this
message in order to avoid unnecessary controversy. In
order to provide a more positive message—though it is not
clear for whom or why the message needs to be more positive—these recommendations have advocated increased
consumption of fruits, vegetables, and grains (a positive
message) and generally have not advocated decreased consumption of foods of animal origin (a negative message).
And when these guidelines are implemented or taught,
attempts usually are made to preserve the traditional
intake of animal foods, in the aggregate, by swapping one
kind of animal food for another, e.g., chicken for beef, or
"low fat" dairy products for whole fat dairy products. This
strategy of exchanging one kind of animal food for another
clearly discourages an increased consumption of foods of
plant origin because it gives a false sense of confidence to the
value of low fat animal foods. Moreover, total calorie intake,
almost entirely determined by calorie expenditure, will

remain about the same before and after the exchange, particularly during the prolonged periods of time when these diseases are forming. Not only are long-term reductions in calorie consumption notoriously unsuccessful, but also, recent investigations on the implementation of the dietary guidelines, although indicating a significant reduction in energy intake, probably resulted, according to the authors *(39)*, from the disciplined effects of record keeping.

With the more holistic option C, dietary fat would only be an *indicator* of the extent to which diets are enriched in foods of animal origin and, as such, it would be much less likely to become the primary object of dietary change.* Indeed, many other factors could also be used to indicate the relative amounts of foods derived from plants (e.g., dietary fiber, various antioxidant micronutrients) or from animals (e.g., heme iron, retinoids, vitamin D steroids exclusively found in animal foods). Some of these factors even may be better indicators of food exchange than dietary fat. Thus, I contend that the assertion that the public is not likely to decrease their dietary fat levels much below 30% sets in motion a series of decisions that is derived from highly reductionist research methodology and that trivializes and confuses the evidence on the diet–disease relationship. Moreover, in my view, this recommendation on fat discourages serious research investigation into the more holistic effects of major dietary change (option C).

Dietary and nutritional effects on these diseases need to be rearticulated. New questions are needed to reflect the more comprehensive effects of the total diet (option C)

---

*Dietary fat is a very good indicator of proportional amounts of animal food in population studies because of the very high correlations between the intakes of total fat and animal protein (see ref. *13*); dietary fat may be a less reliable indicator of dietary enrichment with foods of animal origin when investigating individuals because fat may be added to diets regardless of the relative amounts of foods of plant or animal origin.

upon the development of all degenerative diseases. These questions need to be cognizant of the extraordinarily complex biology that lies at the interface of nutrient exposure and disease processes, yet these same options also need to encourage the search for a few simple truths, if possible. Instead of asking the commonly heard question—What level of dietary fat should be recommended?—a different question is now needed to address the effects of the entire diet and, in doing so, to remove the emphasis given to dietary fat and to its specified level of intake. The question needs to speak to the comprehensive effects of all nutrients and countless other dietary constituents now known to affect degenerative disease risk, and to do it in a way that includes both the direct and indirect (interactive, adaptive) effects of these individual constituents within the context of whole foods—and preferably within the context of a population fully adapted to the desired diet. Thus, a more appropriate first question might be: How rich in plant foods should a diet be to prevent degenerative diseases—while optimizing overall health? And then, *and only then, secondary* questions might be posed to reflect the nutritional and food characteristics of this revised diet, such as: How much fat is present? How much fiber? How much vitamin C? What types of plant foods should be used? Are there special plant foods with more value than others? How much variety is needed? How should variety be measured? Then there are questions concerning other "dietary" approaches, such as: Are there alternative nutritional means of preventing these diseases for those who choose to continue consuming a 30% fat diet? For these individuals, could individual nutrient supplements be used? This latter question, however, prompts more troubling questions such as: Which of these compounds should be used? At what dosage? Under what conditions? By whom? Can possible adverse reactions be controlled? How important are in vivo interactive and adaptive responses when interpreting the narrowly constrained experimental activities of these singly tested chemicals? And most important, how do the benefits and risks of using the combination of nutrient supplements (option B2) and a 30% fat diet (option A) compare with the benefits and risks of using a truly low fat diet solely based on plant foods (option C)?

To summarize the foregoing discussion, two broad strategies already appear to exist among consumers, producers, and scientific researchers for the dietary prevention of degenerative diseases. The first strategy, an aggregation of options A, B1, and B2, is reductionist in concept and is focused on the role of single food constituents, i.e., reduce fat and/or use nutrient supplements. The second strategy (option C) is much more holistic in concept and is directed to an exchange of food of animal origin for food of plant origin. The first strategy clearly enjoys far more popularity than the second, thus "proving" the assumption that the public may not be willing to accept dietary proscription or perhaps satisfying the self-fulfilling prophecy. This strategy also is encouraged commercially because pills and low fat foods generate profits. The second more holistic strategy will require substantial change in food selection, particularly if 10% fat, total vegan diets are indicated as the ultimate goal. In effect, I believe that the dietary recommendation that is focused on fat has set in motion a very simplistic view that discriminates against the additional recommendation in these same guidelines to increase the consumption of fruits, vegetables, and grains.

The foregoing discussion, mostly on theoretical considerations, prompts the question whether there is any empirical evidence showing how consumers respond to the present dietary guidelines and, furthermore, whether such changes lead to meaningful change in disease risk. Namely, how do consumers respond to dietary advice when they attempt to reduce disease risk after being accustomed to a diet rich in animal food and total fat?

To this author's knowledge, the best available evidence on this question may be results of the Women's Health Trial (WHT) feasibility and run-in studies designed to determine whether the experimental subjects would switch to a lower fat diet and maintain the practice in order to test the hypothesis that breast cancer incidence would be reduced (40,41). There were two study groups: the initial "Vanguard" intervention study group of 184 women (39–42) and a second, larger intervention group of 448 women (40) who were to be used in the full-scale study that was never funded (37,40). The following describes the results of the Vanguard group who were followed for three years.

The dietary choices made by the study participants were largely conditioned both by the study objectives and by the research investigators themselves (37,39). During the dietary coaching period, every effort was made by the project managers "… to interfere as little as possible with participants' customary diets …" (p. 316) (40). The participants were provided a "Fat Counter" that listed the fat content of 1500 food items, thus emphasizing the main objective of reducing fat intake. In fact, the investigators made it clear that they intended to focus on a reduction in dietary fat, even though they acknowledged (41) that other investigators (43) had cautioned against such emphasis. Much of the justification for this focus was discussed by Prentice (37).

Henderson et al. (39) reported that these subjects, at 24 months after study initiation, were consuming a 22.6% fat diet compared with a control group consuming a 36.8% fat diet. Consumption of calories from red meat (-48%) and dairy food (–16%) decreased, whereas consumption of calories from "poultry and fish" increased (+17.9%). Surprisingly, aggregate consumption of vegetables, grains, and legumes, as a percent of total calories, did not change. These authors (39) pointed out that there seemed to be resistance to eating more legumes, grains, and starchy vegetables. In a followup study conducted one year after termination of the intervention period *(40)*, there was some evidence that the control and intervention groups were beginning to merge in their intakes of fat (27.7% for intervention group vs 35.1% for the control group). Although fat intakes were lower both at 24 *(39)* and 36 mo *(40)* for the intervention group, the intake of dietary protein, which will be considered below, was significantly increased at both times (16.8–19.2% at 24 mo and 17.3–19.2% of calories at 36 mo). Other nutrient intakes generally were quite similar between the intervention and control groups. In the followup study at 36 mo, the intervention group diet actually exhibited a lower polyunsaturated to saturated fat ratio (0.53 vs 0.60, $p < .001$), which suggests, if anything, a greater intake of fats from foods of animal origin by the intervention group. Although the intervention group participants also reported lower total calorie intake, it was not clear how real this change was, as the investigators believed *(39)* that this may have been because of an "... adaptation of study participants to the process of keeping food records" (p. 120).

Although the investigators *(40)* were encouraged that these findings suggested that the study participants were able to maintain a "low fat" diet for as long as one year after dietary coaching had ceased, this difference did not translate into an altered balance of foods of animal and plant origin. The reduction of fat intake was statistically significant, but protein intake increased *(39–41)* and many of the other constituents, particularly the antioxidant micronutrients mostly found in vegetables, appear to have remained relatively unaltered. For example, 12 mo after coaching had begun and when the difference between control and intervention group fat intakes was the greatest, the proportional aggregate intakes of foods of plant and animal origin remained about the same (e.g., the percent of total protein as animal protein remained at 69–70% [41]). Blood cholesterol, a good indicator of dietary lifestyle over a broad range of dietary fat, also remained relatively high at 24 mo among the intervention subjects, at 5.43 mmol/L (210 mg/dL) (39), at least when compared with populations of truly low breast cancer risk, such as the rural Chinese, whose level is about 3.23 mmol/L (125 mg/dL) (44).

A similar response to the dietary guidelines also was observed in another group of women studied by Boyd et al. (45). These investigators taught 148 women, with evidence of mammographic dysplasia, how to use the dietary guidelines, as part of an investigation to test the effect of a low fat diet upon the risk of breast cancer. After 12 mo, dietary fat intake was 24% in the intervention group compared with 36% in the control group. However, the "low fat" intervention group, like the WHT participants (39–42), still consumed a high protein diet (18% of calories) when compared to the control group (17% of calories). Although food intakes were not reported in this study, the unchanged P/S ratio suggests that the proportional intakes of foods of plant and animal origin remained relatively unchanged, thus compromising the intakes of the disease preventive constituents of plants.

Even free-living individuals appear to make the same type of changes when implementing the dietary guidelines. A sample of NHANES II male and female adults who claimed to be dieters, apparently to reduce their risk of coronary heart disease, were consuming diets containing 32.8% fat, only slightly lower than a level of 36.5% fat for the diets of nondieters (46). Protein intake, as a percent of calories, was 18.3% for the dieters and 15.6% for the nondieters. Thus the dieters were consuming only 10% less fat but 17% more protein. Dieters reported decreased intakes of whole milk products (–45%), meat (–15%), organ meats (–19%), and eggs (–30%), but increased intakes of skim milk products (+60%), poultry (+27%), and fish (+39%). Using the servings/wk measurements in this report *(46)*, I estimate that total and animal protein intakes, as a percent of calories, were about 10% higher for the dieters.

These experiences strongly suggest that, under most circumstances and with or without professional coaching, use of the dietary guidelines to implement dietary change will result in dietary practices that can only have insignificant and/or inconsistent effects upon risk for most degenerative diseases. In large measure this is because of the unsubstantiated emphasis initially given to the independent effect of excess fat on disease risk. Nonetheless, there are many who believe that these modest dietary changes, particularly those focused on fat, may be sufficient to produce important changes in disease risk, obviously more change for some degenerative diseases than for others. This is a widely held view among researchers, medical practitioners, and consumers and was, perhaps, most thoroughly discussed with reference to breast cancer by Prentice et al. *(37)*, who have organized a new WHT, now known as the Women's Health Initiative (WHI) *(38)*. These authors began by citing the strong correlation between breast cancer incidence rates and per capita intake of dietary fat in 21 countries, then undertook "... several regression analyses relating national estimates of caloric intake (for fat and no-fat calories) to breast cancer incidence ..." They suggested that there is "... a specific effect of fat beyond its contribution to total calories," although they added that this suggestion is not necessarily firmly established. Nonetheless they go on to show, in further regression analysis, that "calories from protein ($p = .76$), alcohol ($p = .80$), and carbohydrate ($p = .99$) show no marginal association with breast cancer incidence in this age range, whereas calories from fat continue to show a sig-

nificant association (p = .02)." They then conclude that "...
these data point to fat calories, rather then to those from
protein and carbohydrates, for an explanation of international variations in breast cancer incidence rates" (all
quoted material from p. 805 of ref. *37*). Following this analysis, the authors then cited data from other types of studies to support their contention that fat is capable of
independently enhancing breast cancer risk.

Unfortunately, in my view, this analysis may be fun-
damentally flawed. The potential risk attributed to dietary protein is incorrectly discounted because the authors
used total protein in the analysis and failed to take into
account the vastly different biological and nutritional properties of animal protein when compared with plant protein. For example, animal proteins, *within traditional
levels of intake*, enhance a wide variety of converging physiological responses. This results in an upward bias in the
risks for diverse degenerative diseases, including elevation of blood cholesterol levels in humans *(47–49)* (an effect
that appears to be even more significant than saturated
fat intake *[47]*), enhancement of initiation and promotion
of experimental carcinogenesis induced by different carcinogens in experimental animals *(13,50–54)* by widely
varying converging and supportive mechanisms, enhanced
bone resorption of calcium *(55,56)* that accords with higher
rates of osteoporosis in countries with higher intakes of
animal protein *(57)*, acceleration and maximization of contraindicated body growth rates *(58)*, and stimulation of
caloric retention through enhancement of caloric intake
(59) and diminution of thermogenesis *(60–62)*, among many
other similar effects. Reduction of intake of animal proteins
and/or replacement with equal amounts of plant proteins
generally exhibit the opposite effects. And, of particular
significance to the dietary causes of breast cancer, animal
protein enhances in experimental animals the development

of chemically-induced mammary tumors *(50,52–53)*,* elevates estral surges for progesterone, estradiol, and prolactin *(63)*, increases mammary gland growth *(64)*, duct
proliferation and morphologic development *(65)*, and accelerates body growth rate and shortens the time for sexual
maturation *(50,65)*, all of which phenomena are consistent
with an enhancing effect of animal protein intake on breast
cancer development. That this is relevant to traditional
levels of animal protein intake by humans is indicated by
the very tight coupling that exists between the upper range
of intake required for maximized (but contraindicated) body
growth rate and the lower range of intake required for
promotion of chemically induced liver tumors in the laboratory rat, a phenomenon made all the more relevant
because of the almost identical protein requirement reference points for humans and rats *(66,67)*.** Furthermore,
Armstrong and Doll (18), who investigated virtually the
same countries as Prentice *(37)*, observed a very strong
correlation ($r = .93$) between the intakes of total fat and
animal protein.

Given the internal consistency and biological plausibility of these findings, I hypothesize that the intake of
animal protein, particularly during the perimenarchial
period of life, is a major but almost totally ignored, dietary
cause of breast cancer in humans. Thus a modest decrease
in dietary fat without a compensatory decrease in animal
protein, particularly when limited to the adult age period,
very likely will fail to reduce breast cancer incidence. And

*Caution must be used in the interpretation of studies on the
effect of protein on experimental carcinogenesis, in that in some of
these studies, the level of protein may not include a level sufficiently low to approach growth rate retardation (not attained body
size), and/or an adequately designed feeding protocol, possibly one
that includes the period of sexual maturation.

**A similar analogy can be drawn for other degenerative dis-
ease events and animal protein intake.

furthermore, the failure to take into account an animal protein effect also leads to a sequelae of events that leave intact the balance of foods of animal and plant origin, thus diminishing the effects of the virtually countless constituents present in vegetable and leguminous plants (e.g., antioxidants, complex carbohydrates and fibers, and protease inhibitors, to name but a few) that are capable of preventing the development of a multitude of other degenerative diseases. Thus, the justification that has been used for testing the effects of the so-called low fat diet on breast cancer risk and that now is being applied to many other degenerative diseases may be a serious oversimplification and overgeneralization of the much more substantial, and true, relationship between dietary habits and disease occurrence.

The foregoing discussion suggests that, although application of the dietary guidelines may lead to a prolonged decrease in the intake of total fat, albeit a relatively small decrease, there is little or no change in the intake of most of the other disease modifying nutrients and food constituents. Minimal and fragmentary changes focused on dietary fat cannot be expected to cause either the magnitude or the consistency of change in disease risk inferred by the international correlation studies on various degenerative studies. Perhaps because of this uncertainty (even subliminally?) great efforts are being made to test the abilities of individual nutrients and other constituents to prevent these diseases. Either of these minimalist strategies, the one focused on the subtraction of dietary fat, the other on the addition of individual nutrients or other constituents, are highly reductionist in concept. Most importantly, neither of these strategies represents a test of the ranges of comprehensive dietary experiences that characterize the international correlation studies. Nonetheless, both of these minimalist strategies now are being tested, the single constituents in randomized human trials originally designed for drug studies, and the "low fat" diet in a massive new trial called the Women's Health Initiative (WHI) costing over $100 million.*

The "nutrient supplement" trials, about 17 in number (68), presently are testing about half a dozen candidate compounds (33) in well over 100,000 subjects (68), who are being allowed to continue consuming their customary diets. In order to save time and some of the huge costs of these trials, most of these trials are relying on the use of predisease lesions as endpoint predictors of fully developed diseases. However, there are many questions concerning the efficacy and marketability of these intervention agents yet to be answered, some of which were discussed previously. In addition to those questions, even if a specific supplement is shown to be reliably effective for a specific predisease lesion and a proper dosage schedule is established and the predisease lesion reliably predicts disease outcome (three big "ifs"), will this mean that an individual will be able to continue exposure to the offending cause, e.g., an extravagant diet, without concern for other undesirable health outcomes? Or will it be necessary to identify and test an effective agent for each and every adverse health effect that may result from the consumption of the offending diet? If so, this strategy strikes me as an unusually naive understanding of biology, particularly of nutritional biology, and I therefore do not hold out much hope for its success or for research in this field.

In the new WHI project, the "low fat" diet being consumed is virtually the same as that used in the WHT discussed earlier. However, in this new version, both breast cancer and colon cancer events are being monitored. Once

---

*The full trial, which will include 75 clinical centers and which also will test hormone replacement therapy and supplementation with calcium plus vitamin D in 150,000 postmenopausal women, will cost $625 million (38).

again, this is a minimalist approach, if these women subjects follow the same diet as their WHT predecessors.
Indeed, there now are substantial findings already indicating the futility of this strategy to effect change in breast
cancer risk. Even the projected 17% reduction of breast
cancer incidence rates for women subjects consuming diets
containing 20% fat *(37,69),** is likely to be severely compromised, to say nothing of the many other degenerative
disease risks remaining either unaffected or, perhaps,
adversely affected. In a very large cohort of 89,494 nurses
being carefully followed by Willett and colleagues *(70)*,
there is no evidence, after 8 yr, that this type of diet is
associated with a reduction in breast cancer risk. The level
of fat consumption among this cohort is typically high, with
nine of the ten deciles of fat intake being higher than 29%
of calories. These new findings must be taken seriously,
however, both because they extend and confirm an earlier
report after four years of followup *(43)* and because three
other studies *(71–73)* with similar ranges of fat intake
reported essentially the same results. Although no data
on food and nutrient intakes were given for this cohort
(70), the range of dietary experience undoubtedly is similar to that found for the WHT *(39–42)* and NHANES II
(46) cohorts.

The impressiveness of these prospective study findings *(43,70–73)* is very much in agreement with the negative results of most case-control studies on dietary fat and
breast cancer (reviewed in refs. *43,74*). Although several
criticisms of these case-control studies have been proffered,
as reviewed elsewhere *(74,75)*, the most recent report on

---

*Both of these estimates are likely to be overly optimistic. The
predicted 17% reduction in breast cancer incidence is based on the
linear response from the international correlation studies, which
are not commensurate with the dietary changes actually adopted
by these women. Moreover, the 20% fat diet was not reached in the
WHT trial.

the Nurses' Health Study *(70)* appears to have addressed most of these concerns, except for that concerning the homogeneity of dietary experience. The mutually supportive findings of these prospective and case-control studies therefore contrast with the impressive linear relationship observed for dietary fat and breast cancer in the international correlation studies (reviewed by Goodwin and Boyd *[74]*). Interpreting the latest observations of the Nurses' Health Study, Willett et al. *(70)* suggested that their study may have failed to show a dietary fat effect on breast cancer risk because fat levels below about 20% of calories may be required. But they then discount this possibility because it would imply a nonlinear relationship between dietary fat intake and breast cancer, a finding not in agreement with the international correlation studies. Alternatively, these contrasting kinds of evidence between international correlation studies, on the one hand, and case-control/prospective studies, on the other, could be reconciled by the fact that the international correlation studies reflect not only a linear relationship of breast cancer incidence with dietary fat intake but, *even more important, a linear relationship of breast cancer incidence with the proportional intakes of foods of animal and plant origin,* as evidenced by the very high correlation between the intakes of dietary fat and animal protein among these same countries *(18)*. In contrast, the case-control/prospective cohort studies (of subjects consistently consuming diets rich in foods of animal origin) do not reflect a change in the proportional intakes of these foods over the ranges of fat intake being investigated *(39–42,45)*. As fat intake is decreased in response to the dietary guidelines, the intake of animal protein (thus the proportional intakes of foods of animal and plant origin) either remains unaltered or is increased. As a result, breast cancer risk also remains unaltered or, if anything, might be slightly increased with decreasing fat intake, an observation that is consistent with the results of the Nurses' Health Study (43,70). This interpretation is not meant to suggest that animal protein is the major cause of this disease, even when consumed during and following the perimenarchial age period (63) and even given the experimental animal data (50,52,53). Instead, I hypothesize that, notwithstanding the singular impressiveness of the animal protein data, the focus still should be placed on the proportional intakes of "plant" and "animal" foods as the most consistent indicator of disease risk because this more holistic concept captures the beneficial effects of all nutrients and all degenerative diseases.

A suggestion was made by Willett et al. (70) that consuming the low fat diet, as conventionally used, may have redeeming value by reducing colon cancer risk, as evidenced by the results of the Nurses' Health Study. However, this is not a sufficient redemption because, even if proven to be true, a dietary recommendation that selectively—and only partially—ameliorates one disease risk while leaving others unaltered, and perhaps even increased, is not an acceptable public health policy, particularly if there are alternative strategies that have the potential of far more consistent effects.

3. Summary and Research Recommendations

If a dietary guideline based strategy focused on dietary fat is relatively ineffective—and is certainly not consistently effective—then is there an alternative strategy to prevent these diseases? The hypothesis offered here is that the responsible cause both for breast cancer and to varying degrees for other degenerative diseases, is the proportional intakes of foods of animal and plant origin. This also accords with the international correlation studies to the same extent as does dietary fat *(18)*.* Thus, this hypothesis states that it is the intake of foods of animal origin, at the expense of foods of plant origin, which causes these degenerative diseases. As was pointed out by Willett et al. *(70)*, an hypothesis of this kind is, of course, very difficult to test when using conventional epidemiological procedures. This would be especially true if the desire was to test, at one end of the spectrum, the ability of a 100% "plant food" diet to cause *consistent* decreases in degenerative disease risks or, at the other end of the spectrum, the ability of a 100% "animal food" diet to cause *consistent* increases in degenerative disease risks. Not only are there infinite combinations of food and nutrient intakes to be taken into consideration, but also there are seemingly large numbers of different degenerative diseases. Moreover, a sufficient number of experimental subjects seldom are available for such a study, at least in industrialized societies.

Such a study would be a realistic possibility if there were a few common exposure and outcome indices, although it is doubtful if single indices could be adequate for all purposes. For a biochemical marker index for disease outcome, a measurement that comprehensively reflects at the tissue level for example oxidation damage *(76)* or cell turnover might be useful. Both of these activities, when maximized, represent excessive pressure on the pace of biological change, thus enhancing risk of degenerative diseases. For an exposure index, the proportional intakes of plant and animal protein, expressed or perhaps further modified by total protein intake, as calories or as protein, may be useful.

---

*The relationships of these proportional intakes to risks of various degenerative diseases, although undoubtedly dependent on a wide variety of food constituents provided by these foods, will essentially be of the same sign, differing only in the strength of the correlations.

Short of having access to comprehensive indices, however, a more practical approach, based on present knowledge, is to examine in one study a large number of parameters that reflect diets rich in plant material and that also represent age-related "degenerative type" diseases. This, in fact, became one of the attractive features of our study in rural China *(44)* reported in the previous chapter.

Unlike the usual cross-sectional ecologic surveys where considerable focus is given only to one or a few diet–disease associations, which may in fact be fallacious *(44)*, this study provided considerable opportunity to determine consistency of effect among all relevant associations. With the addition of each relevant correlation to the overall analysis, hypothesis validation exponentially becomes more likely, whereas the possibility of ecologic fallacies exponentially becomes less likely. Thus, what is commonly considered to be a weakness in other cross-sectional studies becomes a strength in this one.

Another unusual strength of a study such as this one *(44)* was the inclusion of a population who already are adapted to the range of nutrient intakes being tested in the hypothesis. Nutritional adaptation, as was discussed previously, is seldom available for consideration in studies of individuals accustomed to nutritionally rich Western diets. Even to the extent that adaptation can be considered, it is usually ignored. Parenthetically, it should be noted that virtually all principles, concepts, and reference standards available in nutrition textbooks and other reference documents in use around the world have been obtained from research investigations on individuals who were accustomed to nutrient exposure levels that may represent—in fact, are likely to represent—only a partial adaptation to adverse health effects. As such, these reference standards and concepts may not represent a superior nutritional strategy for the longer term.

It would be highly desirable to be rigorous in the testing and validation of this broad and unique dietary hypothesis. Rigorous investigation is required for several reasons. First, it would encourage more discriminating research in the future. Second, the extraordinary societal implications of this hypothesis demand its careful analysis. Third, clarification of any apparent inconsistencies in the hypothesis would foster more constructive and less frivolous debate. In light of this view, I suggest that it is necessary to show that all constituent intakes uniquely found in plant foods are either inhibitory or are neutral in their effects on the development of these diseases.* Conversely, all constituent intakes uniquely present in animal foods either should promote or remain neutral in their effects on disease development.

There are several subhypotheses within this broad hypothesis that are inferred in this discussion and I acknowledge that these need empirical validation. Nonetheless, it is this broad hypothesis and this rationale for using a cross-sectional study in China, which I hope will be useful in furthering the debate on which kind of diet is best for the prevention of degenerative diseases.

Acknowledgments

This research was supported in part by NIH Grant 5RO1 CA33638, the Chinese Academy of Preventive Medicine, the United Kingdom Imperial Cancer Research Fund, the United States Food and Drug Administration, the

---

*This requirement is not as difficult as it seems because the biological activities of the constituents in plants known theoretically to favor degenerative disease formation (e.g., mutagens, alkaloids) will be controlled, either largely or entirely, by the other plant constituents simultaneously being consumed—if consumed in the proper amounts.

American Institute for Cancer Research, and several American industry groups.

References

1. Bennett JH. *On Cancerous and Cancroid Growths*. Edinburgh: Sutherland and Knox, 1849.
2. Bennett JH. *Clinical Lectures on the Principals and Practice of Medicine*. Fourth Edition. Edinburgh: Adam and Charles Black, 1865.
3. Tannenbaum A. The genesis and growth of tumors II. Effects of caloric restriction per se. *Cancer Res* 1942; **(2):**460–467.
4. Tannenbaum A, Silverstone H. Nutrition and the genesis of tumours. *Cancer* 1957; **1:**306–334.
5. Keys A, Anderson JT, Grande F. Serum-cholesterol response to dietary fat. *Lancet* 1957; **272:**787.
6. Keys A, Anderson JT, Grande F. Serum cholesterol response to changes in diet. *Metabolism* 1965; **14:**747–787.
7. United States Department of Health and Human Services. *The Surgeon General's Report on Nutrition and Health*. Washington, DC: Superintendent of Documents, US Government Printing Office, 1988.
8. Trowell HC, Burkitt DP. *Western Diseases: Their Emergence and Prevention*. London: Edward Arnold, 1981.
9. Dumont R. *Mes Combats*. Paris: Plon Publishers, 1989.
10. Walshe WH. *The Nature and Treatment of Cancer*. London: Taylor and Walton, 1846.
11. Waterlow JC. Metabolic adaptation to low intakes of energy and protein. *Ann Rev Nutr* 1986; **6:**495–526.
12. American Heart Association. Committee report, rationale for the diet-heart statement of the American Heart Association. *Circulation* 1982; **65:**839A–854A.
13. Committee on Diet Nutrition and Cancer. *Diet, Nutrition and Cancer*. Washington, DC: National Academy Press, 1982.
14. National Research Council, Committee on Diet and Health. *Diet and Health: Implications for Reducing Chronic Disease Risk*. Washington, DC: National Academy Press, 1989.
15. O'Connor TP, Campbell TC. Dietary guidelines. In: Ip C, Birt D, Mettlin C, Rogers A, eds. *Dietary Fat and Cancer*. New York: Liss, 1986; 731–771.
16. Keys A. Coronary heart disease in seven countries. *Circulation* 1979; **41(Suppl. I):**1–19.

17. Drasar BS, Irving D. Environmental factors and cancer of the colon and breast. *Br J Cancer* 1973; **27:**167–172.
18. Armstrong B, Doll R. Environmental factors and cancer incidence and mortality in different countries, with special reference to dietary practices. *Int J Cancer* 1975; **15:**617–631.
19. Carroll KK. Experimental evidence of dietary factors and hormone-dependent cancers. *Cancer Res* 1975; **35:**3374–3383.
20. Keys A. Seven countries. *A Multivariate Analysis of Death and Coronary Heart Disease*. Cambridge, MA: Harvard University Press, 1980.
21. Rous P. The influence of diet on transplanted and spontaneous mouse tumors. *J Exp Med* 1914; **20:**433–451.
22. Tannenbaum A, Silverstone H. Nutrition in relation to cancer. *Adv Cancer Res* 1953; **1:**451–501.
23. Ross MH, Bras G. Tumor incidence patterns and nutrition in the rat. *J Nutr* 1965; **87:**245–260.
24. Freedman LS, Clifford C, Messina M. Analysis of dietary fat, calories, body weight, and the development of mammary tumors in rats and mice: a review. *Cancer Res* 1990; **50:**5710–5719.
25. Rose DP. Dietary factors and breast cancer. *Cancer Surveys* 1986; **3:**671–687.
26. Carroll KK, Braden LM, Bell JA, Kalamegham R. Fat and cancer. *Cancer* 1986; **58:**1818–1825.
27. Cohen LA. Diet and cancer. *Sci Am* 1987; **257:**42–48.
28. US Department of Agriculture ARS. *Composition of Foods: Raw, Processed, Prepared*. American Handbook No. 8, 1968, Expansion (March, 1972). Washington, DC: Government Printing Office, 1978.
29. Greenwald P. Manipulation of nutrients to prevent cancer. *Hosp Pract* 1984; **19:**119–134.
30. Allen RE. *The Concise Oxford Dictionary of Current English*. Oxford, UK: Clarendon Press, 1990.
31. LaChance P. Dietary intake of carotenes and the carotene gap. *Clin Nutr* 1988; **7:**118–122.
32. Malone WF, Kelloff GJ, Boone C, Nixon DW. Chemoprevention and modern cancer prevention. *Prev Med* 1989; **18:**553–561.
33. Boone CW, Kelloff GJ, Malone WE. Identification of candidate cancer chemopreventive agents and their evaluation in animal models and human clinical trials: a review. *Cancer Res* 1990; **50:**2–9.
34. Block G. Vitamin C and cancer prevention: the epidemiologic evidence. *Am J Clin Nutr* 1991; **53:**270S–282S.

35. Chisholm GM. Antioxidants and atherosclerosis: a current assessment. *Clin Cardiol* 1991; **14(2 Suppl. I):** 25–30.
36. Block G. The data support a role for antioxidants in reducing cancer risk. *Nutr Revs* 1992; **50:** 207–213.
37. Prentice RL, Kakar F, Hursting S, Sheppard L, Klein R, Kushi LH. Aspects of the rationale for the Women's Health Trial. *J Natl Cancer Inst* 1988; **80:** 802–814.
38. Anonymous. Fred Hutchinson Center wins $140 million contract to coordinate NIH Women's Health Initiative. *Cancer Lett* 1992; **18:** 1–2.
39. Henderson MM, Kushi LH, Thompson DJ, Gorbach SL, Clifford CK, Insull W, Jr, Moskowitz M, Thompson RS. Feasibility of a randomized trial of a low-fat diet for the prevention of breast cancer: dietary compliance in the Women's Health Trial Vanguard Study. *Prev Med* 1990; **19:** 115–133.
40. White E, Shattuck AL, Kristal AR, Urban N, Prentice RL, Henderson MM, Insull W, Jr, Moskowitz M, Goldman S, Woods MN. Maintenance of a low-fat diet: follow-up of the Women's Health Trial. *Cancer Epidemiol Biomarkers Prev* 1992; **1:** 315–323.
41. Gorbach SL, Morrill-LaBrode A, Woods MN, Dwyer JT, Selles WD, Henderson M, Insull W, Jr, Goldman S, Thompson D, Clifford C, Sheppard L. Changes in food patterns during a low-fat dietary intervention in women. *J Am Diet Assoc* 1990; **90:** 802–809.
42. Insull WI, Jr., Henderson MM, Prentice RL, Thompson DJ, Clifford C, Goldman S, Gorbach S, Moskowitz M, Thompson R, Woods M. Results of a randomized feasibility study of a low-fat diet. *Arch Int Med* 1990; **150:** 421–427.
43. Willett WC, Stampfer MJ, Colditz GA, Rosner BA, Hennekens CH, Speizer FE. Dietary fat and the risk of breast cancer. *N Engl J Med* 1987; **316:** 22–28.
44. Chen J, Campbell TC, Li J, Peto R. *Diet, Life-Style and Mortality in China. A Study of the Characteristics of 65 Chinese Counties.* Oxford, UK; Ithaca, NY; Beijing, PRC: Oxford University Press; Cornell University Press; People's Medical Publishing House, 1990.
45. Boyd NF, Cousins M, Beaton M, Fishell E, Wright B, Fish B, Kriukov V, Lockwood G, Tritchler D, Hanna W, Page DL. Clinical trial of low-fat, high-carbohydrate diet in subjects with mammographic dysplasia: report of early outcomes. *J Natl Cancer Inst* 1988; **80:** 1244–1248.

46. Schectman G, McKinney WP, Pleuss J, Hoffman RG. Dietary intake of Americans reporting adherence to a low cholesterol diet (NHANES II). *Am J Public Health* 1990; **80**:698–703.

47. Sirtori CR, Noseda G, Descovich GC. Studies on the use of a soybean protein diet for the management of human hyperlipoproteinemias. In: Gibney MJ, Kritchevsky D, eds. *Animal and Vegetable Proteins in Lipid Metabolism and Atherosclerosis*. New York: Liss, 1983; 135–148.

48. Carroll KK. Dietary proteins and amino acids—their effects on cholesterol metabolism. In: Gibney MJ, Kritchevsky D, eds. *Animal and Vegetable Proteins in Lipid Metabolism and Atherosclerosis*. New York: Liss, 1983; 9–17.

49. Gaddi A, Ciarrocchi A, Matteucci A, Rimondi S, Raviglia G, Descovich GC, Sirtori CR. Dietary treatment for familial hypercholesterolemia—differential effects of dietary soy protein according to the apolipoprotein E phenotypes. *Am J Clin Nutr* 1991; **53**:1191–1196.

50. Hawrylewicz EJ, Huang HH, Kissane JQ, Drab EA. Enhancement of the 7,12-dimethylbenz(a)anthracene (DMBA) mammary tumorigenesis by high dietary protein in rats. *Nutr Reps Int* 1982; **26**:793–806.

51. Appleton BS, Campbell TC. Dietary protein intervention during the post-dosing phase of aflatoxin B1-induced hepatic preneoplastic lesion development. *J Natl Cancer Inst* 1983; **70**:547–549.

52. Hawrylewicz EJ, Huang HH, Liu J. Dietary protein enhancement of N-nitrosomethylurea-induced mammary carcinogenesis, and their effect on hormone regulation in rats. *Cancer Res* 1986; **46**:4395–4399.

53. Hawrylewicz EJ, Huang HH, Blair WH. Dietary soybean isolate and methionine supplementation affect mammary tumor progression in rats. *J Nutr* 1991; **121**:1693–1698.

54. Youngman LD, Campbell TC. Inhibition of aflatoxin B1-induced gamma-glutamyl transpeptidase positive (GGT+) hepatic preneoplastic foci and tumors by low protein diets: evidence that altered GGT+ foci indicate neoplastic potential. *Carcinogenesis* 1992; **13**:1607–1613.

55. Margen S, Chu J-Y, Kaufmann NA, Calloway DH. Studies in calcium metabolism. *Am J Clin Nutr* 1975; **27**:584–589.

56. Kerstetter JE, Allen LH. Dietary protein increases urinary calcium. *J Nutr* 1990; **120**:134–136.

57. Abelow BJ, Holford TR, Insogna KL. Cross-cultural association between dietary animal protein and hip fracture: a hypothesis. *Calcif Tissue Int* 1992; **50**:14–18.

58. Youngman LD. *The Growth and Development of Aflatoxin B1-Induced Preneoplastic Lesions, Tumors, Metastasis, and Spontaneous Tumors as They are Influenced by Dietary Protein Level, Type, and Intervention.* Ithaca, NY: Cornell University, PhD Thesis, 1990.

59. Dole VP, Dahl LK, Schwartz GC, Cotzias GC, Thaysen JH, Harris C. Dietary treatment of hypertension. The effect of protein on appetite and weight. *J Clin Invest* 1953; **32**:185–191.

60. Miller DS, Payne PR. Weight maintenance and food intake. *J Nutr* 1962; **78**:255–262.

61. Rothwell NJ, Stock MJ, Tyzbir RS. Mechanisms of thermogenesis induced by low protein diets. *Metabolism* 1983; **32**:257–261.

62. Horio F, Youngman LD, Bell RC, Campbell TC. Thermogenesis, low-protein diets, and decreased development of AFB1-induced preneoplastic foci in rat liver. *Nutr Cancer* 1991; **16**:31–41.

63. Huang HH, Hawrylewicz EJ, Kissane JQ, Drab EA. Effect of protein diet on release of prolactin and ovarian steroids in female rats. *Nutr Rpts Int* 1982; **26**:807–820.

64. Pyska H, Styczynski H. Effect of various protein levels in the diet on mammary gland growth in rats. *J Dairy Res* 1979; **46**:551–554.

65. Sanz MCA, Liu J-M, Huang HH, Hawrylewicz EJ. Effect of dietary protein on morphologic development of rat mammary gland. *J Natl Cancer Inst* 1986; **77**:477–487.

66. Subcommittee on Laboratory Animal Nutrition. *Nutrient Requirements of Laboratory Animals.* Second revised edition, number 10. Washington, DC: National Academy Press, 1972.

67. Dunaif GE, Campbell TC. Dietary protein level and aflatoxin B1-induced preneoplastic hepatic lesions in the rat. *J Nutr* 1987; **117**:1298–1302.

68. Meyskens FL. Coming of age—the chemoprevention of cancer. *N Engl J Med* 1990; **323**:825–827.

69. Self S, Prentice R, Iverson D, Henderson M, Thompson D, Byar D, Insull W, Gorbach S, Clifford C, Goldman S, Urban N. Statistical design of the Women's Health Trial. *Controlled Clin Trials* 1988; **9**:119–136.

70. Willett WC, Hunter DJ, Stampfer MJ, Colditz G, Manson JE, Spielgelman D, Rosner B, Hennekens CH, Speizer FE. Dietary fat and fiber in relation to risk of breast cancer. *JAMA* 1992; **268**:2037–2044.

71. Jones DY, Schatzkin A, Green SB, Block G, Brinton LA, Ziegler RG, Hoover R, Taylor PR. Dietary fat and breast can-

cer in the National Health and Nutrition Examination Survey I Epidemiologic follow-up study. *J Natl Cancer Inst* 1987; **79**:465–471.

72. Mills PK, Beeson WL, Phillips RL, Fraser GE. Dietary habits in breast cancer incidence among Seventh-Day Adventists. *Cancer* 1989; **64**:582–590.

73. Knekt P, Albanes D, Seppanen R, Aromaa A, Järvinen R, Hyvönen L, Teppo L, Pekkala E. Dietary fat and risk of breast cancer. *Am J Clin Nutr* 1990; **52**:903–908.

74. Goodwin PJ, Boyd NF. Critical appraisal of the evidence that dietary fat intake is related to breast cancer risk in humans. *J Natl Cancer Inst* 1987; **79**:473–485.

75. Willett W. *Nutritional Epidemiology*. New York: Oxford University Press, 1990.

76. Ames B. Dietary carcinogens and anticarcinogens. *Science* 1983; **221**:1256–1264.

Diet and Western Disease

Fat, Energy, and Cancer

Norman J. Temple

1. The Central Importance of Diet

It is difficult to believe that any reader can be unaware that a wide body of authoritative opinion holds that diet is a major cause of the diseases discussed in this book. The main focus of this chapter is the role of fat, particularly saturated fat, and energy intake in the causation of Western disease. Most attention is paid to cancer, but coronary heart disease and obesity are also examined.

However, as a preliminary it is important to consider key aspects of the supporting evidence. First, in order to properly appreciate the modern Western diet as a cause of Western disease, there is a discussion of non-Western diets. These include both the paleolithic diet and the diets eaten by preindustrial agricultural societies. This is followed by a discussion of the problems that arise with studies investigating the role of diet in disease.

2. A Brief History of Diet

2.1. Lessons from Evolution

In the Western world we have greatly altered our lifestyle in comparison with that lived by our distant ancestors. As a result it is easy to lose sight of the fact that

From: *Western Diseases: Their Dietary Prevention and Reversibility*
Edited by: N. J. Temple and D. P. Burkitt Copyright ©1994 Humana Press, Totowa, NJ

biologically we are virtually unchanged since paleolithic times. Twentieth century humans are the evolutionary product of that long epoch—the hundreds of thousands of years between the emergence of humans and the advent of agriculture. The laws of biology dictate that we must be well adapted to the environment of that period. For all practical purposes we can ignore any changes that may have occurred in the last 10,000 years.

The above concept and its profound relevance to our understanding of disease was repeatedly stressed by Cleave *(1,2)* and later by Boyden *(3,4)*.

Where the environment is little changed, we can still see clear evidence of how well we are adapted to the natural world: Our lungs and blood are perfectly suited to air containing 21% oxygen, our anatomy and physiology to a force of gravity of 1 g, and our eyes to daytime light. To me the most remarkable of these adaptations is the one to temperature extremes. A few years ago I moved from Puerto Rico to Edmonton, Canada. A January night in Puerto Rico is seldom below 19°C, whereas in this region it sometimes drops to below −40°C. Yet the native people, who share a common Siberian origin, managed to thrive in both regions.

Needless to say, the concept that we are highly adapted to the natural environment extends to the diet. This can be demonstrated by considering the close relationship between the supply and requirement of nutrients. The requirement for the 30 or so essential nutrients covers a range of seven orders of magnitude, from 2 μg of vitamin B_{12} to 50 g of protein. Yet for each we consume roughly what we require. This leads to the obvious conclusion that we have become adapted to the diet eaten during human evolution, namely the paleolithic diet. It is therefore highly instructive to examine that diet and to compare it with the modern Western one.

There is no doubt that the paleolithic diet contained appreciable amounts of meat, the fat content of which was much lower than that present in modern meat. Crawford et al. *(5,6)* analyzed meat from wild herbivorous mammals from Uganda and compared it with typical butcher's shop meat purchased in London. The Ugandan meat had a fat content of only 2–5% whereas in London the figure is around 20–40% (before removal of visible fat). In support of this, a major information source on the nutrient content of foods indicates that British meat typically has 16–31% fat (5–11% in major organs) *(7)*. Not only is African meat much leaner but a far larger proportion of the fat is in the form of polyunsaturated fats: 30% in Ugandan wild meat *(5,6)*, whereas for London the values are 2–4% *(5,6)* or 4.5% (9–15% in major organs) *(8)*. (The fat content of meat in Western countries may now be somewhat lower than the above figures indicate.)

Eaton and Konner *(9)* provided further information on the fat content of wild game, including many North American species. This confirms the above data. For instance, such animals as moose, caribou, deer, and bison generally produce a meat with a fat content of no more than 4%.

Eaton and Konner *(9)* also made a detailed study of the paleolithic diet. Assuming that the diet contains 35% meat by weight, which they consider to be a representative figure, then fat provides 21% of energy, and the P:S (polyunsaturated:saturated fat) ratio is 1.41. Even if the diet is 60% meat, the fat content is still only 23%, with a P:S ratio of 1.08. By contrast the modern American diet is around 38% fat with a P:S ratio of about 0.4. This surfeit of fat comes not only from meat but also from food items unavailable in paleolithic times: dairy foods, oils, and margarine. The paleolithic diet also contains at least twice as much dietary fiber. A consideration of the micronutrient content of this diet is given in the following chapter by Temple.

The above evidence strongly indicates that humans are adapted to a diet containing far less fat, especially saturated fat, and far more fiber than the modern diet. This does not prove that a high fat/low fiber diet causes disease, but it is certainly supportive of such a view.

Two further revelations from Eaton and Konner's paper require comment. The paleolithic diet had a high content of protein: At 35% meat it provided 34% of energy as protein, most of which was of animal origin. This is more than twice that of the current American diet. It is difficult to reconcile this with Campbell's evidence that even the level of animal protein in the American diet is too high (*see* preceding chapter by Campbell). The paleolithic diet contained 590 mg of cholesterol/12,600 kJ (3000 kcals), a level similar to the current American level. If the meat intake was 60%, then the cholesterol content rises to 1080 mg. At first glance this appears high and suggests that the diet was potentially atherosclerotic. However, it must be remembered that the major risk factors for coronary heart disease were absent: The diet was low in both fat and saturated fat; hypertension would have been rare (since the diet was low in sodium and high in potassium); cigarets were unavailable; and people were probably physically active. The high cholesterol intake would therefore have been innocuous.

2.2. The Diets of Non-Western Populations

The diets of various non-Western populations was documented in Burkitt and Trowell's book, *Refined Carbohydrates and Disease: Some Implications of Dietary Fibre (10)*. In addition, Eaton and Konner's study of the paleolithic diet, described earlier, was based on the diets eaten by hunter-gatherer societies who either still survive or survived until recently *(9)*. The key features of the great majority of such diets are their low content of fat and their high content of dietary fiber. Carbohydrates are mainly in the form of starch rather than sugar. What little sugar is eaten comes in the form of unprocessed fruit rather than highly refined sugar.

We must at this point draw a distinction between the small number of hunter-gatherer societies and the much larger number of preindustrial agricultural societies. Hunter-gatherers eat little or no cereal foods and no dairy foods at all *(9)*. Their vegetable foods mainly comprise roots, beans, nuts, tubers, and fruit. On the other hand, agricultural societies live mainly on cereals and often produce modest amounts of dairy foods. Traditionally, their cereals are unrefined: They comprise such foods as whole meal flour, brown rice, and unprocessed corn rather than white flour, white rice, or cornflakes.

In the first chapter of this book, Burkitt described the pioneering role of Cleave *(1)* in documenting the relationship between the consumption of refined carbohydrates and Western disease (or the "saccharine disease" as Cleave called it). This work directly led to the key role of dietary fiber becoming recognized and is therefore of tremendous importance. Unfortunately, Cleave dismissed any possible role for excess dietary fat as a cause of Western disease. He did this because of two errors in his reasoning. First, he failed to realize that the meat eaten during paleolithic times had a low fat content. Second, he placed much over-reliance on the availability in recorded history of meat and dairy food with a high fat content. The availability of these foods was a consequence of the Agricultural Revolution, which occurred only 8–10,000 years ago, and which was therefore too recent to be of evolutionary significance. Moreover, it is doubtful whether more than a small minority of humanity were able to eat a fat-rich diet *(11)*. Cleave therefore (wrongly) concluded that humans had eaten a fat-rich diet for a sufficiently long time to have become adapted to it.

The consumption of refined carbohydrates has a deleterious impact on the micronutrient content of the diet. The significance of this is discussed in the chapter "Vitamins and Minerals in Cancer, Hypertension, and Other Diseases" by Temple later in this volume.

3. Problems with Diet Studies

There are several major types of study for investigating the role of fat in disease causation. Comparisons of different populations (interpopulation studies) is the basis for so much in this area, as indicated in the chapter by Burkitt, "Western Diseases and What They Encompass," and the chapter by Walker, "Diet-Related Disease Patterns in South African Interethnic Populations." Studies on individuals within a population are also of tremendous importance. This can take the form of either a prospective study (also known as a cohort study; diet assessment or blood collection is done before onset of the disease in question) or a retrospective study (persons with the disease are compared to those without it). The term case-control study can be applied to either type, but is most often associated with retrospective studies and that term will be used here. Valuable indications of the causes of disease are also provided by clinical trials. For instance, if dietary fiber is an effective treatment for a disorder, then this suggests that lack of fiber is also a cause. Last, there is the evidence from animal investigations. There are now animal models for a wide variety of human disorders, including numerous cancers.

Those of us who have been following the multifaceted debate on the relationship between diet and disease will, no doubt, have been endlessly confused and frustrated by the many contradictions in the evidence. This permeates virtually every aspect of the field. Some of the responsible causes are now briefly discussed.

Studies on individuals are the ones most prone to generating contradictory findings. There appear to be four major reasons for this:

1. It is well established that dietary assessment is subject to serious error. An excellent example of this is the role of salt in hypertension, as discussed in the "Vitamins and Minerals" chapter by Temple. Frost et al. *(12)* demonstrated that many of the reports concerning the relationship between individual intake of salt and blood pressure have failed to detect a positive association simply because of error in measuring salt intake.

2. Many dietary factors are highly correlated. This creates obvious difficulties in distinguishing between true causal associations and spurious ones. For instance, vegetarians often have a high intake of fruit and vegetables. As a result they not only have a low intake of fat but also a high intake of particular nutrients, such as β-carotene. Thus any disease that is uncommon in vegetarians will likely be positively associated with fat and negatively associated with β-carotene. Similarly, a person who has a high energy intake will almost certainly also have a high intake of fat, carbohydrate, and protein; this will inevitably lead to further diet–disease associations. This type of spurious relationship can be characterized as guilt (or protection) by association.

3. The great majority of people within a population tend to have a broadly similar diet. As a result dietary differences may not be large enough to expose diet–disease relationships in prospective and case-control (retrospective) studies. Factors such as genetics and error in diet assessment may generate too much "noise." This problem was explored by Hebert and Kabat *(13)*. First, they noted that differences in fat intake are far larger in animal experiments and

population comparisons than in studies of individuals and therefore the latter type of study is the one least likely to show a relationship between fat intake and cancer. They illustrated this with a case-control study of cigarets and lung cancer. Among men a comparison of never smokers and those smoking 1–19 cigarets/d gave a risk ratio of 3.84. They then repeated the analysis but with the exclusion of all subjects with a low consumption of cigarets. As a result, the comparison groups (15–20 vs 21–30 cigarets/d) became as homogeneous for smoking as are most populations for fat intake. This reduced the risk ratio to a mere 1.04!

4. The critical period for a dietary component to influence the development of a disease may occur ten or more years before the disease actually appears. Clearly, this is more likely to be a problem with case-control studies rather than with prospective ones.

It is easy to imagine these four factors acting in concert, causing several bits of confusion to become a thick fog. Generally, these problems create erratic errors, so particular diet–disease associations are observed in some studies but not others. The high correlation between diet variables is a partial exception to this: It often reduces the strength of genuine diet–disease relationships, but it also generates many false diet–disease relationships, and these can appear consistently.

Let us now approach the problem in reverse. Suppose several studies indicate that factor X is associated with disease Y, whereas other studies show no such relationship. Does the relationship actually exist? We may infer that it probably does, but that the various errors listed cause it to be undetectable for much of the time. We must, however, add two qualifiers. First, the relationship may be truly present but noncausal. Second, large, carefully conducted studies are obviously much more reliable than small, poorly done ones.

4. Cancer

4.1. Interpopulation Studies

In the chapter "Western Diseases and What They Encompass," Burkitt categorized several cancers as Western diseases based on their worldwide distribution. These are colorectal, breast, prostate, lung, and endometrial cancer.

In their landmark study, Armstrong and Doll *(14)* correlated various dietary factors with the international variation in cancer incidence and mortality. Their data indicate that the factors most strongly related to cancer of the colon, rectum, breast, prostate, and ovary are total fat intake and animal protein. Meat also figures prominently, particularly for cancer of the colon and rectum.

More recently, Prentice and Sheppard *(15)* carried out a similar study, focusing on people aged 30–44 and 55–69. They too reported a strong relationship, particularly in the older age group, between fat intake and cancer of the colon, breast, and ovary. In the older group, rectal cancer had a weak relationship with fat, although prostate cancer had a strong one, but neither cancer was related to fat in the younger group. In both this study and the one by Armstrong and Doll *(14)*, fat was more closely correlated with cancer of the colon than of the rectum. Prentice and Sheppard also did a time trend analysis for 10 countries for the 15 years up to 1978–1982. This generally supported the first analysis. However, rectal cancer now showed as much association with fat as did colon cancer. As before, relationships were stronger in the older group. Neither type of analysis indicated a significant role for energy intake.

In a review on breast cancer, Goodwin and Boyd *(16)* assembled additional evidence incriminating dietary fat. Two time trend studies revealed a significant correlation between fat intake and breast cancer. Four out of six national or regional studies that examined total fat intake also saw a significant association with the disease.

At this point it is pertinent to refer to a skeptical viewpoint of the evidence against fat. Willett and Stampfer (17) wrote a critique of Prentice and Sheppard's above epidemiological study (15). In particular, they stressed that data on national food intake may be seriously flawed because of a failure to allow for food wastage. They believed that it is extremely difficult to distinguish between fat intake and energy intake. Other limitations of international studies are well known, most notably that it is never possible to properly allow for confounding variables.

A small study of vegetarian British nuns could find no evidence that they had a lower rate of colon or breast cancer in comparison with other unmarried women (18). On the other hand, studies of Californian Seventh-Day Adventists, a religious group who habitually eat little meat, reveal that they have a low incidence of cancer of the colon and breast (as well as of the prostate) (19).

Clearly, the various types of population studies cannot provide wholly consistent and reliable evidence. Nevertheless, the weight of evidence points to an important role for a high fat diet in certain cancers, particularly colon (i.e., colorectal), breast and prostate, and perhaps ovary. The fat–cancer association is stronger in older than in younger age groups.

4.2. Studies on Individuals

The focus of this section is on whether a diet with a high content of fat or energy increases the risk of certain cancers.

4.2.1. Cancer of the Colon and Rectum

Numerous studies have investigated the dietary histories of patients with cancer of the colon or rectum. Sometimes the two types of cancer have been separately examined;

this permits a comparison of their risk factors. Several studies have observed essentially no difference between the relationship of fat and energy to the two cancers (20–24). Fat and energy intake appear to be more closely associated with rectal cancer in four studies (25–28), with colon cancer in one study (29) and to have no particular trend in three studies (30–32). The studies finding no difference between the two cancers have generally been more reliable based on size and the quality of the diet assessment. This evidence therefore indicates that colon and rectal cancer have the same risk factors, at least with respect to fat and energy. Accordingly, no distinction will henceforth be made between the two cancers. (International studies, it will be recalled, have indicated that fat is more closely associated with cancer of the colon than of the rectum. The cause of this discrepancy clearly requires further investigation.)

It has often been pointed out that there are many apparent inconsistencies concerning whether dietary fat is a risk factor for colon (i.e., colorectal) cancer. There appear to be two main causes of this. First, there is the important question of the type of population being studied. Often studies that have failed to detect at least a weak relationship between fat and colon cancer have been carried out in Mediterranean or East Asian locations. These include Marseilles (33), Northern Italy (22), Israel (34), Japan (35,36), China (25), and Singapore (37). Similarly, a study from Majorca reported a stronger relationship for protein and carbohydrate intake than for fat (31). Mediterranean and East Asian populations typically consume appreciably less fat and saturated fat than populations from northern Europe or North America.

The second important source of negative findings is that many of the studies are based, not on the actual fat intake, but on the intake of fat-rich foods (21,22,28,34–36,38,39). Moreover, in seven of these eight studies the diet assessment was made during the 1960s or 1970s, at which time the methodology was considerably less accurate than it is today. Where populations from northern Europe (i.e., Britain and Sweden), North America, or Australia have been studied using reliable methodology, an association between fat and colon cancer has been demonstrated in nearly all cases. In some studies, the relationship has been weak (20,23,30,40–42), whereas in others it has been fairly clear (24,25,43–45). Taken as a whole these studies indicate that persons whose fat intake is in the top quantile have a risk ratio of typically 1.4 to 2.5 compared to those in the lowest group. In addition, six studies have succeeded in demonstrating a relationship, albeit a weak one in some cases, between the consumption of fat-rich foods and colon cancer (26,29,46–49). Only one study appears to conflict with this conclusion: Essentially no relationship was observed between fat intake and colon cancer in a large, carefully conducted case-control study done in Belgium (32).

Two other negative studies were recently published. A case-control study in Argentina observed no association between fat intake and colon cancer (50). People in this population have a high fat intake, since they eat meat nearly every day. A large prospective study in the United States failed to detect an association between fat intake and colon cancer (51). However, the estimation of fat intake was crude, and energy intake was not measured.

Several studies have reported on the relationship between different types of fat and the risk of colon cancer. In nearly all cases, saturated fat rather than polyunsaturated fat stands out as the type most associated with the disease (20,24,25,30,45,52). In only one study was polyunsaturated fat clearly associated with colon cancer (41). Three studies have suggested that there may also be an association for monounsaturated fat (20,24,45).

The next question to consider is that of energy intake. Eight studies reported a clear association between energy intake and colon cancer (20,24,25,30,31,40,43,50), three studies saw a weaker association *(41,42,44)*, whereas five studies saw essentially no association *(23,27,33,45,53)*. In one study energy intake was associated with rectal cancer but not colon cancer *(32)*. No study has reported a lower energy intake in cases than controls. Clearly, therefore, the batting average points to a relatively high energy intake as a risk factor for the disease. However, since three of the studies that reported no association were large and well conducted, one in Australia *(23)*, one in the United States *(45)*, and one in Marseilles *(33)*, it would be premature to draw firm conclusions.

Diets that are rich in fat also tend to be high in energy, because fat is a concentrated source of energy. This makes it difficult to determine which has the stronger relationship with colon cancer. For instance, Lyon et al. *(40)* concluded that in their case-control study it is not possible to distinguish between the relative roles of fat and energy.

Despite these problems, a number of studies have presented data on the relative strengths of the associations that fat and energy have with colon cancer. Two studies found the association clearly stronger for energy *(31,50)*, three found it somewhat stronger for energy *(30,32,43)*, four found essentially no difference *(20,24,41,42)*, two found it somewhat stronger for fat *(23,44)*, and one found it clearly stronger for fat *(45)*. The conclusion that best fits the evidence is that fat and energy are both important risk factors for the disease and that they are highly interrelated.

4.2.2. Breast Cancer

Howe et al. *(54)* recently carried out a combined analysis of 12 case-control studies, comprising 4427 cases and 6095 controls. A consistent association was demonstrated between the intake of fat, saturated fat, and monounsaturated fat and the risk of breast cancer in postmenopausal women; highly significant odds ratios of 1.4–1.5 were observed when comparing the top and bottom quintiles.

These associations were largely absent in premenopausal women. However, in a critical comment, Willett and Stampfer (17) argued that these odds ratios are exaggerated because of an inappropriate statistical analysis.

Several other studies have also indicated that fat is a risk factor for breast cancer (55–57). In one of these the association was strongest for saturated fat (55). Similarly, a Japanese prospective study reported that meat consumption is related to the disease in postmenopausal women (58). A Finnish prospective study reported that fat is associated with breast cancer but only after adjusting for energy intake (59).

The evidence linking dietary fat with breast cancer is far from consistent. In particular, Willett et al. (60) carried out a large prospective study on American nurses (in their latest report based on eight years of followup, there were 1439 cases in 89,494 subjects). Fat and saturated fat had a weak negative (i.e., apparently protective) relationship with both pre- and postmenopausal breast cancer. No association between fat intake and risk was observed even among women at extremes of fat intake (49% or more vs 29% or less of energy). Several other studies have also failed to detect an association for fat or saturated fat and the disease (19,61–64).

Kushi et al. (65) suggested that these conflicting data may to some extent reflect differing methods of adjusting for energy intake. They carried out a prospective study in Iowa and reported that the presence of an association between fat intake and breast cancer depends on the method used to control for energy intake.

There is mixed evidence that a high energy intake is a risk factor for breast cancer. The combined case-control study reported a clear relationship between energy intake and breast cancer, but only for postmenopausal cases (54). This relationship was entirely accounted for by the higher fat intake in cases. Another case-control study gave similar results *(56)*. On the other hand, the study of American nurses saw no such relationship *(58)* and neither did four other studies *(59,63–65)*.

The role of fat and energy intake in breast cancer is therefore far from clear. This appears to be at variance with interpopulation studies that have pointed to a major role for fat, particularly in older age groups. One possible explanation for these seemingly contradictory observations is that the critical period for a high intake of fat to increase the risk of breast cancer occurs before adulthood. Another possibility is that dietary fat increases breast cancer risk up to a fat intake of about 30% of energy, but beyond that level a plateau occurs.

4.2.3. Prostate Cancer

Prostate cancer resembles colon cancer in that a diet with a high content of fat or of fat-rich foods appears to predispose to the disease. The association is strongest for animal fat or saturated fat. Some studies have demonstrated this with reasonable clarity *(66–69)*, whereas in others the association is weak *(70–75)*. In one of the latter studies, the correlation was stronger for unsaturated than for saturated fat *(72)*, an unusual observation for a Western cancer. In some of these studies the relationship is seen only in particular age groups but no overall pattern is apparent *(66,73–75)*. Three studies failed to detect an association between dietary fat or fat-rich foods and prostate cancer *(76–78)*. The weight of evidence indicates, therefore, that dietary fat is an important risk factor, but further clarification is required.

Two studies reported essentially no difference for energy intake between cases and controls *(76,78)*, whereas a third showed a higher energy intake in cases but only in one age subgroup *(75)*.

4.2.4. Pancreatic Cancer

Pancreatic cancer is not categorized as a Western disease, but it will be so considered here, since studies on individuals have generated data highly relevant to this discussion. Certain other cancers will also be included for the same reason.

Some studies have indicated that subjects with cancer of the pancreas have a relatively high intake of fat *(79)* or of fat-rich animal foods *(36,80)*. In one study, the association was for bacon and fried or grilled meat, suggesting that the causative agent may be mutagens *(81)*. However, a positive association for fat could not be demonstrated in the majority of studies, particularly the more reliable ones *(82–85)*. Some studies have indicated that cholesterol has a strong association with the cancer *(82)*. Overall, the role of fat in pancreatic cancer is far from clear.

A large pooled study (802 cases) linked a high energy intake with cancer of the pancreas *(82)*. In marked contrast, one study observed a strong negative relationship between energy intake and risk *(79)*. However, the latter study was small (only 69 cases). On balance, therefore, energy could well be an important factor.

4.2.5. Other Cancers

Lung cancer cases are reported to show a weak trend toward an above average intake of fat *(86,87)*, fat-rich foods *(88,89)*, and of saturated fat *(87)*. This could not be confirmed in two other studies *(90,91)*, but in these the diet assessment was made around 1960 and in one of them there were only 33 cases *(91)*. There is no evidence that energy intake plays a role *(86,91,92)*. A high intake of fat and saturated fat has also been linked to oral and pharyngeal cancer *(93)*.

Ovarian cancer cases have also been reported to have a relatively high fat intake, particularly of animal fat. The risk ratio (high vs low intake) is about 1.6–2.2. This has been seen in three studies *(94–96)*, but not in two others *(97,98)*. No differences for energy intake have been reported in case-control studies *(96,98)*. A study of cancer of the cervix reported no differences for intake of fat or energy *(99)*. A significant association between bladder cancer and saturated fat intake has been reported *(100)*. Gallbladder cancer has been linked to energy intake but not to fat intake *(101)*.

4.3. Conclusions from Human Studies

Taken as a whole, a large body of evidence indicates that fat consumption is strongly related to colon (i.e., colorectal) cancer and, to a lesser extent, to prostate cancer. Lung, oral-pharyngeal, ovarian, and bladder cancer may also be related to fat intake, but evidence for this is weak. Saturated rather than polyunsaturated fat clearly has the primary responsibility for the fat–cancer association. Numerous studies have indicated that energy consumption is also linked to various cancers, but it is difficult to differentiate between this and fat consumption. The relationship of fat intake to breast and pancreatic cancer is still unresolved.

4.4. Studies on Animals

There has been much controversy over the respective roles of fat and energy intake in experimental carcinogenesis. Dietary fat exerts a promoting action on chemically induced pancreatic tumors in rats and hamsters *(102)*. Similarly, a recent review documented the fact that a high fat diet promotes mammary tumors in rats and mice *(103)*. This is based mainly on chemically induced tumors but includes some reports on spontaneous tumors in mice. However, in a massive study (5700 female rats and mice), which was not referred to in that review, no effect was observed on the incidence of spontaneous mammary adenocarcinomas when diets low or high in fat were given for 2 years *(104)*.

In a series of experiments, Kritchevsky and colleagues
(105–108) demonstrated that energy restriction has a strong
inhibiting effect on mammary tumor formation in rats.

In a detailed analysis, Freedman et al. (103) concluded
that energy and dietary fat have independent effects on
mammary tumors. The effect of fat is two-thirds the magnitude of the energy effect in both mice and Sprague-
Dawley rats.

Dietary fat reportedly increases colon tumor incidence
in Fischer 344 rats but not in Sprague-Dawley rats (109).
Energy restriction has an inhibitory effect on colon tumor
development in rats (106,108), but energy seems unrelated
to the disease when the animals are fed ad libitum (109).

In summary, a high fat diet generally increases the
yield of pancreatic, colon, and mammary tumors in experimental animals, whereas an energy restricted diet protects against colon and mammary tumor development.
These effects appear to be distinct (i.e., fat increases tumor
yield independently of any increase in energy intake).
However, the influence of fat and energy greatly depends
on such factors as the type of tumor, carcinogenic agent,
and species and strain of animal.

For colon tumors, the animal and human data are reasonably consistent. For mammary tumors, the animal data
infer that fat and energy may also be factors in human
breast cancer. This clearly supports the interpopulation
studies rather than the studies on individuals.

5. Coronary Heart Disease

This disease is discussed in more detail in later chapters. There is little doubt that a high intake of saturated fat
and cholesterol has a causal relationship with coronary
heart disease (CHD). The most convincing evidence for this
comes from three key sets of observations. First, the blood
cholesterol level is strongly influenced by the dietary intake

of saturated fat and cholesterol *(110)*. Second, studies in non-human primates demonstrate that atherosclerosis develops in response to a high blood cholesterol level and regresses when the cholesterol level is lowered *(111–113)*. Similarly, growing evidence coming from clinical studies employing angiography indicates that this also holds true in humans *(114)*. Third, clinical trials show that lowering the blood cholesterol level prevents CHD *(115)*.

The above evidence provides strong support for the lipid hypothesis. However, it still has important inadequacies. In a previous article I pointed to numerous inconsistencies in the epidemiological association between fat consumption and CHD and argued that the disease actually has a closer association with refined carbohydrates *(116)*. That evidence has not been refuted. The possible role of refined carbohydrates in CHD therefore merits further attention.

One factor that has been relatively neglected in the etiology of CHD is that of energy intake. Numerous prospective studies have demonstrated that a relatively high energy intake is protective *(117–122)*. However, this probably reflects the benefits of physical activity. The question of whether an excess energy intake is an independent risk factor is almost impossible to measure, since it is overshadowed by variations in energy intake owing to physical activity and other factors. An excess intake of energy may well be a common link between several related diseases, particularly obesity, diabetes (type II), and CHD. It could also explain why both fat and refined carbohydrates are associated with all three diseases: Both tend to result in a surfeit of calorie. Similarly, the protection afforded by exercise against these diseases can be explained, at least in part, by the fact that exercise dissipates excess energy.

With respect to the total intake of fat, the weight of evidence indicates that this is of much less importance than the type of fat *(110,125)*. It has become increasingly apparent in recent years that monounsaturated fat has a beneficial effect on the blood lipid profile. Like polyunsaturated fat it lowers the serum cholesterol, but unlike that fat it does not lower the high-density lipoprotein cholesterol *(123)*.

In summary, a vital component of any strategy aimed at the prevention and treatment of CHD is a major reduction in intake of saturated fat and cholesterol *(124)*. Polyunsaturated and monounsaturated fats, on the other hand, are relatively safe. However, since all dietary fats tend to cause excess energy intake *(see* following section), unsaturated fat should be viewed as a substitute for saturated fat, not as a dietary supplement.

ˎ 6. Obesity

The key question that must be addressed is what aspect of the diet causes an excessive energy intake. Here we focus on the role of fat. However, the crucial role of refined carbohydrates, including lack of fiber, should not be forgotten *(1)*.

Experimental studies have repeatedly demonstrated that a high fat diet induces obesity in rats, mice, and hamsters *(125–128)*. Although some conflicting observations have been reported *(129,130)*, the majority of human studies have indicated that a high fat diet does induce excessive energy intake and thence weight gain. For instance, in a trial 13 women were given unlimited amounts of diets supplying 22 or 37% of energy as fat for 11 wk. The low fat diet caused a spontaneous reduction in energy intake (and consequent weight loss; both changes were significant) *(131)*. A similar trial carried out over 2 years on 303 mildly overweight women also observed significant weight loss with a low fat diet *(132)*.

Since dietary fiber tends to decrease energy intake *(133)* it is predictable that it will work synergistically with a low fat diet. In one experiment in which this was tested subjects were given a high fiber/low fat diet or a low fiber/ high fat one and were instructed to eat to satiety *(134)*. The energy intake was little more than half on the high fiber/low fat diet. Unfortunately, the trial only lasted 5 d. In another study, obese Hawaiians were switched from a typical American diet to one low in fat and high in fiber *(135)*. Average energy intake fell 40% and average weight fell by 7.8 kg in 21 d. Evidence supporting the value of high fiber/low fat diets for the long-term treatment of obesity is reviewed in the chapter by Anderson and Akanji.

On balance, the above evidence indicates that a high fat diet induces a high energy intake, which infers, therefore, that it is an important causative factor of obesity. Support for this conclusion comes from studies showing that dietary fat correlates with the degree of adiposity *(136–139)*. A high fat diet may also cause weight gain independently of excess energy intake. This was demonstrated in a study in which women were fed a diet supplying 20 or 37% of energy as fat *(140)*. They were requested to eat sufficient food so as to maintain their weight. Women on the low fat diet had a significantly higher energy intake but, nevertheless, still lost weight (2.8% over 20 wk).

7. Summary

Let us now return to a theme discussed early in this chapter, namely the paleolithic diet. That diet supplied about 21% of energy as fat, which suggests that this is the fat intake for which evolution has left us best suited. The examination of the evidence linking fat to cancer, CHD, and obesity strongly strengthens the case that the high fat content of the Western diet is a major cause of these diseases. In their chapter, Anderson and Akanji discuss the benefits of a high fiber/high carbohydrate/low fat diet in the treatment of diabetes, and this infers that a high fat diet may have an important role in that disease also.

Studies of the paleolithic diet indicate that saturated fat is the type of fat whose intake has expanded most since evolutionary times. This is entirely consistent with the evidence showing that cancer and CHD are far more closely associated with saturated fat than with other fats. Polyunsaturated and monounsaturated fats play no causative role in CHD, but they are probably as harmful as saturated fat in obesity. Monounsaturated fat appears to be more closely associated with cancer than is polyunsaturated fat.

Closely associated with a high fat diet is an excess intake of energy. There is good evidence that this is an important factor in certain cancers. Though it is speculative it may also be involved in CHD. The obvious role of overconsumption in obesity and the association of that condition with CHD, diabetes, hypertension, and some cancers also supports the role of excess energy in all of these diseases. It follows, therefore, that an excess intake of energy may be a common denominator in a variety of Western diseases.

References

1. Cleave TL, Campbell GD. *Diabetes, Coronary Thrombosis, and the Saccharine Disease*. Bristol, UK: Wright, 1966.
2. Cleave TL. The neglect of natural principles in current medical practice. *J Royal Naval Med Serv* 1956; **42**:55–83.
3. Boyden S. Evolution and health. *Ecologist* 1973; **3**:304–309.
4. Boyden S. *Western Civilization in Biological Perspective. Patterns in Biohistory*. Oxford: Oxford University Press, 1987; 17–21.
5. Crawford MA. Fatty-acid ratios in free-living and domestic animals. Possible implications for atheroma. *Lancet* 1968; **i**:1329–1333.
6. Crawford MA, Gale MM, Woodford MH, Casped NM. Comparative studies on fatty acid composition of wild and domestic meats. *Int J Biochem* 1970; **1**:295–305.

7. Paul AA, Southgate DAT. *McCance and Widdowson's The Composition of Foods*. London: HMSO, 1978.
8. Paul AA, Southgate DAT. *McCance and Widdowson's The Composition of Foods*. Supplement. London: HMSO, 1980.
9. Eaton SB, Konner M. Paleolithic nutrition. A consideration of its nature and current implications. *N Engl J Med* 1985; **312**:283–289.
10. Burkitt DP, Trowell HC. *Refined Carbohydrate Foods and Disease. Some Implications of Dietary Fibre*. London: Academic Press, 1975.
11. Harris M. *Cannibals and Kings. The Origins of Cultures*. New York: Random House, 1977.
12. Frost CD, Law MR, Wald NJ. By how much does dietary salt reduction lower blood pressure? II. Analysis of observational data within populations. *Br Med J* 1991; **302**:815–818.
13. Hebert JR, Kabat GC. Distribution of smoking and its association with lung cancer: implications for studies on the association of fat with cancer. *J Natl Cancer Inst* 1991; **83**:872–874.
14. Armstrong B, Doll R. Environmental factors and cancer incidence and mortality in different countries, with special reference to dietary practices. *Int J Cancer* 1975; **15**:617–631.
15. Prentice RL, Sheppard L. Dietary fat and cancer: consistency of the epidemiological data, and disease prevention that may follow from a practical reduction in fat consumption. *Cancer Causes Contr* 1990; **1**:81–97.
16. Goodwin PJ, Boyd NF. Critical appraisal of the evidence that dietary fat is related to breast cancer risk in humans. *J Natl Cancer Inst* 1987; **79**:473–485.
17. Willett WC, Stampfer MJ. Dietary fat and cancer: another view. *Cancer Causes Contr* 1990; **1**:103–109.
18. Kinlen LJ. Meat and fat consumption and cancer mortality: a study of strict religious orders in Britain. *Lancet* 1982; **i**:946–949.
19. Phillips RL, Snowdon DA. Association of meat and coffee use with cancers of the large bowel, breast, and prostate among Seventh-Day Adventists: preliminary results. *Cancer Res* 1983; **43**:2403S–2408S.
20. Jain M, Cook GM, Davis FG, Grace MG, Howe GR, Miller AB. A case-control study of diet and colorectal cancer. *Int J Cancer* 1980; **26**:757–768.
21. Phillips RL, Snowdon DA. Dietary relationships with fatal colorectal cancer among Seventh-Day Adventists. *J Natl Cancer Inst* 1985; **74**:307–317.

22. La Vecchia C, Negri E, Decarli A, D'Avanzo B, Gallotti L, Gentile A, Franceschi S. A case-control study of diet and colorectal cancer in Northern Italy. *Int J Cancer* 1988; **41:**492–498.
23. Kune S, Kune GA, Watson LF. Case-control study of dietary etiological factors: the Melbourne colorectal cancer study. *Nutr Cancer* 1987; **9:**21–42.
24. Gerhardsson De Verdier M, Hagman U, Steineck G, Rieger A, Norell SE. Diet, body mass and colorectal cancer: a case-referent study in Stockholm. *Int J Cancer* 1990; **46:**832–838.
25. Whittemore AS, Wu-Williams AH, Lee M, Shu Z, Gallagher RP, Deng-ao J, Lun Z, Xianghui W, Kun C, Jung D, Teh C, Chengde L, Yao XJ, Paffenbarger RS, Henderson BE. Diet, physical activity, and colorectal cancer among Chinese in North America and China. *J Natl Cancer Inst* 1990; **82:**915–926.
26. Pickle LW, Greene MH, Ziegler RG, Toledo A, Hoover R, Lynch HT, Fraumeni JF. Colorectal cancer in rural Nebraska. *Cancer Res* 1984; **44:**363–369.
27. Stemmermann GN, Nomura AMY, Heilbrun LK. Dietary fat and the risk of colorectal cancer. *Cancer Res* 1984; **44:**4633–4637.
28. Graham S, Dayal H, Swanson M, Mittlemar A, Wilkinson G. Diet in the epidemiology of cancer of the colon and rectum. *J Natl Cancer Inst* 1978; **61:**709–714.
29. Tajima K, Tominaga S. Dietary habits and gastrointestinal cancers: a comparative case-control study of stomach and large intestinal cancers in Nagoya, Japan. *Jpn J Cancer Res* 1985; **76:**705–716.
30. Potter JD, McMichael AJ. Diet and cancer of the colon and rectum. A case-control study. *J Natl Cancer Inst* 1986; **76:**557–569.
31. Benito E, Stiggelbout A, Bosch FX, Obrador A, Kaldor J, Mulet M, Nunoz N. Nutritional factors in colorectal cancer risk: a case-control study in Majorca. *Int J Cancer* 1991; **49:**161–167.
32. Tuyns AJ, Haelterman M, Kaaks R. Colorectal cancer and the intake of nutrients: oligosaccharides are a risk factor, fats are not. A case-control study in Belgium. *Nutr Cancer* 1987; **10:** 181–196.
33. Macquart-Moulin G, Riboli E, Cornee J, Charnay B, Berthezene P, Day N. Case-control study on colorectal cancer and diet in Marseilles. *Int J Cancer* 1986; **38:**183–191.
34. Modan B, Barell V, Lubin F, Modar M, Greenberg RA, Graham S. Low-fiber intake as an etiologic factor in cancer of the colon. *J Natl Cancer Inst* 1975; **55:**15–18.
35. Haenszel W, Locke FB, Segi M. A case-control study of large bowel cancer in Japan. *J Natl Cancer Inst* 1980; **64:**17–22.

36. Hirayama T. A large-scale cohort study on the relationship between diet and selected cancers of the digestive organs. In: Bruce WR, Correa P, Lipkin M, Tannenbaum SR, eds. *Gastrointestinal Cancer: Endogenous Factors*. Cold Spring Harbor, NY: Cold Spring Harbor Laboratory, 1981; 409–426.

37. Lee HP, Gourley L, Duffy SW, Esteve J, Lee J, Day NE. Colorectal cancer and diet in an Asian population—A case-control study among Singapore Chinese. *Int J Cancer* 1989; **43**:1007–1016.

38. Higginson J. Etiologic factors in gastrointestinal cancer in man. *J Natl Cancer Inst* 1966; **37**:527–545.

39. Bjelke E. Epidemiology of colorectal cancer, with emphasis on diet. In: Davis W, Harrap KR, Stathopoulos G, eds. *Human Cancer: Its Characterization and Treatment*. Amsterdam: Excerpta Medica, 1980; 158–174.

40. Lyon JL, Mahoney AW, West DW, Gardner JW, Smith KR, Sorenson AW, Stanish W. Energy intake: its relationship to colon cancer risk. *J Natl Cancer Inst* 1987; **78**:853–861.

41. West DW, Slattery ML, Robison LM, Schuman KL, Ford MH, Mahoney AW, Lyon JL, Sorensen AW. Dietary intake and colon cancer: sex- and anatomic site-specific associations. *Am J Epidemiol* 1989; **130**:883–894.

42. Freudenheim JL, Graham S, Marshall JR, Haughey BP, Wilkinson G. A case-control study of diet and rectal cancer in western New York. *Am J Epidemiol* 1990; **131**:612–624.

43. Bristol JB, Emmett PM, Heaton KW, Williamson RCN. Sugar, fat, and the risk of colorectal cancer. *Br Med J* 1985; **291**:1467–1470.

44. Graham S, Marshall J, Haughey B, Mittelman A, Swanson M, Zielezny M, Byers T, Wilkinson G, West D. Dietary epidemiology of cancer of the colon in western New York. *Am J Epidemiol* 1988; **128**:490–503.

45. Willett WC, Stampfer MJ, Colditz GA, Rosner BA, Speizer FE. Relation of meat, fat, and fiber intake to the risk of colon cancer in a prospective study among women. *N Engl J Med* 1990; **323**:1664–1672.

46. Haenszel W, Berg JW, Segi M, Kurihara M, Locke FB. Large-bowel cancer in Hawaiian Japanese. *J Natl Cancer Inst* 1973; **51**:1765–1779.

47. Manousos O, Day NE, Trichopoulos D, Gerovassilis F, Tzonou A, Polychronopoulou A. Diet and colorectal cancer: case-control study in Greece. *Int J Cancer* 1983; **32**:1–5.

48. Morgan JW, Fraser GE, Phillips RL, Andress MH. Dietary factors and colon cancer incidence among Seventh-Day Adventists. *Am J Epidemiol* 1988; **128:**918.

49. Bidoli E, Franceschi S, Talamini R, Barra S, La Vecchia C. Food consumption and cancer of the colon and rectum in northeastern Italy. *Int J Cancer* 1992; **50:**223–229.

50. Iscovich JM, L'Abbe KA, Castelleto R, Calzona A, Bernedo A, Chopita NA, Jmelnitzsky AC, Kalder J, Howe GR. Colon cancer in Argentina. II: Risk from fibre, fat and nutrients. *Int J Cancer* 1992; **51:**858–861.

51. Thun MJ, Calle EE, Namboodiri MM, Flanders WD, Coates RJ, Byers T, Boffetta P, Garfinkel L, Heath CW. Risk factors for fatal colon cancer in a large prospective study. *J Natl Cancer Inst* 1992; **84:**1491–1500.

52. Dales LG, Friedman GD, Ury HK, Grossman S, Williams SR. A case-control study of relationships of diets and other traits to colorectal cancer in American blacks. *Am J Epidemiol* 1979; **109:**132–144.

53. Garland C, Shekelle RB, Barrett-Connor E, Criqui MH, Rossot AH, Paul O. Dietary vitamin D and calcium and risk of colorectal cancer: a 19-year prospective study in men. *Lancet* 1985; **i:**307–309.

54. Howe GR, Hirohata T, Hislop TG, Iscovich JM, Yuan J, Katsouyanni K, Lubin F, Marubini E, Modan B, Rohan T, Toniolo P, Shunzhang Y. Dietary factors and risk of breast cancer: combined analysis of 12 case-control studies. *J Natl Cancer Inst* 1990; **82:**561–569.

55. Richardson S, Gerber M, Cenee S. The role of fat, animal protein and some vitamin consumption in breast cancer: a case control study in Southern France. *Int J Cancer* 1991; **48:**1–9.

56. Van't Veer P, Kok FJ, Brants HA, Ockhuizen T, Sturmans F, Hermus RJJ. Dietary fat and the risk of breast cancer. *Int J Epidemiol* 1990; **19:**12–18.

57. Ewertz M, Gill C. Dietary factors and breast-cancer risk in Denmark. *Int J Cancer* 1990; **46:**779–784.

58. Hirayama T. Epidemiology of breast cancer with special reference to the role of diet. *Prev Med* 1978; **7:**173–195.

59. Knekt P, Albanes D, Seppanen R, Aromaa A, Jarvinen R, Hyvonen L, Teppo L, Pakkala E. Dietary fat and risk of breast cancer. *Am J Clin Nutr* 1990; **52:**903–908.

60. Willett WC, Hunter W, Stampfer MJ, Colditz G, Manson JE, Spiegelman D, Rosner B, Hennekens CH, Speizer FE. Dietary

fat and fiber in relation to risk of breast cancer. An 8-year follow-up. *JAMA* 1992; **268:** 2037–2044.

61. Graham S, Marshall J, Mettlin C, Rzepka T, Nemoto T, Byers T. Diet in the epidemiology of breast cancer. *Am J Epidemiol* 1982; **116:**68–75.

62. Lee HP, Gourley L, Duffy SW, Esteve J, Lee J, Day NE. Dietary effects on breast cancer risk in Singapore. *Lancet* 1991; **337:** 1197–1200.

63. Graham S, Hellmann R, Marshall J, Freudenheim J, Vena J, Swanson M, Zielezny M, Nemoto T, Stubbe N, Raimondo T. Nutritional epidemiology of postmenopausal breast cancer in Western New York. *Am J Epidemiol* 1991; **134:**552–566.

64. Jones DY, Schatzkin A, Green SB, Block G, Brinton LA, Ziegler RG, Hoover R, Taylor PR. Dietary fat and breast cancer in the National Health and Nutrition Examination Survey. I. Epidemiologic Follow-up Study. *J Natl Cancer Inst* 1987; **79:**465–471.

65. Kushi LH, Sellers TA, Potter JD, et al. Dietary fat and postmenopausal breast cancer. *J Natl Cancer Inst* 1992; **84:** 1092–1099.

66. Kolonel LN, Yoshizawa CN, Hankin JH. Diet and prostatic cancer: a case-control study in Hawaii. *Am J Epidemiol* 1988; **127:**999–1012.

67. Ross RK, Shimizu H, Paganini-Hill A, Honda G, Henderson BE. Case-control studies of prostate cancer in blacks and whites in Southern California. *J Natl Cancer Inst* 1987; **78:**869–874.

68. Snowdon DA, Phillips RL, Choi W. Diet, obesity, and risk of fatal prostatic cancer. *Am J Epidemiol* 1984; **120:**244–250.

69. Walker ARP, Walker BF, Tsotetsi NG, Sebitso C, Siwedi D, Walker AJ. Case-control study of prostate cancer in black patients in Soweto, South Africa. *Br J Cancer* 1992; **65:** 438–441.

70. Mills PK, Beeson WL, Phillips RL, Fraser GE. Cohort study of diet, lifestyle, and prostate cancer in Adventist men. *Cancer* 1989; **64:**598–604.

71. Mettlin C, Selenskas S, Natarajan N, Huben R. Beta-carotene and animal fats and their relationship to prostate cancer risk. A case-control study. *Cancer* 1989; **64:**605–612.

72. Rotkin ID. Studies in the epidemiology of prostate cancer: expanded sampling. *Cancer Treat Rep* 1977; **61:** 173–180.

73. Heshmat MY, Kaul L, Kovi J, Jackson MA, Jackson AG, Jones GW, Edson M, Enterline JP, Worrell RG, Perry SL. Nutri-

tion and prostate cancer: a case-control study. *Prostate* 1985; **6:**7–17.

74. Graham S, Haughey B, Marshall J, Priore R, Byers T, Rzepka T, Mettlin C, Pontes JE. Diet in the epidemiology of cancer of the prostate gland. *J Natl Cancer Inst* 1983; **70:**687–692.

75. West DW, Slattery ML, Robison LM, French TK, Mahoney AW. Adult dietary intake and prostate cancer risk in Utah: a case-control study with special emphasis on aggressive tumors. *Cancer Causes Contr* 1991; **2:**85–94.

76. Ohno Y, Yoshida O, Oishi K, Okada K, Yamabe H, Schroeder FH. Dietary beta-carotene and cancer of the prostate: a case-control study in Kyoto, Japan. *Cancer Res* 1988; **48:** 1331–1336.

77. Hsing AW, McLaughlin JK, Schuman LM, Bjelke E, Gridley G, Wacholder S, Chien HTC, Blot WJ. Diet, tobacco use, and fatal prostate cancer: results from the Lutheran Brotherhood Cohort Study. *Cancer Res* 1990 **50:**6836–6840.

78. Severson RK, Nomura AM, Grove JS, Stemmermann GN. A prospective study of demographics, diet, and prostate cancer among men of Japanese ancestry in Hawaii. *Cancer Res* 1989; **49:**1857–1860.

79. Durbec JP, Chevillotte G, Bidart JM, Berthezene P, Sarles H. Diet, alcohol, tobacco and risk of cancer of the pancreas: a case-control study. *Br J Cancer* 1983; **47:**463–470.

80. Olsen GW, Mandel JS, Gibson RW, Wattenberg LW, Schuman LM. A case-control study of pancreatic cancer and cigarettes, alcohol, coffee and diet. *Am J Public Health* 1989; **79:** 1016–1019.

81. Norell SE, Ahlbom A, Erwald R, Jacobson G, Lindberg-Navier I, Olin R, Tornberg B, Wiechel KL. Diet and pancreatic cancer: a case-control study. *Am J Epidemiol* 1986; **124:**894–902.

82. Howe GR, Ghadirian P, Bueno de Mesquita HB, Zatonski WA, Baghurst PA, Miller AB, Simard A, Baillargeon J, DeWaard F, Przewozniak K, McMichael AJ, Jain M, Hsieh CC, Maisonneuve P, Boyle P, Walker AM. A collaborative case-control study of nutrient intake and pancreatic cancer within the Search Programme. *Int J Cancer* 1992; **51:**365–372.

83. Mills PK, Beeson WL, Abbey DE, Fraser GE, Phillips RL. Dietary habits and past medical history as related to fatal pancreas cancer risk among Adventists. *Cancer* 1988;**61:**2578–2585.

84. Gold EB, Gordis L, Diener MD, Seltser R, Boitnott JK, Bynum TE, Hutcheon DF. Diet and other risk factors for cancer of the pancreas. *Cancer* 1985; **55:**460–467.

85. Mack TM, Yu MC, Hanisch R, Henderson BE. Pancreas cancer and smoking, beverage consumption, and past medical history. *J Natl Cancer Inst* 1986; **76:**49–60.

86. Byers TE, Graham S, Haughey BP, Marshall JR, Swanson MK. Diet and lung cancer risk: findings from the Western New York Diet Study. *Am J Epidemiol* 1987; **125:**351–363.

87. Knekt P, Seppanen R, Jarvinen R, et al. Dietary cholesterol, fatty acids, and the risk of lung cancer among men. *Nutr Cancer* 1991; **16:**267–275.

88. Mettlin C. Milk drinking, other beverage habits, and lung cancer risk. *Int J Cancer* 1989; **43:**608–612.

89. Kvale G, Bjelke E, Gart JJ. Dietary habits and lung cancer risk. *Int J Cancer* 1983; **31:**397–405.

90. Byers T, Vena J, Mettlin C, Swanson M, Graham S. Dietary vitamin A and lung cancer risk: an analysis by histological subtypes. *Am J Epidemiol* 1984; **120:**769–776.

91. Shekelle RB, Lepper M, Liu, S, Maliza C, Raynor WJ, Rossof AH, Paul O, Shryock AM, Stamler, J. Dietary vitamin A and risk of cancer in the Western Electric study. *Lancet* 1981; 2:1185–1190.

92. Kalandidi A, Katsouyanni K, Voropoulou N, Bastas G, Saracci R, Trichopoulos D. Passive smoking and diet in the etiology of lung cancer among non-smokers. *Cancer Causes Contr* 1990; 1:15–21.

93. McLaughlin JK, Gridley G, Block G, Winn DM, Preston-Martin S, Schoenberg JB, Greenberg DS, Stemhagen A, Austin DF, Ershow AG, Blot WJ, Fraumeni JF. Dietary factors in oral and pharyngeal cancer. *J Natl Cancer Inst* 1988; **80:**1237–1243.

94. Cramer DW, Welch WR, Hutchison GB, Wilett W, Scully RE. Dietary animal fat in relation to ovarian cancer risk. *Obstet Gynecol* 1984; **63:**833–838.

95. La Vecchia C, Decarli A, Negri E, Parazzini F, Gentile A, Cecchetti G, Fasoli M, Franceschi S. Dietary factors and the risk of epithelial ovarian cancer. *J Natl Cancer* 1987; **79:** 663–669.

96. Shu XO, Gao YT, Yuan JM, Ziegler RG, Brinton LA. Dietary factors and epithelial ovarian cancer. *Br J Cancer* 1989; **59:**92–96.

97. Byers T, Marshall J, Graham S, Mettlin C, Swanson M. A case control study of dietary and nondietary factors in ovarian cancer. *J Natl Cancer Inst* 1983; **71:**681–686.

98. Slattery ML, Schuman KL, West DW, French TK, Robison LM. Nutrient intake and ovarian cancer. *Am J Epidemiol* 1989; **130:**497–502.

99. Brock KE, Berry G, Mock PA, MacLennan R, Truswell AS, Brinton LA. Nutrients in diet and plasma and risk of in situ cervical cancer. *J Natl Cancer Inst* 1988; **80**:580–585.

100. Ribolo E, Gonzalez CA, Lopez-Abente G, Errezola M, Izarzugaza I, Escolar A, Nebot M, Hemon B, Agudo A. Diet and bladder cancer in Spain: a multi-centre case-control study. *Int J Cancer* 1991; **49**:214–219.

101. Zatonski WA, La Vecchia C, Przewozniak K, Maisonneuve P, Lowenfels AB, Boyle P. Risk factors for gallbladder cancer: a Polish case-control study. *Int J Cancer* 1992; **51**:707–711.

102. Rogers AE, Longnecker MP. Dietary and nutritional influences on cancer: a review of epidemiologic and experimental data. *Lab Invest* 1988; **59**:729–759.

103. Freedman LS, Clifford C, Messina M. Analysis of dietary fat, calories, body weight, and the development of mammary tumors in rats and mice: a review. *Cancer Res* 1990; **50**:5710–5719.

104. Appleton BS, Landers RE. Oil gavage effects on tumor incidence in the National Toxicology Program's 2-year carcinogenesis bioassay. In: Poirier LA, Newberne PM, Pariza MW, eds. *Essential Nutrients in Carcinogenesis*. New York: Plenum Press, 1986; 99–104.

105. Kritchevsky D, Welch CB, Klurfeld DM. Response of mammary tumors to caloric restriction for different time periods during the promotion phase. *Nutr Cancer* 1989; **12**:259–269.

106. Klurfeld DM, Weber MM, Kritchevsky D. Inhibition of chemically induced mammary and colon tumor promotion by caloric restriction in rats fed increased dietary fat. *Cancer Res* 1987; **47**:2759–2762.

107. Klurfeld DM, Welch CB, Lloyd LM, Kritchevsky D. Inhibition of DMBA-induced mammary tumorigenesis by caloric restriction in rats fed high-fat diets. *Int J Cancer* 1989; **43**:922–925.

108. Kritchevsky D. Fat, calories and cancer. In: Horwitz C, Rozen P, eds. *Progress in Diet and Nutrition*. Basel: Karger, 1988; 188–198.

109. Zhao LP, Kushi LH, Klein RD, Prentice RL. Quantitative review of studies of dietary fat and rat colon carcinoma. *Nutr Cancer* 1991; **15**:169–177.

110. Kris-Etherton PM, Krummel D, Russell ME, Dreon D, Mackey S, Borchers J, Wood PD. The effect of diet on plasma lipids, lipoproteins, and coronary heart disease. *J Am Diet Assoc* 1988; **88**:1373–1400.

111. St. Clair RW. Atherosclerosis regression in animal models: current concepts of cellular and biochemical mechanisms. *Prog Cardiovasc Dis* 1983; **26**:109–132.

112. Armstrong ML, Warner ED, Connor WE. Regression of coronary atheromatosis in rhesus monkeys. *Circ Res* 1970; **27**:59–67.

113. Clarkson TB, Bond MG, Bullock BC, McLaughlin KJ, Sawyer JK. A study of atherosclerosis regression in *Macaca mulatta*. *Exp Mol Pathol* 1984; **41**:96–118.

114. Waters D, Lesperance J. Regression of coronary atherosclerosis: an achievable goal? Review of results from recent clinical trials. *Am J Med* 1991; **91(Suppl. 1B)**:10S–17S.

115. Little JA. Coronary heart disease prevention trials. In: Carroll KK, ed. *Diet, Nutrition, and Health*. Montreal: McGill-Queen's University Press, 1989; 73–84.

116. Temple NJ. Coronary heart disease—dietary lipids or refined carbohydrates? *Med Hypotheses* 1983; **10**:425–435.

117. Yano K, Rhoads GG, Kagan A, Tillotson J. Dietary intake and risk of coronary heart disease in Japanese men living in Hawaii. *Am J Clin Nutr* 1978; **31**:1270–1279.

118. Garcia-Palmieri MR, Sorlie P, Tillotson J, Costas R, Cordero E, Rodriguez M. Relationship of dietary intake to subsequent coronary heart disease incidence: the Puerto Rico Heart Health Program. *Am J Clin Nutr* 1980; **33**:1818–1827.

119. Medalie JH, Kahn HA, Nuefeld HM, Riss E, Goldbourt U. Five-year myocardial infarction incidence. II. Association of single variables to age and birthplace. *J Chronic Dis* 1973; **26**:329–349.

120. Morris JN, Marr JW, Clayton DG. Diet and heart disease: a postscript. *Br Med J* 1977; **2**:1307–114.

121. Kromhout D, Coulander CD. Diet, prevalence and 10-year mortality from coronary heart disease in 871 middle-aged men. The Zutphen Study. *Am J Epidemiol* 1984; **119**:733–741.

122. Lapidus L, Andersson H, Bengtsson C, Bosaeus I. Dietary habits in relation to incidence of cardiovascular disease and death in women: a 12-year follow-up of participants in the population study of women in Gothenburg, Sweden. *Am J Clin Nutr* 1986; **44**:444–448.

123. Ulbricht TLV, Southgate DAT. Coronary heart disease: seven dietary factors. *Lancet* 1991; **338**:985–992.

124. Connor SL, Connor WE. Coronary heart disease: prevention and treatment by nutritional change. In: Carroll KK, ed. *Diet, Nutrition, and Health*. Montreal: McGill-Queen's University Press, 1989; 33–72.

125. Lemonnier D. Effect of age, sex and site on the cellularity of the adipose tissue in mice and rats rendered obese by a high-fat diet. *J Clin Invest* 1972; **51**:2907–2915.

126. Mickelsen O, Takahashi S, Craig C. Experimental obesity. I. Production of obesity in rats by feeding high-fat diets. *J Nutr* 1955; **57**:541–554.

127. Fenton PF, Dowling MT. Studies on obesity. I. Nutritional obesity in mice. *J Nutr* 1953; **49**:319–331.

128. Birt DF, Higginbotham SM, Patil K, Pour P. Nutritional effects on the lifespan of Syrian hamsters. *Age* 1982; **5**:11–19.

129. Foltin RW, Rolls BJ, Moran TH, Kelly T, McNelis AL, Fischman MW. Caloric, but not macronutrient, compensation by humans for required-eating occasions with meals and snacks varying in fat and carbohydrate. *Am J Clin Nutr* 1992; **55**:331–342.

130. Leibel RL, Hirsch J, Appel BE, Checani GC. Energy intake required to maintain body weight is not affected by wide variation in diet composition. *Am J Clin Nutr* 1992; **55**:350–355.

131. Kendall A, Levitsky DA, Strupp BJ, Lissner L. Weight loss on a low fat diet: consequence of the imprecision of the control of food intake in humans. *Am J Clin Nutr* 1991; **53**:1124–1129.

132. Shepphard L, Kristal AR, Kushi LH. Weight loss in women participating in a randomized trial of low-fat diets. *Am J Clin Nutr* 1991; **54**:821–828.

133. Haber GB, Heaton KW, Murphy D, Burroughs L. Depletion and disruption of dietary fibre. Effects on satiety, plasma-glucose, and serum-insulin. *Lancet* 1977; **2**:679–682.

134. Duncan KH, Bacon JA, Weinsier RL. The effects of high and low energy density diets on satiety, energy intake, and eating time of obese and nonobese subjects. *Am J Clin Nutr* 1983; **37**:763–767.

135. Shintani TT, Hughes CK, Beckham S, O'Connor HK. Obesity and cardiovascular risk intervention through the ad libitum feeding of traditional Hawaiian diet. *Am J Clin Nutr* 1991; **53**:1647S–1651S.

136. Tucker LA, Kano MJ. Dietary fat and body fat: a multivariate study of 205 adult females. *Am J Clin Nutr* 1992; **56**:616–622.

137. Miller WC, Lindeman AK, Wallace J, Niederpruem M. Diet composition, energy intake, and exercise in relation to body fat in men and women. *Am J Clin Nutr* 1990; **52**:426–430.

138. Romieu I, Willett WC, Stampfer MJ, Colditz GA, Sampson L, Rosner B, Hennekens CH, Speizer FE. Energy intake and other determinants of relative weight. *Am J Clin Nutr* 1988; **47**:406–412.

139. Dreon DM, Frey-Hewitt B, Ellsworth N, Williams PT, Terry RB, Wood PD. Dietary fat: carbohydrate ratio and obesity in middle-aged men. *Am J Clin Nutr* 1988; **47**:995–1000.

140. Prewitt TE, Schmeisser D, Bowen PE, Aye P, Dolecek TA, Langenberg P, Cole T, Brace, L. Changes in body weight, body composition, and energy intake in women fed high- and low-fat diets. *Am J Clin Nutr* 1991; **54**:304–310.

Dietary Fiber

K. W. Heaton

The great 17th century physician, Boerhaave, kept an elaborately bound volume which was said to contain all the secrets of medicine. When it was opened after his death, all the pages were found to be blank—except one. Inscribed on this page was but one sentence: "Keep the head cool, the feet warm and the bowels open."

(Sanders N, Paterson C, *Lancet* 1991; **337**:600,601.)

1. Introduction

Dietary fiber is not a single substance but a family of diverse substances, like the vitamins *(1)*. Indeed, it is not really a substance at all but rather a concept—a concept that has thrown the science of nutrition into turmoil and that has generated more new ideas and more changes in eating behavior than any nutritional idea since the vitamins in the 1920s.

2. Concepts

The essence of dietary fiber is plant cell walls, and the essential fact about plant cell walls is that their *raison d'être* is physical or mechanical, not chemical. Cell walls give a plant its solidity, shape, and stiffness, they conduct

From: *Western Diseases: Their Dietary Prevention and Reversibility*
Edited by: N. J. Temple and D. P. Burkitt Copyright ©1994 Humana Press, Totowa, NJ

its sap, and they stop it drying out. Of course, their physical properties depend on their chemical structure and it is interesting to study the chemistry of cell walls. But, from the point of view of human health and disease, it is more important to understand how the presence of cell walls affects the properties of food and the way food is handled by the alimentary tract than it is to understand the chemical structure of cell walls. For this reason, the chemistry of dietary fiber is omitted from this chapter except for a simple table of its components (Table 1). It has been extensively reviewed in many other places, together with its physicochemical properties in vitro (2).

How cell walls affect food and its handling depends on whether they are intact or disrupted. Intact cell walls, as in raw or lightly cooked fruit and vegetables and in whole or coarsely milled cereal grains, give food its shape and texture; indeed its very solidity and appearance. When the cell walls are disrupted, as in extended cooking, homogenizing, or milling, a food often becomes unrecognizable. Humankind is the only living thing that eats food in an unrecognizable form, with its cellular structure destroyed. Loss of solidity and cellularity have important implications for the ease and speed with which nutrients can be ingested and digested and, consequently, for metabolism and gastrointestinal function.

3. Measurement and Its Limitations

There is so far no way of measuring the cellularity of a food or the degree of its disruption, yet this is what we really need to know. It is easy to measure the amount of cell wall material in a food, exploiting the fact that, with many foods, cell walls are all that remains in a solid state when the food is exposed to a process of simulated digestion. With the help of a gas chromatograph, one can mea-

Table 1
The Main Components of Dietary Fiber (1)[a]

Name	Chemical nature	Classification
Cellulose	Straight β-glucan chains	
Noncellulosic polysaccharides		
Glucans	Branched β-glucan chains	Cell wall nonstarch
Hemicelluloses	Galactomannans, xylans, xyloglucans	polysaccharides (NSP)
Pectic substances	Rhamnogalacturonans, arabinogalactans	
Gums and mucilages	Gluco- and glucurono-mannans, galactans, galactomannan, xylans, xyloglucans	Noncell wall NSP
Lignins	Phenylpropane polymers	Noncarbohydrate components of cell walls[b]
Cutin, suberin	Complex esters	

[a]Polysaccharide food additives were included in early definitions of dietary fiber.

[b]Plant cell walls also contain proteins and inorganic components, e.g., calcium, silica.

sure the individual polysaccharides and, even without it, one can separate soluble polysaccharides from insoluble ones. However, these measurements do not distinguish intact cell walls from fragmented ones. Broken up cell walls no longer serve their original function of trapping cell contents. Tough cell wall material that has been reduced to a fine powder no longer has the mechanical effects on the alimentary tract it once had and can lose its physicochemical properties too, like binding water and various chemicals. Finely ground bran is a poor laxative (3).

As Eastwood (4) observed, the fiber content of a food or diet tells you practically nothing about the effects of that food or diet on the body. Beside the reasons already given, there is huge variability between the types of nonstarch polysaccharide (NSP) in different foods and even in a single food grown in different circumstances *(5)*. Even a single polysaccharide, like arabinoxylan or β-glucan, can have completely different properties depending on the details of its chemical structure *(5,6)*. As an added complication, the laxative effects of a diet, indeed all its effects on colonic physiology, are probably owing as much (or more) to its content of starch and the extent to which that starch escapes undigested into the colon as to its content of cell wall material *(7)*.

With all these limitations on the meaning of measurements of dietary fiber it may seem surprising that any epidemiological studies have found significant correlations between fiber intake and the risk of disease. The fact that they have in some cases must not be accepted at face value. The fiber content of a diet may simply be a surrogate for another property of fiber-containing foods, like their starch content (which is rarely measured) or their content of trace elements or antioxidant vitamins, especially with fruits and vegetables.

4. Physiological Properties and Their Significance

The important effects of dietary fiber are of four kinds:

1. Effects of an intact cellular structure in food;
2. Effects of a high content of soluble NSP in food;
3. Effects of intact particulate material reaching the distal intestine; and
4. Effects of fermentable carbohydrate reaching the colon.

The physiological results that matter are:

1. Slower ingestion of solid vs fluid food (because of 1 above);
2. Slower digestion of starch (because of 1 or possibly 2 above);

3. Slower absorption of sugars (because of 1 or possibly 2 above);
4. Laxation (because of 3 and 4 above);
5. Greater production and absorption of short-chain fatty acids (because of 4 above); and
6. Altered bacterial metabolism in the colon with reduced absorption of bacterial metabolites (because of 4 above).

5. Promotion of Disease by Cell-Wall-Depleted Foods

5.1. Metabolic Effects

Rapid digestion of starch and absorption of sugars provoke the pancreatic β-cells to secrete more insulin; there is postprandial hyperinsulinemia. To maintain homeostasis and prevent rebound hypoglycemia, insulin antagonists like adrenaline, cortisol, and growth hormone are released. The body is forced to resist its own insulin, in effect to do battle with itself. The most important recognized causes of insulin resistance are adiposity, especially abdominal adiposity (paunchiness or "middle-age spread") and physical unfitness caused by lack of exercise (8,9). Insulin resistance is dangerous because, if the body fails to overcome it by secreting more insulin all the time, it becomes insulin-deficient and diabetes mellitus supervenes. If the body succeeds, it does so at the price of permanently high circulating insulin levels (hyperinsulinemia). This is even more dangerous because insulin stimulates the synthesis of cholesterol and alters many other metabolic activities. The postprandial hyperinsulinemia that results from rapidly digested carbohydrate has not been proved to have these consequences but, logically, it is likely to have them. Hyperinsulinemia is linked causally with coronary artery disease, hypertension, and probably gallstones (8,10).

The faster ingestion of food and reduced satiation that results from disruption and depletion of cell walls was demonstrated in studies comparing whole fruit with fruit puree and fruit juice *(11,12)*. Other studies prove that, for the most part, sugars ingested in liquid form bypass the body's calorie-controlling mechanisms—they are truly empty calories *(13)*. The implication is that fiber-depleted sugars, now known as nonmilk extrinsic sugars *(14)*, are inherently fattening and contribute to the unwanted weight gain that is experienced by most people in Western countries.

It is a reasonable but untested assumption that any industrial or domestic process that makes food easier and quicker to eat (like making bread soft rather than chewy) will also promote excess calorie intake. Such processes invariably destroy the cellular architecture of plant products.

Milling of cereal grains makes them more digestible because of disruption of the cellular architecture and increased surface area for enzymatic action. Finely milled wheat flour is digested more rapidly than coarsely ground flour and evokes a greater insulin response *(15,16)*. Combinations of finely ground flour and extrinsic sugar are extremely popular as well as palatable (cakes, biscuits, pastries). They must also be very insulinogenic. Their ease and speed of consumption and digestion, coupled with their content of hidden fat (fat that would not be eaten without them), makes them prime candidates as causes of obesity.

5.2. Intestinal Effects

5.2.1. Laxation

Surprisingly, the laxative effects of cell wall material are still not fully understood. Two closely associated phenomena are involved—greater fecal output and faster colonic transit, but it is not known which comes first or how exactly dietary fiber acts in either case. In bulking the feces, the major effects are the binding of water by insoluble NSP and the promotion of bacterial multiplication by fermentation of NSP *(17)*. Transit may be stimulated by distension from increased bulk (including gas) or by a direct mechanical effect of solid particles *(18)*.

The colonic lumen is a highly complex milieu and the motility of the colon and rectum are subject to many hormonal and neural influences as well as to the stimuli they receive from their contents. In particular, defecation is subject to psychological and social factors. It is easily suppressed and, when it is, the function of the whole colon alters and stool output drops *(19)*. In a group of people, fecal output is related as much to personality factors as to fiber intake *(20)*. All the same, almost everyone can alter their bowel habit by altering their fiber intake and, when people with high and low fiber intakes are compared there are marked differences in fecal output, stool consistency and form, and intestinal transit time *(21)*, so there is no doubt that fiber is an important determinant of bowel function.

It is less certain what diseases are caused by low fecal output or the changes in stool form, transit time, and defecatory behavior that go with it. In the early years of this century a wide array of diseases was blamed on constipation (admittedly with scant evidence) *(22)*. By the 1960s the pendulum had swung so far in the opposite direction that it was considered normal to have only three bowel actions a week *(23)* and straining at stool was deemed to be necessary and natural *(24)*. Now the pendulum is swinging back. The link between low stool output and the incidence of colon cancer is so strong *(25)* that the British government has recommended a 50% raise in dietary fiber intake (to a population average of 18 g NSP daily) *(26)*.

5.2.2. Fermentation

When carbohydrate reaches the colon it is fermented by anaerobic bacteria to short-chain fatty acids—mainly the one-, two-, and three-carbon acids acetic, propionic and butyric respectively. At the same time gases are produced—hydrogen and carbon dioxide and, in about half the population, methane. The main carbohydrates reaching the colon are NSP (dietary fiber), starch and the glycoproteins of colonic mucus *(7)*. The acids are mostly absorbed and metabolized—they contribute 1–5% to total energy intake. Butyric acid is metabolized as soon as it enters the colonic mucosa and may be the colonic mucosa's main fuel source. Lack of it is possibly a cause of colitis because, if the colon is bypassed as part of a surgical procedure, it commonly becomes inflamed and this inflammation is cured by infusions containing short-chain fatty acids *(27)*. Butyric acid may also protect against cancer of the colon *(see* the following section).

6. The Role of Dietary Fiber in Individual Diseases

6.1. Constipation

Constipation is the one condition in which everyone agrees there is a role for dietary fiber. However, there are nondietary aspects of modern life that can inhibit defecation, such as having to rush to work in the morning, travel, irregular lifestyle, embarrassment about bowel function, competition for toilet facilities, stress, and unphysiological posture at defecation. Squatting is the "natural" posture and people strain less when they squat than when they sit to defecate *(28)*.

Constipation is important not just because it is common and uncomfortable but because it may be a risk factor for diverticular disease, colorectal cancer, and anal problems (hemorrhoids and anal fissure) and, possibly, breast cancer. It is also possible that prolonged straining at stool impedes the return of blood from the legs and leads to venous dilatation and hence to varicose veins and deep vein thrombosis *(29)*.

It has been firmly stated that "a low intake of dietary fiber is the major causative factor in the constipation which is so prevalent in western communities" *(30)*. However, doubt is creeping in. An expert committee has stated that "constipation should probably be regarded as a disorder of colonic or anorectal motility that may respond to the mild laxative action of complex carbohydrate, rather than simply the result of a 'fiber deficient diet' " *(31)*, and some gastroenterologists believe that the value of bran as a laxative has been exaggerated *(32)*.

What is the evidence relating stool weight and intestinal transit time to dietary fiber intake? Between-country comparisons are not available. Only one within-country study has been published *(33)*. This involved a more-or-less random sample of 62 people aged 18–80 yr from Edinburgh, Scotland. Fecal weight ranged from 19–278 g/24 h and dietary fiber intake (Southgate method) from 4.3–32.6 g/24 h. The correlation between the two was significant but was weak so that only 17% of the variance in stool output was explained by dietary fiber. This was a field study that necessarily limited the accuracy of the measurements and it may have underestimated the strength of the relationship. In a more tightly controlled study of 51 volunteers with a wider range of fiber intakes (10–78 g/d) the correlation between fiber intake and fecal weight was extremely close *(21)*. In this study the consistency of the stools was measured using a penetrometer and there was a good correlation between fiber intake and stool softness.

There is, of course, overwhelming evidence that people can increase their fecal output by increasing their fiber intake *(3)*, though people with constipation may respond less well. Wheat bran is particularly effective and there is a linear relationship between bran intake and fecal output, whether the bran, is taken raw, baked into bread or boiled and processed into a breakfast cereal (when it is

least effective). There is no doubt, therefore, that constipation can be prevented and relieved by dietary means (if
colorectal physiology is intact).

Overall, the bulking effect of fiber from fruit and vegetables is slightly less than that from wheat, the mean
increase in stool wt/g of fiber being 4.9 and 5.7 g, respectively (3). In practice, fruit and vegetables are less useful
for this purpose because the concentration of fiber in fruit
and vegetables (1–3%) is much lower than it is in whole
grain cereals (8–12%).

Cooking and grinding to small particles reduces the
bulking effect of wheat bran and may well do so with other
forms of fiber (3,34).

6.2. Diverticular Disease of the Colon

Colonic diverticular disease was the first disease (outside constipation) in which a high fiber intake was claimed
to have a therapeutic role. Such treatment seemed logical
because the leading theory for the pathogenesis of diverticulosis was and still is that the colon "ruptures itself" in
its struggle to propel the stools and its job is believed to be
harder with small stools (35,36). This theory has survived,
albeit with some anomalies. A key finding is that diverticulosis is common in black Americans, though it is still rare
in Africa (37). In England it is substantially less common
in vegetarians (38). Vegetarians eat more dietary fiber than
meat eaters and pass heavier and softer stools (21,39). In
Japan, fiber intake has fallen steadily since World War II,
and the prevalence of diverticular disease has risen; moreover, case-control studies there and in Greece show that
people with diverticular disease eat less fiber than matched
healthy controls (40). In these studies the subjects with
diverticulosis were all symptomatic. Since only a few people
with diverticulosis get symptoms, the results may be biased
and cannot safely be extrapolated to the generality of cases.
There is only one case-control study based on asympto-

matic cases detected by population screening and the results were not so clear-cut *(38)*.

Experimental studies give some support to the fiber hypothesis. When groups of rats were fed different amounts of wheat fiber for the whole of their lives, the prevalence of colonic diverticulosis was inversely proportional to fiber intake *(41)*. However, even a high fiber intake (17% of diet by weight) allowed 9.4% of rats to develop diverticulosis, so other factors must be operating. Age-related weakening of the bowel wall may be relevant but there must also be factors that allow the pressure in the colon to rise too high.

Bran, a high fiber diet, or bulking agents are widely used to treat symptomatic diverticular disease, but placebo-controlled trials have given mixed results; in surgical patients bran was clearly superior to placebo *(42)*, but in medical patients it was not *(43)*. Symptomatic diverticular disease is just irritable bowel syndrome (that is, intestinal pain with constipation or diarrhea) in a person who happens to have diverticulosis *(44)* and in irritable bowel syndrome bran is superior to placebo only when there is constipation *(45)*. Patients with abdominal symptoms who reach a surgeon are more likely to have constipation than diarrhea because patients with diarrhea are generally sent to physicians, not surgeons.

6.3. Cancer of the Large Bowel

The possible role of dietary fiber as a protective factor in large bowel cancer has been intensively investigated since Burkitt *(46)* put forward his hypothesis. There is no shortage of mechanisms by which fiber (and undigested starch) might be protective (Table 2), but as yet there is no proof that any of them are operative. The epidemiological data are inconsistent. One huge Australian case-control study concluded there was no protective effect—even a trend toward a promoting effect *(47)*—whereas shortly afterward another one concluded there *was* a protective effect *(48)*.

Table 2
Effects of Fiber on the Large Bowel
that Should in Theory Reduce Cancer Risk

Dilution of mutagens (owing to bulkier bowel contents)
Adsorption of mutagens to surface of particulate matter
Faster transit time, so less contact of mutagens with mucosa
Consequences of more fermentation and lower pH
    Greater production of butyrate (which is strongly antineoplastic)
    Reduced microbial production of mutagens including degraded bile acids

---

Recently, Trock et al. *(49)* assessed the strength of evidence from 37 epidemiological studies that they considered methodologically sound, namely 23 case-control studies, one international correlation study, eight within-country correlation studies, two cohort studies, and three time-trend studies. Thirteen studies were deemed to give strong support to the fiber hypothesis in that inverse associations between fiber intake or surrogates thereof and colon cancer incidence/mortality were statistically significant and remained so after adjustment for fats, meat, or energy. Eight studies gave moderate support in that they showed a significant inverse association between fiber intake and colon cancer risk but they did not control for confounding by meats or fats. Fourteen studies were equivocal, neither supporting nor refuting the fiber hypothesis, but eight of them were at least consistent with it. Only two studies were clearly opposed to it, that is, despite being methodologically sound and of adequate power, they found no protective effect of fiber. Altogether, 29 studies (78%) provided some evidence of protection.

Trock et al. *(49)* also pooled the results of the 16 case-control studies from which odds ratios could be calculated and found that the risk of cancer in people eating the most fiber vs those eating the least was 0.57 (95% confidence limits 0.50–0.64).

Some experts believe that the evidence for a protective effect from vegetables is stronger than that from fiber *(50)*. In the meta-analysis of Trock et al. *(49)*, the odds ratio for cancer in people eating the most vegetables was indeed lower than that for fiber (0.48, 95% confidence limits 0.41–0.57). Vegetables may be protective through their content of antioxidant vitamins and this is in line with the finding of low serum levels of vitamin E and selenium in the serum of patients with bowel cancer or adenomatous polyps (the precursor lesion) *(51)*.

If dietary fiber is protective in itself, there should be an inverse relationship between a community's average stool weight and its incidence of large bowel cancer. Cummings and colleagues *(25)* have collated all published data on mean stool output of 23 different communities and have shown that there is indeed a significant inverse relationship between stool weight and colonic cancer mortality in these populations ($r = 0.78$).

Besides dietary fiber, the only important component of the diet that probably contributes to fecal weight is starch. Few studies have examined starch intake in relation to bowel cancer risk. A large case-control study did so and found that a high starch intake reduced the relative risk to 0.82 *(52)*. Undigested starch resembles nonstarch polysaccharides in being fermented by colonic bacteria to short-chain fatty acids. Fermentation of starch produces relatively more butyrate than fermentation of NSP. Butyrate has marked antineoplastic properties and the feces of patients with colorectal cancer produce less butyrate than the feces of controls *(50)*.

Most colorectal cancers probably arise from adenomatous polyps. The few studies that have been done on dietary intake and adenoma formation have reported a protective effect of high polysaccharide intake *(53,54)* and a large daily dose of wheat bran has been shown to delay the recurrence of polyps after polypectomy *(55)*. However, the latter study was done in a very atypical group of patients—people with familial adenomatous polyposis who had undergone colectomy and ileorectal anastomosis.

In conclusion, the evidence for a protective effect from a fiber-rich diet is substantial but not conclusive. Definite answers should come from large prospective trials, which are in progress.

6.4. Breast Cancer

The evidence available is not great but does suggest a link between low intake of fiber-containing foods and breast cancer. In a combined analysis of 12 case-control studies from eight countries, a protective effect was found for several markers of fruit and vegetable intake, at least in postmenopausal women (56). This effect was not strong, the relative risk being reduced to 0.83 with a high intake of dietary fiber (though, subsequently, a much stronger effect has been reported from Moscow [57]), and, again, it is impossible to separate the role of fiber from that of antioxidant vitamins and minerals. However, there are two reasons for suspecting a direct effect from the laxative action of fiber. First, it has been shown that adding wheat bran to the diet of premenopausal women lowers their serum estrogen concentrations (58); this is relevant because breast cancer is linked to high estrogen exposure. Second, women with severe constipation have been found to have precancerous cells in their breast fluid (59). Much more work is needed.

6.5. Gallstones

Gallstones have become commoner in Europe in the last century and are prevalent in all "Westernized" communities but their incidence may now be falling in some European countries (60).

Obesity is a major risk factor in women and abdominal obesity is a risk factor in men (10). The risk of gallstones is also related to the biliary content of deoxycholate—a bacterial metabolite of the bile salt cholate, absorbed from the colon—and biliary deoxycholate levels can be increased by lengthening colonic transit time and decreased by laxative treatment of constipated people (61). This suggests that constipation predisposes to gallstones and preliminary epidemiological findings support this idea (62).

Case-control studies have yielded mixed results (60). Two have found a protective effect of a high fiber intake, two have found a trend in this direction, and one found no effect at all. This inconsistency could be a result of the inherent limitations of case-control diet studies. The published studies give no indication that cereal fiber is especially protective, but wheat fiber has been shown to lower biliary deoxycholate and biliary cholesterol saturation, which should reduce the risk of gallstones. In one study a diet based on naturally fiber-rich foods made bile less saturated with cholesterol without any "bran effect" on bile deoxycholate (63). Indirect support for a protective effect of a naturally fiber-rich diet comes from the relative scarcity of gallstones in vegetarians (64).

6.6. Appendicitis

As long ago as 1920, Rendle Short suggested that appendicitis was caused by lack of fiber in the diet (or cellulose as he called it). More recently, Burkitt (65,66) produced much epidemiological evidence for a connection between appendicitis and westernization, including anecdotal reports of a sudden increase in the disease when white bread was introduced to groups of people. He proposed that on a low fiber diet fecoliths or viscous cecal contents block the mouth of the appendix. However, there are discrepancies in the evidence. For instance, in the middle of this century, there was a substantial fall in appendicitis rates when dietary fiber intakes were static or even falling (67). Also, the higher rate of appendicitis in Ireland compared with Britain is associated with higher, not lower, consumption of potatoes and cereals *(68)*. Case-control studies have not resolved the issue. Three have found a low intake of dietary fiber in cases but two have not *(69)*. The situation is further confused by the finding that across England and Wales appendicitis rates correlate negatively with green vegetable consumption but positively with potato consumption *(70)*.

Different etiological factors may operate at different ages. Barker *(67)* proposed that, in children, good hygiene and freedom from infections in early life allow the lymphoid tissue at the base of the appendix to remain immunologically naive so that when the immune system is eventually challenged by, say, a virus infection in later childhood, it overreacts and the consequent lymphoid hyperplasia occludes the lumen of the appendix. If this is true, it is still possible that, in older people, appendiceal obstruction is caused by a fecolith or viscous cecal contents in association with constipation.

7. Conclusions

Dietary fiber is a much-hyped but poorly understood entity. It is the only major dietary constituent that completely escapes digestion, but undigested starch may be equally important for colonic function and health. Bulky stools seem to protect against large bowel cancer and probably other diseases, and such stools are safely and reliably obtained by raising dietary fiber intake. The nutrient-trapping function of intact cell walls is an important and neglected role of dietary fiber. By eating most of their plant food with its cellular structure intact, people can protect themselves from over-nutrition and hyperinsulinemia and thereby lower their risk of diabetes, hypertension, coronary heart disease, and gallstones. Some of the benefits of such foods may be because of their content of micronutrients such as trace elements and antioxidant vitamins. The safest diet is the most natural one, namely, a wide variety of whole foods.

Acknowledgment

The author's interest in nutrition was kindled by Surgeon-Captain T. L. Cleave (1906–1983)—a true pioneer.

References

1. Southgate D, Englyst H. Dietary fiber: chemistry, physical properties and analysis. In: Trowell H, Burkitt D, Heaton K, eds. *Dietary Fibre, Fibre-Depleted Foods and Disease*. London: Academic Press, 1985; 31–55.
2. Southgate DAT, Waldron K, Johnson IT, Fenwick GR, eds. *Dietary Fibre: Chemical and Biological Aspects*. Cambridge, UK: Royal Society of Chemists, 1990.
3. Cummings JH. The effect of dietary fiber on fecal weight and composition. In: Spiller GA, ed. *CRC Handbook of Dietary Fiber in Human Nutrition*. Boca Raton, FL: CRC Press, 1986; 211–280.
4. Eastwood MA. What does the measurement of dietary fibre mean? *Lancet* 1986; i:1487–1488.
5. Selvendran RR. Chemistry of plant cell walls and dietary fibre. *Scand J Gastroenterol* 1987; **22(Suppl. 129)**:33–41.
6. Åman P, Graham H. Mixed-linked β-(1→3), (1→4)-D-glucans in the cell walls of barley and oats—chemistry and nutrition. *Scand J Gastroenterol* 1987; **22(Suppl. 129)**:42–51.
7. Cummings JH, Englyst HN. Fermentation in the human large intestine and the available substrates. *Am J Clin Nutr* 1987; **45**:1243–1255.
8. Reaven GM. Role of insulin resistance in human disease. *Diabetes* 1988; **37**:1595–1607.
9. Houmard JA, Wheeler WS, McCammon MR, Holbert D, Israel RG, Barakat HA. Effect of fitness level and the regional distribution of fat on carbohydrate metabolism and plasma lipids in middle- to older-aged men. *Metabolism* 1991; **40**:714–719.

10. Heaton KW, Braddon FEM, Emmett PM, Mountford RA, Hughes AO, Bolton CH, Ghosh S. Why do men get gallstones? Roles of abdominal fat and hyperinsulinaemia. *Eur J Gastroenterol Hepatol* 1991; **3:**745–751.

11. Haber GB, Heaton KW, Murphy D, Burroughs L. Depletion and disruption of dietary fibre. Effects on satiety, plasma-glucose, and serum-insulin. *Lancet* 1977; **ii:**679–682.

12. Bolton RP, Heaton KW, Burroughs LF. The role of dietary fiber in satiety, glucose, and insulin: studies with fruit and fruit juice. *Am J Clin Nutr* 1981; **34:**211–217.

13. Porikos KP, Hesser MF, van Itallie TB. Caloric regulation in normal-weight men maintained on a palatable diet of conventional foods. *Physiol Behav* 1982; **29:**293–300.

14. Committee on medical aspects of food policy. *Dietary Sugars and Human Disease.* Department of Health and Social Security Report 37. London: HMSO, 1989.

15. Heaton KW, Marcus SN, Emmett PM, Bolton CH. Particle size of wheat, maize and oat test meals: effects on plasma glucose and insulin responses and on the rate of starch digestion in vitro. *Am J Clin Nutr* 1988; **47:**675–682.

16. O'Donnell LJD, Emmett PM, Heaton KW. Size of flour particles and its relation to glycaemia, insulinaemia, and colonic disease. *Br Med J* 1989; **298:**1616–1617.

17. Stephen A. Constipation. In: Trowell H, Burkitt D, Heaton K, eds. *Dietary Fibre, Fibre-Depleted Foods and Disease.* London: Academic Press, 1985; 133–144.

18. Tomlin J, Read NW. Laxative effects of indigestible plastic particles. *Br Med J* 1988; **297:**1175,1176.

19. Klauser A, Voderholzer WA, Heinrich CA, Schindlbeck NE, Müller-Lissner SA. Behavioral modification of colonic function. Can constipation be learned? *Dig Dis Sci* 1990; **35:**1271–1275.

20. Tucker DM, Sandstead HH, Logan GM, Klevay LM, Mahalko J, Johnson LK, Inman L, Inglett GE. Dietary fiber and personality factors as determinants of stool output. *Gastroenterology* 1981; **81:**879–883.

21. Davies GJ, Crowder M, Reid B, Dickerson JWT. Bowel function measurements of individuals with different eating patterns. *Gut* 1986; **27:**164–169.

22. Hertz AF. *Constipation and Allied Intestinal Disorders.* Oxford: Oxford Medical Publications, 1909.

23. Connell AM, Hilton C, Irvine G, Lennard-Jones JE, Misiewicz JJ. Variation of bowel habit in two population samples. *Br Med J* 1965; **2:**1095–1099.

24. Mendeloff AI. Defecation. In: Code CF, ed. *Handbook of Physiology, Section 6: Alimentary Canal. Vol IV. Motility.* Washington: American Physiological Society, 1968; 2140–2143.

25. Cummings JH, Bingham SA, Heaton KW, Eastwood MA. Fecal weight, colon cancer risk and dietary intake of non-starch polysaccharides (dietary fiber). *Gastroenterology* 1992; **103:**1783–1789.

26. Committee on medical aspects of food policy. *Dietary Reference Values for Food Energy and Nutrients for the United Kingdom.* Department of Health Report on Health and Social Subjects 41; London: HMSO, 1991.

27. Harig JM, Soergel KH, Komorowski RA, Wood CM. Treatment of diversion colitis with short-chain-fatty acid irrigation. *N Engl J Med* 1989; **320:**23–28.

28. Fedail SS, Harvey RF, Burns-Cox CJ. Abdominal and thoracic pressures during defaecation. *Br Med J* 1979; **1:**91.

29. Burkitt D. Varicose veins, haemorrhoids, deep-vein thrombosis and pelvic phleboliths. In: Trowell H, Burkitt D, Heaton K, eds. *Dietary Fibre, Fibre-Depleted Foods and Disease.* London: Academic Press, 1985; 317–329.

30. Trowell H, Burkitt D, Heaton K. *Dietary Fibre, Fibre-Depleted Foods and Disease.* London: Academic Press, 1985; 421.

31. BNF Task Force on Complex Carbohydrates in Foods. *Report of the Task Force.* London: Chapman and Hall, 1990.

32. Müller-Lissner SA. Effect of wheat bran on weight of stool and gastrointestinal transit time: a meta-analysis. *Br Med J* 1988; **296:**615–617.

33. Eastwood MA, Brydon WG, Baird JD, Elton RA, Helliwell S, Smith JH, Pritchard JL. Fecal weight and composition, serum lipids, and diet among subjects aged 18 to 80 years not seeking health care. *Am J Clin Nutr* 1984; **40:**628–634.

34. Wyman JB, Heaton KW, Manning AP, Wicks ACB. The effect on intestinal transit and the feces of raw and cooked bran in different doses. *Am J Clin Nutr* 1976; **29:**1474–1479.

35. Painter NS, Truelove SC, Ardran GM, Tuckey M. Segmentation and the localisation of intraluminal pressures in the human colon, with special reference to the pathogenesis of colonic diverticula. *Gastroenterology* 1965; **49:**169–177.

36. Painter NS, Burkitt DP. Diverticular disease of the colon: a deficiency disease of Western civilisation. *Br Med J* 1971; **2:**450–454.

37. Burkitt DP, Clements JL, Eaton SB. Prevalence of diverticular disease, hiatus hernia, and pelvic phleboliths in black and white Americans. *Lancet* 1985; **ii:**880,881.

38. Gear JSS, Ware A, Fursdon P, Mann JI, Nolan DJ, Brodribb AJM, Vessey MP. Symptomless diverticular disease and intake of dietary fibre. *Lancet* 1979; **i:**511–514.
39. Davies GJ, Crowder M, Dickerson JWT. Dietary fibre intakes of individuals with different eating patterns. *Hum Nutr: Appl Nutr* 1985; **39A:**139–148.
40. Heaton KW. Dietary fibre in the prevention and treatment of gastro-intestinal disorders. In: Schweitzer TF, Edwards CA, eds. *Dietary Fibre—A Component of Food: Nutritional Function in Health and Disease.* London: Springer, 1992; 249–263.
41. Fisher N, Berry CS, Fearn T, Gregory JA, Hardy J. Cereal dietary fiber consumption and diverticular disease: a lifespan study in rats. *Am J Clin Nutr* 1985; **42:**788–804.
42. Brodribb AJM. Treatment of symptomatic diverticular disease with a high-fibre diet. *Lancet* 1977; **i:**664–666.
43. Ornstein MH, Littlewood ER, Baird IM, Fowler J, North WRS, Cox AG. Are fibre supplements really necessary in diverticular disease of the colon? A controlled clinical trial. *Br Med J* 1981; **282:**1353–1356.
44. Thompson WG, Patel DG, Tao H, Nair R. Does uncomplicated diverticular disease cause symptoms? *Dig Dis Sci* 1982; **27:**605–608.
45. Cann PA, Read NW, Holdsworth CD. What is the benefit of coarse wheat bran in patients with the irritable bowel syndrome? *Gut* 1984; **25:**168–173.
46. Burkitt DP. Epidemiology of cancer of the colon and rectum. *Cancer* 1971; **28:**3–13.
47. Potter JD, McMichael AJ. Diet and cancer of the colon and rectum; a case-control study. *J Natl Cancer Inst* 1986; **76:**557–569.
48. Kune S, Kune GA, Watson LF. Case-control study of dietary etiologic factors; the Melbourne Colorectal Cancer Study. *Nutr Cancer* 1987; **9:**21–42.
49. Trock B, Lanza E, Greenwald P. Dietary fiber, vegetables, and colon cancer: critical review and meta-analysis of the epidemiological evidence. *J Natl Cancer Inst* 1990; **82:**650–661.
50. Bingham SA. Mechanisms and experimental and epidemiological evidence relating dietary fibre (non-starch polysaccharides) and starch to protection against large bowel cancer. *Proc Nutr Soc* 1990; **49:**153–171.
51. O'Sullivan KR, Mathias PM, Tobin A, O'Morain C. Risk of adenomatous polyps and colorectal cancer in relation to serum antioxidants and cholesterol status. *Eur J Gastroenterol Hepatol* 1991; **3:**775–779.

52. Tuyns AJ, Kaaks R, Haelterman M. Colorectal cancer and the consumption of foods: a case-control study in Belgium. *Nutr Cancer* 1988; **11:**189–204.

53. Macquart-Moulin G, Riboli E, Cornée J, Charnay B, Berthezène P, Day N. Case-control study on colorectal cancer and diet in Marseilles. *Int J Cancer* 1986; **38:**183–191.

54. Hoff G, Moen IE, Trygg K, Frølich W, Souar J, Vatn M, Gjone E, Larsen S. Epidemiology of polyps in the rectum and sigmoid colon. Evaluation of nutritional factors. *Scand J Gastroenterol* 1986; **21:**199–204.

55. DeCosse JJ, Miller HH, Lesser ML. Effect of wheat fiber and vitamins C and E on rectal polyps in patients with familial adenomatous polyposis. *J Natl Cancer Inst* 1989; **81:**1290–1297.

56. Howe GR, Hirohata T, Hislop TG, Iscovich JM, Yuan J-M, Katsouyanni K, Lubin F, Marubini E, Modun B, Rohan T, Toniolo P, Shunzhang Y. Dietary factors and risk of breast cancer: combined analysis of 12 case-control studies. *J Natl Cancer Inst* 1990; **82:**561–569.

57. Zaridze D, Lifanova Y, Maximovitch D, Day NE, Duffy SW. Diet, alcohol consumption and reproductive factors in a case-control study of breast cancer in Moscow. *Int J Cancer* 1991; **48:**493–501.

58. Rose DP, Goldman M, Connolly JM, Strong LE. High-fiber diet reduces serum estrogen concentrations in premenopausal women. *Am J Clin Nutr* 1991; **54:**520–525.

59. Petrakis NL, King EB. Cytological abnormalities in nipple aspirates of breast fluid from women with severe constipation. *Lancet* 1981; **ii:**1203–1205.

60. Heaton KW. Gallstone prevention: clues from epidemiology. In: Northfield T, Jazrawi R, Zentler-Munro P, eds. *Bile Acids in Health and Disease*. Lancaster, UK: MTP Press, 1988; 157–169.

61. Marcus SN, Heaton KW. Intestinal transit, deoxycholic acid and the cholesterol saturation of bile—three inter-related factors. *Gut* 1986; **27:**550–558.

62. Heaton KW, Emmett PM, Symes CL, Braddon FEM. An exxplanation for gallstones in normal-weight women: slow intestinal transit. *Lancet* 1993; **341:**8–10.

63. Thornton JR, Emmett PM, Heaton KW. Diet and gall stones: effects of refined and unrefined carbohydrate diets on bile cholesterol saturation and bile acid metabolism. *Gut* 1983; **24:**2–6.

64. Pixley F, Wilson D, McPherson K, Mann J. Effect of vegetarianism on development of gall stones in women. *Br Med J* 1985; **291**:11,12.
65. Burkitt DP. The aetiology of appendicitis. *Br J Surg* 1971; **58**:695–699.
66. Burkitt DP. Appendicitis. In: Burkitt DP, Trowell HC, eds. *Refined Carbohydrate Foods and Disease: Some Implications of Dietary Fibre*. London: Academic Press, 1975; 87–97.
67. Barker DJP. Acute appendicitis and dietary fibre; an alternative hypothesis. *Br Med J* 1985; **290**:1125–1127.
68. Morris J, Barker DJP, Nelson M. Diet, infection, and acute appendicitis in Britain and Ireland. *J Epidemiol Comm Health* 1987; **41**:44–49.
69. Larner AJ. The aetiology of appendicitis. *Br J Hosp Med* 1988; **June**:540–542.
70. Barker DJP, Morris J, Nelson M. Vegetable consumption and acute appendicitis in 59 areas in England and Wales. *Br Med J* 1986; **292**:927–930.

Vitamins and Minerals in Cancer, Hypertension, and Other Diseases

Norman J. Temple

So much precise research has been done in the laboratories and so many precise surveys have been made that we know all we need to know about the food requirements of the people. ... The position is perfectly clear-cut [with respect to Britain].

Drummond* and Wilbraham,
in "The Englishman's Food" (1939)

1. Evolution and the Modern Diet

It was argued in a preceding chapter ("Diet and Western Disease") that humans are most likely adapted to the paleolithic diet. Eaton and Konner *(1)* estimated that such a diet provides a daily intake of 1580 mg calcium, 392 mg vitamin C, and 690 mg sodium. Compared with the typical Western diet the paleolithic one provides approximately twice as much calcium, over four times as much vitamin C, but has a sodium content only about 18% as high. The

---

*Jack Drummond was a major nutrition authority in the 1920s and 1930s. He coined the term "vitamin."

From: *Western Diseases: Their Dietary Prevention and Reversibility*
Edited by: N. J. Temple and D. P. Burkitt Copyright ©1994 Humana Press, Totowa, NJ

potassium:sodium ratio was 16.1, whereas in the modern diet it is around 0.7.

Eaton and Konner's study is of considerable importance. However, it does have limitations. First, it is based on an assumption that the typical diet was 35% meat by weight. Using the data in their paper I have recalculated the content of selected nutrients. If the meat content was as low as 10% or as high as 80% by weight (but with energy constant at 12,600 kJ or 3000 kcal), calcium content then becomes 2153 mg (10% meat) or 617 mg (80% meat), vitamin C becomes 5569 mg or 116 mg, whereas the potassium:sodium ratio becomes 33.5 or 7.4. Thus, when the diet contains little meat, it provides a large quantity of calcium and vitamin C. If, though, the diet is predominantly meat, then calcium falls to a level similar to the modern diet, although vitamin C is still rather higher than in that diet. Regardless of the meat content, there is always many times more potassium than sodium, which contrasts sharply with today's diet.

A serious limitation of Eaton and Konner's study is the lack of information on many micronutrients. The typical modern diet has a large content of refined carbohydrates. These undergo substantial nutrient losses during the refining process. In the case of sugar, the content of vitamins and minerals is essentially zero. With cereal foods, such as wheat and rice, refining causes substantial losses of micronutrients (2,3). These includes vitamins B_6 and E, pantothenic acid, biotin, folate, potassium, magnesium, manganese, copper, zinc, chromium, selenium, and molybdenum (and also choline). The high consumption of these nutrient-depleted foodstuffs suggests that the modern diet may be relatively low in various nutrients when compared to more traditional human diets.

What we need to know is the impact of refining carbohydrates on the diet as a whole. Valuable information in this regard was provided by Heaton et al. (4). Patients suffering from gallstones or diabetes were placed on a refined or an unrefined diet for periods of six weeks each. On the refined diet the patients avoided whole grain products and ate unlimited amounts of white bread. The gallstone patients were also instructed to consume plenty of sugar and to limit their intake of fruit and vegetables. On the unrefined diet the patients avoided sugar and refined cereals but were permitted unlimited amounts of unrefined carbohydrates such as whole meal bread, brown rice, whole grain breakfast cereals, fruit and vegetables. Neither diet placed any restriction on food quantity or on the amount of animal products.

The unrefined diet provided significantly more vitamins B_6 and E, pantothenic acid, folate, biotin, potassium, magnesium, copper and zinc, as well as more vitamin C. It also resulted in a doubling of the intake of dietary fiber. However, calcium was higher on the refined diet, since white flour in Britain has added calcium.

A similar study was carried out by Temple (5), except that this one was an abstract exercise. Two diets were formulated. The first was a British diet and contained refined carbohydrates in the quantities typically consumed. The other was the same diet with the same energy content, except that refined carbohydrates were replaced by unrefined foods, namely whole meal bread, unrefined breakfast cereals, potatoes, apples, peanuts, and beans. It is assumed that this would be a typical dietary adaptation if refined carbohydrates were no longer available. The total content of nutrients in each diet was then calculated. Using the unrefined diet as the reference diet, the results show that refined carbohydrates lower the intake of nutrients by from 31% (zinc and potassium) to 72% (manganese). The other nutrients (vitamins B_6 and E, folate, magnesium, and copper, as well as dietary fiber) had losses of from 45 to 65%. There is also a substantial loss of selenium and chromium, but the extent of this cannot be accurately quantified. Tea is a rich source of manganese and the substantial loss of this mineral (72%) would have been even higher had the diet not contained a generous amount of this beverage. The loss of cations (zinc, potassium, manganese, magnesium, and copper) and of vitamin B_6 is counterbalanced, to a greater or lesser extent, by the fact that these nutrients are better absorbed from a low fiber diet.

There has been an unfortunate tendency to equate refined carbohydrates solely with lack of fiber. Based on the above evidence the effects of refining go much further.

Clearly, the possibility must be considered that the modern Western diet provides an insufficient amount of various vitamins and minerals, particularly in comparison with more traditional diets. In the remainder of this chapter evidence will be considered that indicates that a relatively low intake of particular nutrients, as well as an excess of sodium, is strongly associated with disease, particularly with certain cancers and with hypertension.

2. Vitamins, Minerals, and Cancer

2.1. General Considerations

In the chapter "Western Diseases and What They Encompass," Burkitt highlighted the fact that Western diseases include only certain cancers, notably those of the lung, colon, breast, and prostate. Other cancers are far more common in particular nonindustrialized countries. The explanation for this is the worldwide distribution of the causative factors for the various cancers. However, there are many other factors that are not true causes of particular cancers but do increase or decrease its incidence. Several vitamins and minerals come under this heading. As a rule each one acts on several cancers that may be unrelated to primary cause (and geographical distribution).

Progress in this area has been a case of three steps forward, two steps back. This is because of the intrinsic difficulties in unraveling the relationships between diet and disease. The causes of this were discussed in the chapter "Diet and Western Disease," by Temple.

2.2. β-Carotene

In 1981, Peto et al. (6) proposed that β-carotene is protective against certain cancers. They based this hypothesis mainly on a careful analysis of case-control (retrospective) and prospective studies. Few previous studies had considered β-carotene. A far more common study parameter was "vitamin A"; that is, the sum of preformed vitamin A (retinol and certain related chemicals) plus β-carotene (provitamin A). This paper sparked a great deal of interest in β-carotene. Over the last decade the results from dozens of human studies have appeared.

The evidence concerning the β-carotene hypothesis was previously reviewed by Temple and Basu (7). It will now be briefly reassessed, including the considerable amount of new evidence that has appeared in the 5 years since that review was written.

All human studies on the relationship between β-carotene status and cancer risk suffer from a lack of precision. First, dietary assessment is well known to be fraught with inaccuracies. More importantly, when case-control or prospective studies reveal a negative association between dietary β-carotene intake and particular cancers, this does not prove that β-carotene itself is protective; β-carotene may be a mere marker for the true protective factor. If half the population ate a carrot daily, the richest source of β-carotene, the task of the investigator would be greatly simplified. Instead, people eat highly variable amounts of vegetables, leading to significant correlations between intakes of numerous dietary components, including β-carotene, other carotenoids, vitamins C and E, and noncereal dietary fiber. This creates obvious problems in identifying the true protective factor. For this reason when a dietary study indicates that vegetables are protective against various types of cancer, a more specific conclusion is often not possible.

An additional source of error is that people change their diets over the course of time. As a result, the intake of β-carotene when the dietary assessment was made may not be indicative of the intake at the stage of carcinogenesis when β-carotene is effective.

Measurement of blood β-carotene level is an alternative strategy, since it is an indicator of β-carotene intake in the preceding days and weeks. Nevertheless, it too may give rise to misleading results. First, as with dietary assessment, the blood may have been collected at the "wrong" time. Second, blood β-carotene may be a mere marker for the intake of the true protective factor. Third, if several years have elapsed between blood collection and analysis of plasma or serum (as is often the case in prospective studies), then errors can arise because of loss of β-carotene in storage.

The evidence that β-carotene protects against certain cancers is strongest for cancer of the lung. This was found in both the earlier studies (7) as well as in more recent ones (8–13). Overall, persons with a relatively high β-carotene intake have about half the risk of the disease, compared to those with a relatively low intake. Of course, appropriate corrections are made for confounding variables, particularly smoking.

There are numerous reports linking a relatively poor β-carotene status to several other common cancers in addition to lung. This includes cancer of the stomach (7,14–16), breast (postmenopausal) (17–19), cervix (7,20–22), bladder (7), and colon (23–25). In each of these cases the association lacks consistency and the possibility cannot be excluded that it is spurious. With stomach cancer, for instance, important confounding variables are vitamin C, vegetables, and smoking.

Several studies have indicated that β-carotene is protective against experimental tumors *(7)*. Temple and Basu *(26)* showed that β-carotene is strongly protective against chemically induced colon tumors in mice. This is probably the study most relevant to humans, since it used an animal model of a tumor common in humans and a nutritionally relevant dose of β-carotene (equivalent to about 150–300 g carrots/12,600 kJ or 3000 kcals).

2.3. Vitamin C

Block *(27)* recently reviewed the human evidence concerning vitamin C and cancer prevention. In many investigations the study parameter has been fresh fruit rather than vitamin C itself. She concluded that for nonhormone-dependent cancers there is strong evidence that vitamin C is protective; those with a relatively high intake having approximately half the risk of cancer than those with a low intake. The evidence for this seems strongest for cancer of the oral cavity, larynx, esophagus, stomach, and pancreas. It is also strong for cancer of the breast, rectum, and cervix, as well as for cervical dysplasia, a precancerous condition. Somewhat weaker evidence indicates that vitamin C may also be protective in cancer of the colon. In the short time since the above review was written, another handful of studies have confirmed that a low intake of vitamin C is associated with an increased risk of several cancers *(18–20,28,29)*.

Animal experiments have indicated that vitamin C protects against several tumor types *(30)*. However, it would be an oversimplification to describe vitamin C as simply anticarcinogenic; in some experiments it has resulted in an increased tumor yield.

2.4. Vitamin E

Animal studies provide few clear clues as to whether vitamin E is anticarcinogenic. Depending on the tumor model and the experimental details, vitamin E prevents, has no effect on, or, on occasion, even enhances tumor formation *(31–33)*.

Case-control studies have been inconsistent in showing that a low intake of vitamin E is associated with cancer *(15,16,19,25,34–37)*. Any such association could well be secondary to other nutrients.

Several prospective studies have related blood vitamin E level to future cancer risk. Though there is a lack of consistency, these studies suggest that a relatively low blood level of the vitamin may increase the risk of cancer, particularly colorectal *(10,31,33,38)*.

The relationship between vitamin E and cancer risk presents a confusing picture. Dietary and blood analyses have sometimes indicated that a relatively poor status of the nutrient is a risk factor for certain types of cancer. But is the association real? To answer that question let us ask what we would expect to find in the literature if vitamin E is indeed protective but only to a minor extent and only with a minority of cancers. We would expect that only some studies would report a protective association. In others, because of such problems as poor selection of controls, errors in dietary assessment, random error and the inclusion of patients with cancers not protected against by vitamin E, no association would be demonstrable. This appears to be the situation with which we are faced. Clearly, firm conclusions are premature but the findings to date are consistent with the view that vitamin E offers limited protection against some types of cancer.

2.5. Selenium

Dozens of animal experiments have shown an anticarcinogenic action in several tumor models *(39)*. This has been readily demonstrated using a selenium level of greater than approx 1 ppm in the diet or approx 0.7 ppm in drinking water, a level that is near toxic. However, experiments using lower levels of selenium supplementation, and that are more akin to human nutrition, have been less likely to show an anticarcinogenic action and, in fact, often result in an increased tumor yield *(39)*.

Growing evidence from human studies indicates that selenium is protective against various forms of cancer. Epidemiological studies have demonstrated an inverse association between the geographical distribution of selenium and cancer mortality *(40,41)*. Prospective studies over the last few years have shown several times that a low tissue selenium level (as measured by analysis of blood or toenails) poses an increased risk of cancer. The mineral appears to protect against cancer in several organs *(42–49)*. Typically, risk ratios of two to three have been reported when comparing the groups with the highest and lowest selenium levels.

2.6. Calcium

In 1984 Newmark et al. *(50)* hypothesized that calcium has a preventive action against colon cancer. Reviewing the results of several human studies Sorenson et al. *(51)* found the evidence generally supportive of the hypothesis. A recent case-control retrospective study also gave qualified support *(23)*. However, five other studies, three case-control retrospective *(36,52,53)* and two prospective *(54,55)*, revealed no evidence that calcium is protective against cancer of the colon or rectum. Taken as a whole, the evidence suggests that calcium is protective, but clearly this requires confirmation.

2.7. Sodium

Various evidence indicates that an excessive salt (sodium) intake increases the risk of stomach cancer *(56)*. Joossens and Geboers *(56)* have pointed to marked parallels in the mortality rates from stroke and stomach cancer. This occurs both between and within countries. In addition, the two mortality rates often fall in parallel. It must be stressed that there are important epidemiological differences between the two diseases: Stroke is a Western disease, whereas stomach cancer is not. Nevertheless, the similarities are strong enough to suggest that there must be an etiological factor that plays an important role in both. Salt is by far the best candidate. Case-control retrospective studies have also indicated that salt is a risk factor for stomach cancer (57–61). Supporting evidence has come from experimental studies using rats: Salt acts as a cocarcinogen and as a promoter (62–65).

2.8. Summary

We now have strong evidence that the intake of several vitamins and minerals plays an important role in determining who within a population succumbs to cancer. It is still premature to draw firm conclusions, but the balance of evidence is that an above-average intake of β-carotene and vitamin C leads to a significant lowering of the risk for a variety of cancers. This probably also applies to selenium. For vitamin E the evidence is more hazy. The balance of probabilities indicates that calcium is protective against colon cancer, and, conversely, excess salt increases the risk of stomach cancer.

It is intellectually satisfying to identify the specific nutrients that modify cancer risk, but from a practical point of view it is far more important to characterize the diet that will be most effective at reducing the total burden of cancer. The evidence is strong that a diet rich in fruits and vegetables will help achieve this. Such a diet has, in many of the studies referred to, been associated with a reduced cancer risk. In a recent review Block et al. (66) reported that a significant protective association with fruits and vegetables against cancer has been observed in 132 of 170 studies. This protection extends to most major cancers. They conclude, "For most cancer sites, persons with low fruit and vegetable intake ... experience about twice the risk of cancer compared with those with high intake, even after control for potentially confounding factors. ... It would appear that major public health benefits could be achieved by substantially increasing consumption of these foods." Numerous studies published since that review was written have lent additional support to these conclusions (67–77).

There may be much uncertainty as to which substances in fruits and vegetables are anticarcinogenic (β-carotene, other carotenoids, vitamin C, nonnutrients, and the like), but there is little doubt that these foods are generally protective against cancer. In addition, recommending a generous intake of fruits and vegetables is entirely consistent with all other dietary recommendations made in this book.

In 1927 Heisenberg, the great physicist, formulated his uncertainty principle as follows: "The more precisely we determine the position, the more imprecise is the determination of velocity in this instant." The relationship between nutrients and cancer follows an analogous uncertainty principle: "The more precisely we study specific nutrients, the more imprecisely can we determine their role in cancer."

3. Minerals and Blood Pressure

3.1. Sodium

An excess salt intake has long been linked to hypertension (78). Thirty years ago Meneely and Dahl (79) stated:

> Evidence has been presented which indicates that, among animals and humans consuming large amounts of salt, hypertension will be common. It is suggested that the chronic intake of salt in amounts well in excess of requirements may play a primary role in the pathogenesis of hypertension in man.

Individual susceptibility will determine which one individual in a group develops the disease. It is suggested that the excessive consumption of salt by man may be as subtly lethal as, it appears, is the excessive consumption of calories, tobacco or alcohol.

Subsequent research has confirmed the correctness of this statement.

In most Western populations the daily intake of salt is typically 8–12 or more g/d. This contrasts with populations eating a traditional diet, where the intake is typically 0.5 to 2 g/d and rarely exceeds 4 g/d. In such populations, hypertension is virtually unknown, as are its cardiovascular complications *(80)*, and, moreover, blood pressure does not increase with age *(81)*.

On a population basis, sodium intake has a highly significant correlation with blood pressure *(81,82)*. However, studies of individuals within a population have generally found the correlation to be weak or nonexistent. Simpson *(82)* pointed out that such a correlation has been reported several times in East Asian populations but not in Western ones. In a recent reanalysis of published data, Frost et al. *(83)* drew attention to the cause of this apparent hole in the salt hypothesis. It is owing to random error in the measurement of each person's sodium intake. There is considerable day-to-day variation in sodium intake and therefore a single measurement, which is the typical procedure used in such studies, is inaccurate. Frost et al. *(83)* concluded that when this source of error is corrected, studies of individuals within a population collectively show a highly significant association between blood pressure and sodium intake ($p < 0.001$).

Many clinical trials have investigated whether salt restriction is beneficial in hypertension, but results have been mixed. Law et al. *(84)*, in a reanalysis of 70 published studies, concluded that salt restriction does indeed lower blood pressure, particularly in hypertensives. The main reason why many studies have failed to observe this is that they have been of insufficient duration (four weeks or less). (*Note:* The above two reanalyses of published findings, refs. *81,83,84,* are part of a recent single comprehensive study of salt and blood pressure.)

Based on the above evidence as a whole, Law et al. *(84)* concluded that in people aged 50–59 a reduction in the daily sodium intake of 50 mmol (about 3 g of salt) would, after a few weeks, lower systolic blood pressure by an average of 5 mm Hg, and by 7 mm in hypertensives. Diastolic blood pressure would be lowered by about half as much. They predict that such a change by a whole Western population would reduce the incidence of stroke by 22% and of CHD (coronary heart disease) by 16%. This is appreciably better than could be achieved by treating all hypertensives. A reduction in sodium consumption of 100 mmol/d (about 6 g/d of salt), which would necessitate major changes in salt use in all areas of food production and consumption, would reduce stroke mortality by 39% and CHD mortality by 30%.

3.2. Potassium

Interpopulation studies indicate that hypertension is also associated with a relatively low potassium intake *(85).* However, within a population numerous studies have failed to detect a correlation between potassium and either blood pressure or the presence of hypertension. On the other hand, such an association was reported in the INTERSALT study *(86),* and using data from the first National Health and Nutrition Evaluation Survey (NHANES I) *(87).* Quite possibly the same methodological flaws are responsible for these inconsistencies as have dogged testing of the salt–blood pressure hypothesis.

Cappucio and MacGregor *(88)* carried out a meta-analysis of published trials performed on hypertensive patients and concluded that oral potassium supplements have a small but statistically significant blood pressure lowering effect. The data are consistent with the hypothesis that potassium helps counteract the effects of an excessive sodium intake.

The benefits of a raised potassium intake may extend beyond its hypotensive action. Evidence from studies on both humans (89) and rats (90) indicate that supplemental potassium affords protection from stroke independent of its effects on blood pressure.

It is still unclear whether blood pressure relates best to sodium or to the sodium:potassium ratio. There are good grounds for believing that sodium and potassium are two sides of the same coin. Whether one is dealing with entire Western populations or with hypertensives only, there are strong reasons to recommend a cut in salt intake and an increase in potassium.

3.3. Calcium

Dietary calcium is another factor that has become increasingly recognized as being related to blood pressure. An inverse relationship between calcium intake and blood pressure has been consistently seen in studies done within a population (91–93). Moreover, trials of calcium supplementation have generally found a blood pressure lowering effect, albeit minor (92).

4. Other Examples of the Role of Vitamins and Minerals in Disease

4.1. Calcium

Quite apart from its possible protective role against colon cancer and hypertension, calcium appears to be protective against osteoporosis, particularly in postmenopausal women (94). Studies have indicated that bone loss is more rapid in postmenopausal women whose calcium intake is relatively low (95,96) and is slowed by calcium supplementation *(97)*. A prospective study on elderly men and women revealed an inverse association between calcium intake and subsequent risk of hip fracture *(98)*. It must be stressed that epidemiological evidence demonstrates that a low calcium intake is not a primary cause of the condition *(99)*; therefore it is only a contributing factor.

4.2. Vitamin C

There is suggestive evidence that vitamin C status is inversely associated with blood pressure *(100)* and is positively associated with HDL-cholesterol *(101,102)*. These associations infer a protective benefit against cardiovascular disease. It is still unclear whether vitamin C reduces the total blood cholesterol level.

Another health problem apparently related to vitamin C status is male fertility. Supplementation with the nutrient leads to qualitative and quantitative improvements in human sperm *(103)*.

4.3. Chromium

According to R. A. Anderson *(104)*:

The dietary chromium intake of most individuals is considerably less than the suggested safe and adequate intake. Consumption of refined foods, including simple sugars, exacerbates the problem of insufficient dietary chromium since these foods are not only low in dietary chromium but also enhance additional chromium losses.

A low chromium status appears to be a significant factor in impaired glucose tolerance. This association was stumbled on in 1853 when it was noticed that yeast, a rich source of chromium, alleviates diabetes *(105)*. Several clinical studies have tested the effect of supplemental chromium in subjects with an elevated glucose level following a glucose load or with maturity-onset diabetes. The treatment improves the glucose tolerance and insulin levels (i.e., blood glucose and insulin levels both fall) *(104,106)*. This indicates an improvement in tissue insulin sensitivity as a result of correcting a marginal chromium deficiency. Supplemental chromium also increases the glucose level in subjects with hypoglycemia *(104)*. This is one of the rare instances where the same treatment corrects two conditions that appear to be the opposite of each other.

Chromium may also be protective against atherosclerosis and CHD but this still needs clarification *(107)*.

5. Discussion

Traditionally, understanding a disease has consisted of identifying its cause. During the last century microbes were recognized as the cause of infectious disease. Earlier this century vitamin deficiencies were identified as the cause of such diseases as beri-beri and rickets. The concept that each disease has one or, at most, a few specific causes exerts a powerful influence in the medical sciences including nutrition.

Based on this concept it seemed logical that since each deficiency disease is caused by lack of a specific nutrient, then the converse must also be true: Nutrients function only to prevent specific deficiency diseases. (In the case of vitamin E there is no clear deficiency disease so it has often been described as "a vitamin in search of a disease.") Recommended intakes are established largely on this thinking. For instance, the recommended intake of vitamin C is based on the amount to prevent scurvy. But the hole in this concept is that a low nutrient intake is often not a cause of a disease but merely a *factor* in it. In other words, it does not play a central role in determining whether the disease occurs but rather increases the likelihood of the disease. Furthermore, the intake of the nutrient required to properly perform this role is often appreciably more than is required simply to prevent a deficiency.

This concept is well established in the area of infectious disease. The true cause is, of course, infection by microbes. However, nutrition is an important factor in determining whether an infected person develops the particular disease. When the true cause (microbial infection) is removed, infection cannot occur, regardless of nutrition. If, though, the true cause is present, optimal nutrition will prevent some, perhaps many, cases of the infectious disease but will never prevent all cases. Another good example is fluoride and tooth decay. By no stretch of the imagination is fluoride deficiency a true cause. Yet fluoride supplementation is strongly protective. If cariogenic foods were removed from the diet, then fluoride would automatically cease to be protective.

Of the nutrient–health relationships described in this chapter the only ones where the intake of a nutrient can reasonably be described as a cause is sodium in hypertension and, perhaps also, of potassium in hypertension and chromium in a minority of persons with impaired glucose tolerance. In all other cases the nutrient is only a factor. Accordingly, optimizing the intake of the nutrient will reduce the incidence of the condition but will never entirely prevent it. In such cases the first priority is obviously to deal with the true cause. Optimizing the intake of nutrients such as vitamin C, selenium, and calcium should be viewed as a second line of defense. In the battle against disease, as in other battles, the second line of defense is potentially of great importance. It seems a fair estimate that this can probably reduce incidence rates by anywhere from 10 to 50%.

It is apparent, therefore, that to define recommended nutrient intakes narrowly in terms of preventing deficiency diseases is myopic. For at least a few nutrients, we can define three levels of intake. A low intake is clearly deficient by any definition and will produce clinical symptoms or, at least, subclinical deficiency. A somewhat higher intake will prevent deficiency symptoms but will not fully protect health. It is best described as a suboptimal intake. A yet greater intake is required for maximal health preservation. The optimal intake will often be more than would be required if a person led a completely healthy lifestyle. Indeed, it may be "unnaturally" high in that it is more than the amount that humans became adapted to during evolution. However, this can be justified, since it is an antidote to the true cause of the disease. The optimal intake can be likened to the use of fish as a preventive of CHD. Eating fish obviously does not satisfy a requirement but, rather, may help counteract the true cause of CHD.

We can illustrate this most clearly by considering β-carotene and lung cancer. The cause of most lung cancer is tobacco, but strong evidence indicates that an increased intake of β-carotene can cut the risk of the disease in half. Provided a person is not vitamin A deficient, the requirement for β-carotene to prevent a deficiency is zero. Yet in diehard smokers a prudent recommendation would be for a generous intake (i.e., the optimal intake), so as to minimize the risk of cancer of the lung (and of other organs).

Much the same applies to vitamin C. Diet surveys indicate that the mean intake exceeds the RDA *(108)*. Few people, therefore, are at risk of deficiency. Yet most people would benefit by an increased intake, the main reason being to reduce their risk of various types of cancer.

Important evidence that the optimum intake of vitamin C is much more than that needed merely to prevent scurvy has come from studies on guinea pigs, one of the few mammals to require the nutrient. Levine *(109)* pointed out that guinea pigs need approximately ten times more vitamin C to maintain general health than to merely prevent scurvy. Unless they receive the higher intake, they cannot reproduce efficiently, wounds no longer heal speedily, and their health suffers in other ways.

Selenium provides a similar example. The mean intake exceeds the RDA *(110)*. Indeed, deficiency is virtually unheard of in the Western world. Yet, as with vitamin C, dietary studies in the field of cancer suggest that a suboptimal intake is common.

β-carotene, vitamin C, and selenium provide the clearest examples of the concept of suboptimal intake. With other nutrients a suboptimal intake may be simply a straightforward case of marginal deficiency. The benefits of a supplemental intake of chromium might well be a case of correcting a deficient intake, but this is still unclear. Diet surveys indicate that many women have a low calcium intake *(108,110)*. Its association with colon cancer and hypertension may therefore reflect a widespread state of marginal deficiency, at least in women. Alternately, there is every possibility that even if the entire population consumed the RDA for calcium, increasing its intake would still be beneficial.

Clearly, much work still needs to be done to determine optimal nutrient intakes. Quite likely these will be found to depend on individual features, such as whether the person is hypertensive or smokes.

The evidence is now convincing that vitamins and minerals have roles undreamed of 20 years ago. Nutritionists, for understandable reasons, have paid more attention to fat and dietary fiber. Clearly, in order to minimize the risk of cancer, hypertension, and other lifestyle-related diseases, it is necessary to optimize the intake of many nutrients.

References

1. Eaton SB, Konner M. Paleolithic nutrition. A consideration of its nature and current implications. *N Engl J Med* 1985; **312:**283–289.
2. Paul AA, Southgate DAT. *McCance and Widdowson's The Composition of Foods*. London: HMSO, 1978; 38–53.
3. Schroeder HA. Losses of vitamins and trace minerals resulting from processing and preservation of foods. *Am J Clin Nutr* 1971; **24:**562–573.
4. Heaton KW, Emmett PM, Henry CL, Thornton JR, Manhire A, Hartog M. Not just fibre—the nutritional consequences of

refined carbohydrate foods. *Hum Nutr Clin Nutr* 1983; **37C:**31–35.

5. Temple NJ. Refined carbohydrates—a cause of suboptimal nutrient intake. *Med Hypotheses* 1983; **10:**411–424.

6. Peto R, Doll R, Buckley JD, Sporn MB. Can dietary beta-carotene materially reduce human cancer rates? *Nature* 1981; **290:**201–208.

7. Temple NJ, Basu TK. Does beta-carotene prevent cancer? A critical appraisal. *Nutr Res* 1988; **8:**685–701.

8. Mettlin C. Milk drinking, other beverage habits, and lung cancer risk. *Int J Cancer* 1989; **43:**608–612.

9. Harris RWC, Key TJA, Silcocks PB, Bull D, Wald NJ. A case-control study of dietary carotene in men with lung cancer and in men with other epithelial cancers. *Nutr Cancer* 1991; **15:**63–68.

10. Comstock GW, Helzlsouer KJ, Bush TL. Prediagnostic serum levels of carotenoids and vitamin E as related to subsequent cancer in Washington County, Maryland. *Am J Clin Nutr* 1991; **53:**260S–264S.

11. Connett JE, Kuller LH, Kjelsberg MO, Polk BF, Collins G, Rider A, Hulley SB. Relationship between carotenoids and cancer. The Multiple Risk Factor Intervention Trial (MRFIT). *Cancer* 1989; **64:**126–134.

12. Smith AH, Waller KD. Serum beta-carotene in persons with cancer and their immediate families. *Am J Epidemiol* 1991; **133:**661–671.

13. Le Marchand L, Yoshizawa CN, Kolonel LN, Hankin JH, Goodman MT. Vegetable consumption and lung cancer risk: a population-based case-control study in Hawaii. *J Natl Cancer Inst* 1989; **81:**1158–1164.

14. LaVecchia C, Negri C, Decarli A, D'Avanzo B, Franceschi S. A case-control study of diet and gastric cancer in northern Italy. *Int J Cancer* 1987; **40:**484–489.

15. Risch HA, Jain M, Choi NW, Fodor JG, Pfeiffer CJ, Howe GR, Harrison LW, Craib KJP, Miller AB. Dietary factors and the incidence of cancer of the stomach. *Am J Epidemiol* 1985; **122:**947–959.

16. Buiatti E, Palli D, Decarli A, Amadori D, Avellini C, Bianchi S, Bonaguri C, Cipriani F, Cocco P, Giacosa A, Marubini E, Minacci C, Puntoni R, Russo A, Vindigni C, Fraumeni JF, Blot WJ. A case-control study of gastric cancer and diet in Italy: II. Association with nutrients. *Int J Cancer* 1990; **45:**896–901.

17. Howe GR, Hirohata T, Hislop TG, Iscovich JM, Yuan J, Katsouyanni K, Lubin F, Marubini E, Modan B, Rohan T, Toniolo P, Shunzhang Y. Dietary factors and risk of breast cancer: combined analysis of 12 case-control studies. *J Natl Cancer Inst* 1990; **82**:561–569.

18. Zaridze D, Lifanova Y, Maximovitch D, Day NE, Duffy SW. Diet, alcohol consumption and reproductive factors in a case-control study of breast cancer in Moscow. *Int J Cancer* 1991; **48**:493–501.

19. Graham S, Hellmann R, Marshall J, Freudenheim J, Vena J, Swanson M, Zielezny M, Nemoto T, Stubbe N, Raimondo T. Nutritional epidemiology of postmenopausal breast cancer in Western New York. *Am J Epidemiol* 1991; **134**:552–566.

20. Herrero R, Potischman N, Brinton LA, Reeves WC, Brenes MM, Tenorio F, deBritton RC, Gaitan E. A case-control study of nutrient status and invasive cervical cancer. *Am J Epidemiol* 1991; **134**:1335–1346.

21. Brock KE, Berry G, Mock PA, MacLennan R, Truswell AS, Brinton LA. Nutrients in diet and plasma and risk of in situ cervical cancer. *J Natl Cancer Inst* 1988; **80**:580–585.

22. Palan PR, Mikhail MS, Basu J, Romney SL. Plasma level of antioxidant beta-carotene and alpha-tocopherol in uterine cervix. *Nutr Cancer* 1991; **15**:13–20.

23. Whittemore AS, Wu-Williams AH, Lee M, Shu Z, Gallagher RP, Deng-ao J, Lun Z, Xianghui W, Kun C, Jung D, Teh C, Chengde L, Yao XJ, Paffenbarger RS, Henderson BE. Diet, physical activity, and colorectal cancer among Chinese in North America and China. *J Natl Cancer Inst* 1990; **82**: 915–926.

24. West DW, Slattery ML, Robison LM, Schuman KL, Ford MH, Mahoney AW, Lyon JL, Sorensen AW. Dietary intake and colon cancer: sex- and anatomic site-specific associations. *Am J Epidemiol* 1989; **130**:883–894.

25. Heilbrun LK, Nomura A, Hankin JH, Stemmermann GN. Diet and colorectal cancer with special reference to fiber intake. *Int J Cancer* 1989; **44**:1–6.

26. Temple NJ, Basu TK. Protective effect of beta-carotene against colon tumors in mice. *J Natl Cancer Inst* 1987; **78**:1211–1214.

27. Block G. Vitamin C and cancer prevention: the epidemiologic evidence. *Am J Clin Nutr* 1991; **53**:270S–282S.

28. Van Eenwyk J, Davis FG, Bowen PE. Dietary and serum carotenoids and cervical intraepithelial neoplasia. *Int J Cancer* 1991; **48**:34–38.

29. Boeing H, Frentzel-Beyme R, Berger M, Berndt V, Gores W, Korner M, Lohmeier R, Menarcher A, Mannl HFK, Meinhardt M, Muller R, Ostermeier H, Paul F, Schwemmle K, Wagner KH, Wahrendorf J. Case-control study on stomach cancer in Germany. *Int J Cancer* 1991; **47**:858–864.

30. Carpenter MP. Vitamins E and C in neoplastic development. In: Laidlaw SA, Swendseid ME, eds. *Vitamins and Cancer Prevention*. New York: Wiley-Liss, 1991; 61–90.

31. Knekt P, Aromaa A, Maatela J, Aaran R, Nikkari T, Hakama M, Hakulinen T, Peto R, Teppo, L. Vitamin E and cancer prevention. *Am J Clin Nutr* 1991; **53**:283S–286S.

32. Temple NJ, El-Khatib SM. Cabbage and vitamin E: their effect on colon tumor formation in mice. *Cancer Lett* 1987; **35**:71–77.

33. Mergens WJ, Bhagavan HN. Alpha-tocopherols (vitamin E). In: Moon TE, Micozzi MS, eds. *Nutrition and Cancer Prevention*. New York: Marcel Dekker, 1989; 305–340.

34. Gerber M, Cavallo F, Marubini E, Richardson S, Barbieri A, Capitelli E, Costa A, Crastes De Paulet A, Crastes De Paulet P, De Carli A, Pastorino U, Pujol H. Liposoluble vitamins and lipid parameters in breast cancer. A joint study in Northern Italy and Southern France. *Int J Cancer* 1988; **42**:489–494.

35. Byers TE, Graham S, Haughey BP, Marshall JR, Swanson MK. Diet and lung cancer risk: findings from the Western New York Diet Study. *Am J Epidemiol* 1987; **125**:351–363.

36. Fraudenheim JL, Graham S, Marshall JR, Haughey BP, Wilkinson G. A case-control study of diet and rectal cancer in Western New York. *Am J Epidemiol* 1990; **131**:612–624.

37. Ghadirian P, Simard A, Baillargeon J, Maisonneuve P, Boyle P. Nutritional factors and pancreatic cancer in the Francophone community in Montreal, Canada. *Int J Cancer* 1991; **47**:1–6.

38. Longnecker MP, Martin-Moreno J, Knekt P, Nomura AMY, Schober SE, Stahelin HB, Wald NJ, Gey KF, Willett WC. Serum alpha-tocopherol concentration in relation to subsequent colorectal cancer: pooled data from five cohorts. *J Natl Cancer Inst* 1992; **84**:430–435.

39. Combs GF. Selenium. In: Moon TE, Micozzi MS, eds. *Nutrition and Cancer Prevention*. New York: Marcel Dekker, 1989; 389–420.

40. Shamberger RJ, Frost DV. Possible protective effect of selenium against human cancer. *Can Med Assoc J* 1969; **100**:682.

41. Clark LC. The epidemiology of selenium and cancer. *Fed Proc* 1985; **44**:2584–2589.

42. Helzlsouer KJ, Comstock GW, Morris JS. Selenium, lycopene, alpha-tocopherol, beta-carotene, retinol, and subsequent bladder cancer. *Cancer Res* 1989; **49**:6144–6148.

43. Knekt P, Aromaa A, Maatela J, Alfthan G, Aaran R, Teppo L, Hakama M. Serum vitamin E, serum selenium and the risk of gastrointestinal cancer. *Int J Cancer* 1988; **42**:846–850.

44. Burney PGJ, Comstock GW, Morris JS. Serologic precursors of cancer: serum micronutrients and the subsequent risk of pancreatic cancer. *Am J Clin Nutr* 1989; **49**:895–900.

45. Willett WC, Polk BF, Morris JS, Stampfer MJ, Pressel S, Rosner B, Taylor JO, Schneider K, Hames CG. Prediagnostic serum selenium and risk of cancer. *Lancet* 1983; **ii**:130–134.

46. Salonen JT, Alfthan G, Huttunen JK, Puska P. Association between serum selenium and the risk of cancer. *Am J Epidemiol* 1984: **120**:342–349.

47. Salonen JT, Salonen R, Lappetelainen R, Maenpaa PH, Alfthan G, Puska P. Risk of cancer in relation to serum concentrations of selenium and vitamins A and E: matched case-control analysis of prospective data. *Br Med J* 1985; **290**:417–420.

48. Fex G, Pettersson B, Akesson B. Low plasma selenium as a risk factor for cancer death in middle-aged men. *Nutr Cancer* 1987; **10**:221–229.

49. Kok FJ, de Bruijn AM, Hofman A, Vermeeren R, Valkenburg HA. Is serum selenium a risk factor for cancer in men only? *Am J Epidemiol* 1987; **125**:12–16.

50. Newmark HL, Wargovich MJ, Bruce WR. Colon cancer and dietary fat, phosphate, and calcium: a hypothesis. *J Natl Cancer Inst* 1984; **72**:1323–1325.

51. Sorenson AW, Slattery ML, Ford MH. Calcium and colon cancer: a review. *Nutr Cancer* 1988; **11**:135–145.

52. Negri E, LaVecchia C, D'Avanzo B, Franceschi S. Calcium, dairy products, and colorectal cancer. *Nutr Cancer* 1990; **13**:255–262.

53. Tuyns AJ, Haelterman M, Kaaks R. Colorectal cancer and the intake of nutrients: oligosaccharides are a risk factor, fats are not. A case-control study in Belgium. *Nutr Cancer* 1987; **10**:181–196.

54. Wu AH, Paganini-Hill A, Ross RK, Henderson BE. Alcohol, physical activity and other risk factors for colorectal cancer: a prospective study. *Br J Cancer* 1985; **55**:687–694.

55. Stemmermann GN, Nomura A, Chyou P-H. The influence of dairy and nondairy calcium on subsite large-bowel cancer risk. *Dis Colon Rectum* 1990; **33**:190–194.

56. Joossens JV, Geboers J. Dietary salt and risks to health. *Am J Clin Nutr* 1987; **45**:1277–1288.
57. Coggon D, Barker M P, Cole RB, Nelson M. Stomach cancer and food storage. *J Natl Cancer Inst* 1989; **81**:1178–1182.
58. Graham S, Haughey B, Marshall J, Brasure J, Zielezny M, Freudenheim J, West D, Nolan J, Wilkinson G. Diet in the epidemiology of gastric cancer. *Nutr Cancer* 1990; **13**:19–34.
59. Tuyns A. Salt and gastrointestinal cancer. *Nutr Cancer* 1988; **11**:229–232.
60. Haenszel W, Kurihara M, Segi M, Lee RKC. Stomach cancer among Japanese in Hawaii. *J Natl Cancer Inst* 1972; **49**:969–988.
61. Hu J, Zhang S, Jia E, Wang Q, Liu S, Liu Y, Wu Y, Cheng Y. Diet and cancer of the stomach: a case-control study in China. *Int J Cancer* 1988; **41**:331–335.
62. Shirai T, Fukushima S, Ohshima M, Masuda A, Ito N. Effects of butylated hydroxyanisole, butylated hydroxytoluene, and NaCl on gastric carcinogenesis initiated with N-methyl-N'-nitro-N-nitrosoguanidine in F344 rats. *J Natl Cancer Inst* 1984; **72**:1189–1198.
63. Takahashi M, Kokubo T, Fukukawa F, Kurokawa Y, Hayashi Y. Effects of sodium chloride, saccharin, phenobarbital and aspirin on gastric carcinogenesis with N-methyl-N'-nitro-N-nitrosoguanidine. *Gann* 1984; **75**:494–501.
64. Kim JP, Park JG, Lee MD, Han M, Park S, Lee B, Jung S. Co-carcinogenic effects of several Korean foods on gastric cancer induced by N-methyl-N'-nitro-N-nitrosoguanidine in rats. *Jpn J Surg* 1985; **15**:427–437.
65. Tatematsu M, Takahashi M, Fukushima S, Hananouchi M, Shirai T. Effects in rats of sodium chloride on experimental gastric cancers induced by N-methyl-N'-nitro-N-nitrosoguanidine or 4-nitroquinoline-1-oxide. *J Natl Cancer Inst* 1975; **55**:101–106.
66. Block G, Patterson B, Subar A. Fruit, vegetables, and cancer prevention: a review of the epidemiological evidence. *Nutr Cancer* 1992; **18**:1–29.
67. Walker ARP, Walker BF, Tsotetsi NG, Sebitso C, Siwedi D, Walker AJ. Case-control study of prostate cancer in black patients in Soweto, South Africa. *Br J Cancer* 1992; **65**:438–441.
68. Swanson CA, Mao BL, Li JY, Lubin JH, Yao SX, Wang JZ, Cai SK, Hou Y, Luo QS, Blot WJ. Dietary determinants of lung cancer risk: results from a case-control study in Yunnan Province, China. *Int J Cancer* 1992; **50**:876–880.
69. Tuyns AJ, Kaaks R, Haelterman M, Riboli E. Diet and gastric cancer. A case-control study in Belgium. *Int J Cancer* 1992; **51**:1–6.

70. Howe GR, Ghadirian P, Bueno de Mesquita HB, Zatonski WA, Baghurst PA, Miller AB, Simard A, Baillargeon J, De Waard F, Przewozniak K, McMichael AJ, Jain M, Hsieh CC, Maisonneuve P, Boyle P, Walker AM. A collaborative case-control study on nutrient intake and pancreatic cancer within the search programme. *Int J Cancer* 1992; **51:**365–372.

71. Negri E, LaVecchia C, Franceschi S, D'Avanzo B, Parazzini F. Vegetables and fruit consumption and cancer risk. *Int J Cancer* 1991; **48:**350–354.

72. Zheng W, Blot WJ, Shu XO, Gao YT, Ji BT, Ziegler RG, Fraumeni JF. Diet and other risk factors for laryngeal cancer in Shanghai, China. *Am J Epidemiol* 1992; **136:**178–191.

73. Cheng KK, Day NE, Duffy SW, Lam TH, Fok M, Wong J. Pickled vegetables in the aetiology of oesophageal cancer in Hong Kong Chinese. *Lancet* 1992; **339:**1314–1318.

74. Candelora EC, Stockwell HG, Armstrong AW, Pinkham PA. Dietary intake and risk of lung cancer in women who never smoked. *Nutr Cancer* 1992; **17:**263–270.

75. Forman MR, Yao SX, Graubard BI, Qiao YL, McAdams M, Mao BL, Taylor PR. The effect of dietary intake of fruits and vegetables on the odds ratio of lung cancer among Yunnan tin miners. *Int J Epidemiol* 1992; **21:**437–441.

76. Franceschi S, Barra S, LaVecchia C, Bidoli E, Negri E, Talamini R. Risk factors for cancer of the tongue and the mouth. A case-control study from Northern Italy. *Cancer* 1992; **70:**2227–2233.

77. Thun MJ, Calle EE, Namboodiri MM, Flanders WD, Coates RJ, Byers T, Boffetta P, Garfinkel L, Heath CW. Risk factors for fatal colon cancer in a large prospective study. *J Natl Cancer Inst* 1992; **84:**1491–1500.

78. Ambard Z, Beaujard E. Causes de l'hypertension arterielle. *Arch en Med* 1904; **1:**520–533.

79. Meneely GR, Dahl LK. Electrolytes in hypertension: the effects of sodium chloride. The evidence from animal and human studies. *Med Clin N Am* 1961; **45:**271–283.

80. Trowell HC. Hypertension, obesity, diabetes mellitus and coronary heart disease. In: Trowell HC, Burkitt DP, eds. *Western Diseases: Their Emergence and Prevention*. London: Edward Arnold, 1981;3-32.

81. Law MR, Frost CD, Wald NJ. By how much does dietary salt reduction lower blood pressure? I. Analysis of observational data among populations. *Br Med J* 1991; **302:**811–815.

82. Simpson FO. Blood pressure and sodium intake. In: Laragh JH, Brenner BM, eds. *Hypertension: Pathophysiology, Diagnosis and Management*. New York: Raven Press, 1990; 205–215.

83. Frost CD, Law MR, Wald NJ. (Part II of ref. *81*). II. Analysis of observational data within populations. *Ibid* 815–818.

84. Law MR, Frost CD, Wald NJ. (Part III of ref. *81*). III. Analysis of data from trials of salt reduction. *Ibid* 819–824.

85. Svetkey LP, Klotman PE. Blood pressure and potassium intake. In: Laragh JH, Brenner BM, eds. *Hypertension: Pathophysiology, Diagnosis and Management*. New York: Raven Press, 1990; 217–227.

86. Intersalt Cooperative Research Group. Intersalt: an international study of electrolyte excretion and blood pressure. Results for 24 hour urinary sodium and potassium excretion. *Br Med J* 1988; **297:**319–328.

87. Gruchow HW, Sobocinski KA, Barboriak JJ. Alcohol, nutrient intake, and hypertension in US adults. *JAMA* 1985; **253:**1567–1570.

88. Cappucio FP, MacGregor GA. Does potassium supplementation lower blood pressure? *J Hypertens* 1991; **9:**465–473.

89. Khaw KT, Barrett-Connor E. Dietary potassium and stroke-associated mortality. A 12-year prospective population study. *N Engl J Med* 1987; **316:**235–240.

90. Tobian L. The protective effects of high-potassium diets in hypertension, and the mechanisms by which high-NaCl diets produce hypertension—a personal view. In: Laragh JH, Brenner BM, eds. *Hypertension: Pathophysiology, Diagnosis and Management*. New York: Raven Press, 1990; 49–61.

91. Iso H, Terao A, Kitamura A, Sato S, Naito Y, Kiyama M, Tanigaki M, Iida M, Konishi M, Shimamoto T, Komachi Y. Calcium intake and blood pressure in seven Japanese populations. *Am J Epidemiol* 1991; **133:**776–783.

92. Cutler JA, Brittain E. Calcium and blood pressure. *Am J Hypertens* 1990; **3:**137S–146S.

93. McCarron DA, Morris CD, Young E, Roullet C, Drueke T. Dietary calcium and blood pressure: modifying factors in specific populations. *Am J Clin Nutr* 1991; **54:**215S–219S.

94. Heaney RP. Calcium in the prevention and treatment of osteoporosis. *J Intern Med* 1992; **231:**169–180.

95. Andon MB, Smith KT, Bracker M, Sartoris D, Saltman P, Strause L. Spinal bone density and calcium intake in healthy postmenopausal women. *Am J Clin Nutr* 1991; **54:**927–929.

96. Dawson-Hughes B, Jacques P, Shipp C. Dietary calcium intake and bone loss from the spine in healthy postmenopausal women. *Am J Clin Nutr* 1987; **46:**685–687.

97. Dawson-Hughes B. Calcium supplementation and bone-loss: a review of controlled clinical trials. *Am J Clin Nutr* 1991; **54**:274S–280S.
98. Holbrook TL, Barrett-Connor E, Wingard DL. Dietary calcium and risk of hip fracture: 14-year prospective population study. *Lancet* 1988; **ii**:1046–1049.
99. Walker ARP, Walker BF, Labadarios D. Calcium supplementation in black Africa. *Med J Australia* 1990; **153**:572.
100. Yoshioka M, Matsushita T, Chuman Y. Inverse association of serum ascorbic acid level and blood pressure or rate of hypertension in male adults aged 30–39 years. *Int J Vitam Nutr Res* 1984; **54**:343–347.
101. Jacques PF, Hartz SC, McGandy RB, Jacob RA, Russell RM. Ascorbic acid, HDL, and total plasma cholesterol in the elderly. *J Am Coll Nutr* 1987; **6**:169–174.
102. Jacques PF, Hartz SC, McGandy RB, Jacob RA, Russell RM. Vitamin C and blood lipoproteins in an elderly population. *Ann NY Acad Sci* 1987; **498**:100–109.
103. Dawson EB, Harris WA, Rankin WE, Charpentier LA, McGanity WJ. Effect of ascorbic acid on male fertility. *Ann NY Acad Sci* 1987; **498**:312–323.
104. Anderson RA. Chromium metabolism and its role in disease processes in man. *Clin Physiol Biochem* 1986; **4**:31–41.
105. Rolls R. Brewers's yeast and diabetes. *Br Med J* 1977; **1**:905.
106. Anderson RA, Polansky MM, Bryden NA, Canary JJ. Supplemental-chromium effects on glucose, insulin, glucagon, and urinary chromium losses in subjects consuming controlled low-chromium diets. *Am J Clin Nutr* 1991; **54**:909–916.
107. Liu VJK, Chen XS. Trace elements in the cardiovascular patient: prevention and treatment. In: Watson RR, ed. *Nutrition and Heart Disease*. Vol II. Boca Raton, Florida: CRC Press, 1987; 57–69.
108. Pao EM, Mickle SJ, Burk MC. One-day and 3-day nutrient intakes by individuals—Nationwide Food Consumption Survey Findings, Spring 1977. *J Am Diet Assoc* 1985; **85**:313–324.
109. Levine M. New concepts in the biology and biochemistry of ascorbic acid. *N Engl J Med* 1986; **314**:892–902.
110. Pennington JAT, Young BE. Total diet study nutritional elements, 1982–1989. *J Am Diet Assoc* 1991; **91**:179–183.

PART III

THE POSSIBILITY OF DISEASE REVERSIBILITY

Reversing Coronary Heart Disease

Hans Diehl

They are as sick that surfeit with too much as they that starve with nothing.

Shakespeare, *Merchant of Venice*

1. Atherosclerosis: Portrait of a Killer

1.1. Extent

In the United States, as in many industrialized Western countries, almost every second man and woman dies from degenerative vascular disease related to atherosclerosis, a process that expresses itself most frequently as coronary artery disease, with angina and myocardial infarction as its classical manifestations. The disease is so prevalent that it has often been assumed to be a natural concomitant of the aging process.

1.2. Nature

Despite the fact that at least 90% of typical angina patients have significant atherosclerotic obstructive abnormalities, most coronary artery disease is asymptomatic until the atherosclerotic plaque has narrowed the arterial diameter by 75–85% and begins to interfere with the flow of oxygenated blood to the myocardium (Figs. 1 and 2).

The progressive shortage of oxygen (hypoxia), largely related to atherosclerotic plaque buildup over the years,

From: *Western Diseases: Their Dietary Prevention and Reversibility*
Edited by: N. J. Temple and D. P. Burkitt Copyright ©1994 Humana Press, Totowa, NJ

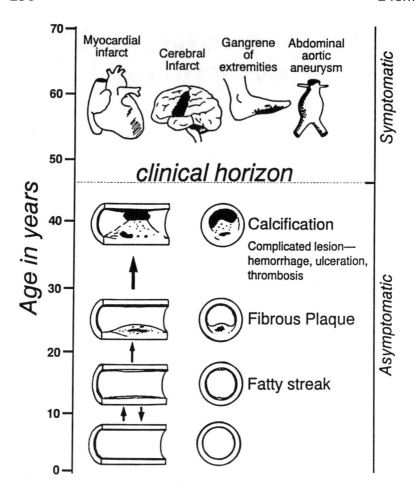

Fig. 1. Diagram of the natural history of atherosclerosis. (Slightly modified and reproduced with permission from McGill HC, Jr, et al. In: Standler M, Bourne GH, eds. *Atherosclerosis and Its Origin*, New York, Academic Press, 1963.)

and its complications, such as thrombus development, is a principal underlying cause of various hypoxic diseases affecting the circulatory system (Fig. 3 and Table 1).

Atherosclerosis is an insidious killer. A coronary artery, for instance, may be 90% narrowed before athero-

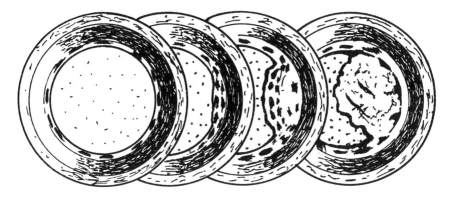

Fig. 2. Atherosclerotic plaque growth causing progressive artery narrowing and closure.

sclerosis becomes symptomatic. Lewis Kuller *(1)*, in his Sudden Death Study in Baltimore, found that 24% of the victims had seen a physician within 7 days prior to the onset of the fatal event, and many had received a clean bill of health. Their first sign of heart disease was also their last one. On the other hand, thrombotic events superimposed on slowly developing atheromatous plaques may suddenly trigger the catastrophic clinical event, even though the artery may not be stenosed more than 50–70%. Regardless of the trigger mechanisms, atherosclerosis is like a loaded gun, where time usually pulls the trigger.

In Western industrialized societies, these atherosclerotic processes usually begin in the preteen years and progress relentlessly to clinical endpoints such as angina and myocardial infarction (when it involves the coronary arteries). Postmortem studies recorded that over three-quarters of American soldiers who died in the Korean conflict already had significantly narrowed coronary arteries. Their average age was 22 years *(2,3)*; and yet atherosclerosis is not natural, nor is it the inevitable result of the aging process. Large populations in the world are clinically unaffected by it *(4)*.

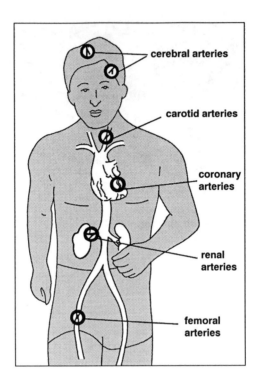

Fig. 3. Atherosclerotic sites. Atherosclerotic plaques most commonly involve the coronary, carotid, and cerebral arteries affecting blood flow and oxygen delivery to the heart muscle and to brain structures. Plaques can also interfere with proper blood flow to the kidneys, the male sex organ, and the legs by affecting the renal, penile, and femoral arteries.

1.3. Causes: Early Findings

The mystery of atherosclerosis began to unravel around the turn of the century when Russian researchers found that rabbits fed a diet of meat and eggs developed high levels of cholesterol and severely clogged arteries (5).

In 1916 a Dutch researcher in Indonesia reported that Javanese people had very low cholesterol levels and were virtually free of atherosclerosis, in contrast to Dutch people living in the Netherlands. When he noticed, however, that

Table 1
Hypoxic Diseases[a]

Clinical syndrome	Arteries	Commentary
Cerebral infarction	Cerebral Carotid	Brittle, narrowed arteries rupture or close causing paralysis or sudden death
Senility	Cerebral Carotid	An inadequate oxygen supply to vital brain tissue is involved in 45% of cases
Hearing loss	Carotid Others	Often owing to atherosclerosis, especially in high frequency range
Visual loss	Retinal	Retinal changes often relate to blood vessel changes and inadequate blood flow
Angina pectoris	Coronaries	Impeded or interrupted blood flow to myocardium, especially during exertion
Myocardial infarction	Coronaries	Portion of myocardium becomes hypoxic and dies
Dissecting aneurysm	Aorta	Atherosclerosed aorta becomes brittle and aneurysm can break much easier
Hypertension	Renal Others	Greater force (higher BP) is needed to push blood through narrowed vessels
Degenerative disc disease	Vertebral	Majority of hypertensives exhibit stenosed vertebral arteries, which may possibly interfere with proper nutrient delivery to discs
Impotence	Penile	In 50%, atherosclerosis interferes with adequate and sustained blood flow
Intermittent claudication	Femoral	Restricted flow of oxygenated blood to leg muscles causing cramps and pain
Gangrene	Tibial	Tissue, usually in the toes and feet, decays and dies often necessitating amputation

[a]Insufficient oxygen due to atherosclerotic processes, which narrow arterial diameter and may promote thrombotic episodes and spasms.

Javanese stewards working—and eating—on Dutch steam-
ships soon developed the same high cholesterol levels as
did the Dutch, he suggested that atherosclerosis might not
be related to race as much as to diet (6).

Additional support for the emerging "diet theory" was
found during war time, particularly during World War II,
when most Europeans were forced to change their eating
habits from their customary diet of meat, eggs, and dairy
products to a more austere diet of potatoes, beans, grains,
roots and vegetables. For nearly a decade the average blood
cholesterol levels plummeted to 4.0 mmol/L (156 mg%) (7).
At the same time, a substantial decrease in atherosclerosis and CHD was documented in various European populations (8). Finland, deprived of meat and butterfat as a
result of the Russian invasion, experienced a sharp drop
in atherosclerosis (9). Norway (10) and Sweden (8) experienced a similar fate because of Nazi occupation. In Belgium, William Castelli and other pathologists had a
difficult time finding atherosclerosed arteries at the time
of autopsy to show medical students.

"There was a marked decline in CHD," Castelli recalls,
"because the Nazis had taken away their livestock, and
most of the people were living on potatoes and bread" (11).

The most unusual evidence came from Nazi concentration camps. Despite subhuman diets and torture, survivors
of the holocaust were surprisingly free of atherosclerosis.
It was the first indication, later confirmed by angiographic
examinations of American POWs in Vietnam, that the process of atherosclerosis was reversible. Those who were held
the longest in captivity had the cleanest arteries.

2. The Risk Factor Concept

In 1947, Ancel Keys and colleagues (12) enrolled 281
businessmen in their 40s and 50s into a 15-yr study of
heart disease at the University of Minnesota. When the

medical records of the men who suffered coronary events during the period were compared with those of the men who did not, Keys found three big differences: The heart attack victims had higher levels of cholesterol, higher blood pressures, and more of them smoked.

The Framingham Study confirmed Keys' *risk factor* concept *(13)*. Some 40 years ago, 5209 men and women between the ages of 30–60 and living in the town of Framingham, MA, were enrolled in this "life and death" study of cardiovascular disease. The participants have lived in a scientific "fish bowl" ever since, where their habits, physical characteristics, histories (medical and otherwise), and laboratory tests have been carefully assessed to see if they were possibly related to various circulatory diseases. Under the present leadership of William Castelli, the researchers in this monumental study found the following:

- Fifty-yr-old men with blood *cholesterol* over 7.6 mmol/ L (295 mg%) are nine times more likely to develop hardening of the arteries than men of the same age with cholesterol under 5.2 mmol/L (200 mg%).
- *Smokers* are ten times more likely to die by the age of 60 than nonsmokers.
- Men 20% *overweight* are five times more likely to develop cardiovascular disease by the age of 60 than men of normal weight.
- Every second *death* is a result of heart disease or stroke.
- By the age of 60, one out of every five men has a *heart attack*.
- By the age of 50, every third person is *hypertensive* and three times more likely to die of cardiovascular disease than a normotensive person.

The Framingham Study confirmed Keys' risk factor concept in determining the likelihood of developing cardiovascular disease and made it as important as the germ theory is to infectious diseases (Fig. 4).

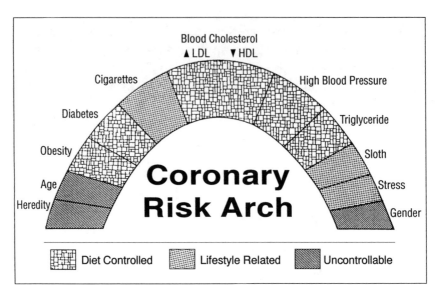

Fig. 4. Coronary risk factors. The higher on the arch, the higher the contribution of the risk factor to heart disease. Five of the eight controllable risk factors are mostly under the control of diet.

The higher the risk factor is located on the risk arch, the more important and consequential is its contribution to cardiovascular disease risk. Therefore, high total serum cholesterol levels (especially high LDL-cholesterol and low HDL-cholesterol), high blood pressure readings, and smoking are *major* risk factors. Any combination of risk factors only magnifies the problem.

Risk factors like age, gender, and heredity are beyond human control. Women are somewhat protected against heart disease until menopause compared with men of the same age. Only about one person in 500 is actually "programmed" genetically for heart disease.

Fortunately, most risk factors in heart disease are controllable. Serum cholesterol, blood pressure, triglycerides, obesity, and type II diabetes can be greatly modified through a change in diet. Stress management skills can be learned, the smoking habit kicked, and physical activity enhanced.

These risk factors play an important role in determining atherogenesis and the subsequent pace of atherosclerosis *(14)*. The blood lipids (total cholesterol, HDL-, and LDL-cholesterol, and triglycerides), hypertension, cigaret smoking, type II diabetes, and a sedentary lifestyle have been repeatedly shown to be independently related to the rates of development of clinical coronary artery disease and to the extent of occlusive coronary atherosclerosis as shown on angiography *(15)*. The presence of multiple risk factors also plays a major role as rate determinants of venous graft closure after coronary bypass surgery *(16–18)*.

By placing risk factor information into a composite cardiovascular risk profile, the *joint* effects can be estimated in the form of multivariate risk functions taking into account the multifactorial elements of cardiovascular risk and the continuous gradient of response *(19)*. These multivariate risk functions make it possible to estimate risk over a wide range and to identify the segment of the asymptomatic population who are prime candidates for developing hypoxic diseases. For these, special lifestyle intervention can be planned and implemented *(20,21)*.

Table 2 displays the risk differential for coronary risk in individuals with different coronary risk factors. It compares high risk male "A" with low risk male "B." At age 35, male "A" is 140 *times* more likely to develop angina or a myocardial infarction within 6 yr than his contemporary with ideal risk factor levels. If "A" was 45 yr of age, he would have 32 times the chance of developing heart disease within 6 yr. And at 55, his chances would be 15 times greater. Understandably, age diminishes the coronary risk differential, since aging itself contributes to the coronary risk in that it measures how long the arteries have been exposed to certain risk factors, such as serum cholesterol.

Table 2
Coronary Risk[a]

Risk factor	Male "A"	Male "B"	Risk differential
Cholesterol (mmol/L)	8.0	4.0	
Blood Pressure	160/95	110/75	
Smoking	yes	no	
Diabetes	yes	no	
ECG-LVH	positive	negative	
Risk of developing heart disease within 6 yr			
at age 35	14%	0.1%	140x
at age 45	32%	1.0%	32x
at age 55	42%	2.8%	15x

[a]Risk of developing heart disease in six years according to risk factors.

The older the person, the longer the exposure, the greater the probability of atherosclerotic buildup.

Multimillion dollar studies funded by the National Institutes of Health have shown that 63–80% of all major coronary events before age 65 could be prevented if the population would lower serum cholesterol levels to at least 4.7 mmol/L (181 mg%), lower systolic blood pressure to less than 125 mmHg, and quit smoking (22).

Evaluating the data from the Multiple Risk Factor Intervention Trial (MRFIT), which followed 356,222 men between the ages of 35–57 for 6 yr, the Stamler group (23) concluded, "In the absence of smoking and diabetes, 87% of all coronary deaths could be prevented if the blood cholesterol level was kept below 181 mg% (4.7 mmol/L) and systolic blood pressure under 120." Many people could accomplish this in less than four weeks if they would conscientiously break with their rich Western diet and their smoking (24).

3. The Etiological Preeminence of Hyperlipidemia

3.1. Dietary Changes (US 1860–1990)

Evidence from epidemiological, clinicopathological, and animal studies has demonstrated that the development and progression of atherosclerotic lesions and their subsequent clinical manifestations in various arterial diseases is not an inevitable consequence of aging or genetics but a matter of lifestyle. These hypoxic diseases develop under the confluence of affluence-related causes: rich diet, smoking, sedentary life patterns, and stress. Stamler, Burkitt, and many others have repeatedly pointed to the far-reaching changes in the dietary composition of industrialized and affluent nations, as causally related to the development of cardiovascular disease and other Western killer diseases (Fig. 5).

Jeremiah Stamler *(25)* in his George Lyman Duff Memorial lecture stated:

> It is further reasonable and sound to designate "rich diet" as a *primary, essential, necessary cause* of the current epidemic of premature atherosclerotic disease raging in western industrialized countries. Cigarette smoking and hypertension are important secondary or complementary causes.

Complex carbohydrates (starches), as found in grains, vegetables, legumes, and potatoes, are no longer the mainstay of the Western diet. Economic development, modern food processing, and animal husbandry relate directly to diets higher in total fat, saturated fat and processed vegetable oil, refined sugar, animal protein, salt, and cholesterol. These increases occurred while the consumption of inexpensive, unrefined starch foods, rich in minerals, vitamins, and fiber, experienced a substantial decrease.

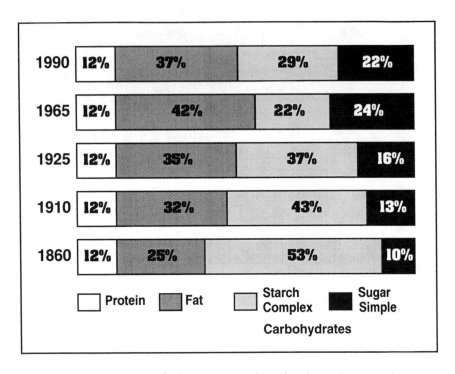

Fig. 5. Dietary trends (in percent of total calories), United States 1860–1990. The American diet has been shifting, resulting in a dramatic change in its composition, which reached its high sugar, high fat culmination in the 1970s. Since then complex carbohydrate foods are gradually increasing at the expense of fat and sugar.

3.2. Effect of Dietary Changes on Atherosclerosis

Extensive research has prominently linked the emergence and expansion of cardiovascular disease and other Western diseases in various countries to these dietary changes. Dietary patterns largely determine national levels of blood lipids (via the amount and type of fat as well as the amount of cholesterol consumed), blood pressures (through sodium and potassium content), triglycerides (by amounts of sugar, fat, alcohol and total calories eaten), and the prevalence of obesity and type II diabetes (related to calorie-dense foods).

Similarly, the substantial decline in CHD mortality in certain Western nations over the last two decades is also concordant with dietary changes. The 1978 Bethesda Conference, exploring the 25% decline in coronary deaths during the period 1968–1978 in the United States, concluded that primary preventive measures, such as dietary changes that lowered the average national serum cholesterol level from 6.1 to 5.4 mmol/L (235 to 210 mg%) and smoking cessation, played dominant roles (22).

It has also been acknowledged that better medical care and technological and surgical interventions have

> probably made an additional contribution. However, by themselves, these advances in care of heart disease patients cannot be the sole or main reason for the long-term decline, since the decline began and unfolded well before the emergence and widespread use of these specific new interventions (26).

3.2.1. Experimental Atherogenesis

Ever since the experiments Ignatowsky (27) and Anitschkow (5) carried out in Russia at the beginning of this century, scientists continue to note that high cholesterol feedings have consistently induced atherosclerosis in susceptible animals. These animals have systematically demonstrated that experimentally induced hypercholesterolemia is the key factor for atherogenesis.

Taylor and colleagues (28) became the first team to feed a monkey to death. After being fed a typically prepared American diet for some 30 months, the rhesus monkey died of a massive myocardial infarction with damage and atherosclerotic plaques similar to that found in human hearts and coronaries after a serious heart attack.

In his landmark studies with rhesus monkeys, Robert Wissler and his group at the University of Chicago demonstrated again and again that the typical American diet causes high serum cholesterol levels and produces significant atherosclerotic plaque buildup. After simulating human plaques in various species for more than 30 years, Wissler *(29)* recently stated, "Animal experiments show that no species is immune to atherosclerosis, once sustained elevation of serum cholesterol has been accomplished."

Armstrong, Warner, and Connor fed a group of 30 adult rhesus monkeys a 40% fat, very high cholesterol diet that increased their serum cholesterol levels from 3.6 (140 mg%) to 18 mmol/L (700 mg%). After 17 mo on this atherogenic diet, 10 of the monkeys were sacrificed. Their coronary lumina had stenosed an average of 60% *(30)*. When the remaining monkeys were fed a cholesterol-lowering diet, their serum cholesterol levels returned to normal, and significant plaque regression was demonstrated as early as 20 mo *(31)* (Fig. 6).

William Connor *(32)*, an eminent cardiologist and nutrition researcher at the University of Oregon, recently said, "If ever a human disease can be produced in animals, it is atherosclerosis. And if ever the requirements for this disease have been isolated, they are fat and cholesterol in the diet."

3.2.2. Human Studies

Evidence from clinicopathological and epidemiological studies overwhelmingly confirms that hyperlipidemia in a human population is the primary prerequisite for atherosclerosis manifested in premature cardiovascular disability and death. Hyperlipidemia is caused by a diet high in fat, especially saturated fat and cholesterol.

The International Atherosclerosis Project, examining 23,000 autopsied arteries from 14 countries, found that the degree of artherosclerosis was directly proportional to the prevalence of CHD and stroke, and that lipid levels were directly related to plaque damage. The higher the serum cholesterol, the greater the plaque buildup *(33)*.

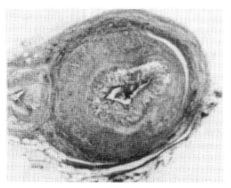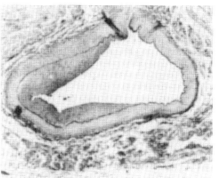

Fig. 6. Atherosclerotic plaque regression. Experiment on regression of coronary atherosclerosis in male rhesus monkeys *(31)*. **Left:** Coronary atherosclerosis after 18 mo on a high-cholesterol diet from egg yolks. **Right:** Regression of atherosclerosis 40 mo after cessation of feeding of the high-cholesterol diet. (Reproduced with permission from the American Heart Association.)

Ancel Keys *(34)*, in a monumental population study, observed more than 12,000 men between the ages of 40–59 in seven countries over a period of 10 years. Countries with diets high in saturated fat (mostly from animal products) and cholesterol (exclusively from animal products) had high serum cholesterol levels and high rates of CHD (Fig. 7).

Conversely, countries with diets relatively low in saturated fat and cholesterol had low serum cholesterol levels and low rates of CHD. Sophisticated data analyses showed that 70% of the variance in CHD incidence was attributable to habitual dietary differences in saturated fat. Similarly, 80% of the variance in the average serum cholesterol levels between these populations could be explained by differences in habitual diet.

During the ten years of the study, Keys found 14 times as many fatal myocardial infarctions in East Finland as in Japan. The Finns' average serum cholesterol level was

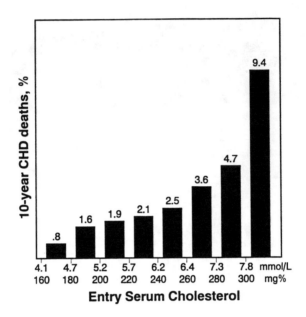

Fig. 7. The Seven Countries Study. Total serum cholesterol and deaths from CHD *(34)*.

6.8 mmol/L (264 mg%), but Japanese men averaged 3.7 mmol/L (144 mg%). During that time, the Finns consumed 22% of their total calories as saturated fat, whereas the Japanese consumed only 3% from animal fat, with their total fat intake below 10% *(34)*.

When Keys and his colleagues *(35)* studied Japanese men who had migrated to California, they found that these men had also left behind their very low-fat, low-cholesterol diets and their apparent immunity to heart disease. Their fat intake had moved from 10 to 40% of total calories, their serum cholesterol had climbed from 3.9 mmol/L (150 mg%) to 5.9 mmol/L (238 mg%), and their CHD death rate had increased tenfold, almost equaling that of American males (Fig. 8).

More than 200 such migrant studies have confirmed that atherosclerosis is not a disease of genetics but is largely

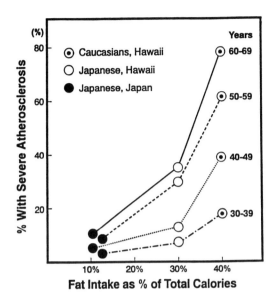

Fig. 8. Fat intake and extent of atherosclerosis. International autopsy studies show that as people eat more fat and cholesterol, the extent of arterial plaque buildup increases. Caucasians living in Hawaii had more artery damage at 30–39 yr than those Japanese living in Japan had at age 60–69 *(35)*.

a disease of dietary lifestyle. As "protected" people move to a coronary-prone culture, with its dietary excesses and its accompanying high serum cholesterol levels, they soon begin to develop the hypoxic diseases of their host country *(36)*.

Angiograms of the coronary arteries of 723 men under the age of 40 admitted to the Cleveland Clinic with slight chest pain showed that the extent of arterial closure related directly and positively to the serum cholesterol levels *(37)* (Fig. 9).

Similarly, Page and colleagues showed that the strongest and most reliable predictor of arterial narrowing is the level of serum cholesterol *(38)*.

More recently, the $115 million MRFIT Study, which followed 356,222 men between the ages of 35–57 for 6 yr,

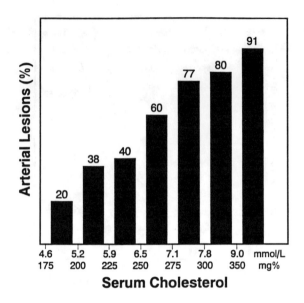

Fig. 9. The Cleveland Study. Total serum cholesterol levels and significant arterial lesions. The higher the cholesterol, the higher the prevalence of significant artery narrowing (37).

showed that the risk of fatal CHD increased steadily as cholesterol levels increased. The relationship between the two was continuous and graded, and increased in a curvilinear fashion. The relative risk of a male (40–44 yr) dying within 6 yr from CHD was 5.8 times higher at a cholesterol ≥6.2 mmol/L (≥245 mg%) when compared to a level of ≤4.7 mmol/L (≤181 mg%) (23).

Actually, William Roberts, an accomplished investigator of cardiovascular disease and the editor of the *American Journal of Cardiology*, recently concluded in one of his editorials that only one true risk factor exists in coronary artery disease, namely the lifetime presence of a serum cholesterol level of over 150 mg%. With a cholesterol level persistently below 150 mg%—regardless of family history, hypertension, obesity, smoking, maleness, and other common risk factors—within the serum enough substrate simply does not exist to initiate or progressively increase atherosclerosis. Roberts suggested that risk factors can accelerate the disease only as serum cholesterol levels rise above 150 mg% *(38a)*.

It has been known for more than 15 yr that serum cholesterol levels below 4.1 mmol/L (160 mg%) protect populations from atherosclerosis. Antonio Gotto *(39)*, then President of the American Heart Association, stated before the US Senate that, "In societies where the blood cholesterol is under 160 mg%, there is virtually no CHD, or atherosclerosis."

William Castelli *(40)*, Director of the Framingham Study, went one step further, "Diet would not only prevent but also reverse coronary artery disease in 90% of patients if we could get everybody's cholesterol level below 150 mg%."

3.3. Defining Hypercholesterolemia

Although some clinicians still operate under the old definition of hypercholesterolemia as being anything ≥7.8 mmol/L (≥300 mg%), the Consensus Conference of the National Institutes of Health and the National Cholesterol Education Program have redefined the levels of risk according to age. Those aged 20–29 yr are at risk with cholesterols above 5.2 mmol/L (200 mg%). Those aged 30–39 are at risk above 5.7 mmol/L (220 mg%); and those over 40 are at risk above 6.2 mmol/L (240 mg%) *(41)*. Many investigators, however, prefer to see ideal cholesterols *below* 5.2 mmol/L (200 mg%), and some even below 4.6 mmol/L (180 mg%). A few, moreover, feel safest putting the ideal cholesterol level at *100 plus age* (in mg%) *(42)*.

These variations are understandable when one considers the sobering fact that more than half of the American adults have cholesterol levels above 5.2 mmol/L (≥200 mg%), which is clearly associated with an increased risk of developing CHD.

How important are lipoprotein fractions? Epidemiological data strongly suggest that average total cholesterol (TC) and low-density lipoprotein (LDL) values correlate more strongly with diet and CHD findings *between* populations than do high-density lipoprotein (HDL) values. Data from affluent Western cultures, however, favor the TC/HDL ratio, in that HDL appears to be protective *within a high lipid environment (43)*. Since in Western society the risk of CHD climbs with TC/HDL ratio over 4.5, Castelli *(44)* recommends as ideal a ratio of under 3.5, but certainly not one over 4.5.

With two-thirds or more of the cholesterol being carried by the LDL fraction, total serum cholesterol measurements provide, from a practical point of view, a reasonable basis on which to build an intervention therapy.

3.4. Possible Mechanism of Atherogenesis

Examining an arterial plaque, the German pathologist Rudolf Virchow found that it consisted of a fibrous, wax-like substance that turned out to be mostly cholesterol. He postulated that an injury to the endothelium, the thin, innermost arterial lining, caused plaque to form.

Almost 100 years later, Ross and Glomset *(45–47)* confirmed Virchow's concept. They conceived the following scenario:

1. The sensitive endothelial cells lining the artery become damaged, either mechanically or biochemically.
2. Excess lipids and LDL particles circulating in the bloodstream are then able to gain entry into the muscle layer of the arterial wall.
3. These foreign substances irritate the smooth muscle cells, which, in turn, engulf the invaders and multiply, eventually forming an intramural lesion that decreases the luminal diameter and interferes with adequate blood flow.

But what causes the endothelial damage in the first place? Studies show that insufficient oxygen in the blood can biochemically alter the endothelial layer by changing the permeability of the endothelial lining. The lining becomes porous, or leaky, as the junctions between endothelial cells widen. This permits fats and LDL-cholesterol particles to enter the muscle layer and set up inflammatory processes (48).

Studies also show that if the serum cholesterol level is below 4.1 mmol/L (160 mg%), the endothelial injury will usually heal and the lesion will shrink. However, with typical American cholesterol levels of over 5.2 mmol/L (200 mg%), the lesions often do not heal and fat and LDL particles continue to leak through the endothelial lining into the muscle layer and further aggravate the inflammatory process. Over time, the result of these processes is a lesion with a fibrous cap of collagen, smooth muscle lipid-filled macrophages and smooth muscle cells, enclosing a disorganized central core of lipids, cholesterol, plasma proteins, and cellular debris—the atheromatous plaque.

In the presence of persistent hypercholesterolemia, the plaque usually enlarges, often leading to ulceration of the luminal surface (endothelial denudation). The lesion now can rupture and discharge its debris, causing thrombi and emboli that, in turn, set the stage for catastrophic clinical events.

Plaques are like tire patches: they are the body's response to arterial wall damage. In responding to continued irritation over the years, the plaque either gradually enlarges until it eventually interferes with blood flow, or it fissures and cracks, leading to hemorrhage into the plaque, which in turn, can rapidly enlarge the plaque volume and degree of stenosis and lead to intraluminal thrombosis (49) (Fig. 10).

If, indeed, subnormal blood oxygen levels undermine the integrity of the endothelial lining, what causes oxygen levels to be lowered?

Smoking can do it. Inhaled carbon monoxide enters the bloodstream and attaches itself to the erythrocytes with

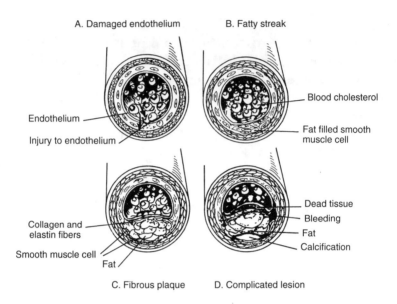

A. Damaged endothelium   B. Fatty streak

Endothelium

Injury to endothelium

Blood cholesterol

Fat filled smooth
muscle cell

Collagen and
elastin fibers

Smooth muscle cell

Fat

Dead tissue

Bleeding

Fat

Calcification

C. Fibrous plaque   D. Complicated lesion

Fig. 10. Progressive steps in plaque formation. **A.** Once the endothelium has been injured, toxic substances like cholesterol (LDL) may enter the arterial wall. **B.** Special muscle cells within the intima become filled with fats forming a fatty streak. **C.** A fibrous plaque develops containing cholesterol, collagen, and calcium. **D.** A complicated lesion builds up that extends into the inner lumen of the artery, eventually leading to "suffocation" of target organs.

an affinity 200 times greater than that of oxygen. When the ability of erythrocytes to carry oxygen is compromised, blood oxygen levels go down, and endothelial permeability goes up *(50)*.

Diet can do it. Excess fats in the blood slows circulation and reduces the oxygen-carrying capacity of the erythrocytes by setting up erythrocytic agglutination *(51)*. These erythrocytic aggregates, no longer able to pass with ease through some of the capillaries, may block as much as 20% of the capillary capacity producing tissue hypoxia *(52)*. In addition, agglutinated erythrocytes are unable to carry their optimal load of oxygen. Normally, each erythrocyte carries usable oxygen over its entire surface. But when agglutinated, only the outside surface can carry oxygen. Kuo and Joyner *(53)* were able to induce angina in resting heart disease patients by having them drink a glass of heavy cream.

Arthur Williams *(54)* reported erythrocytic agglutination in the conjuctival vessels, coronary ischemia, and anginal pain 4 h after feeding his 10 angina patients a breakfast of fried eggs, ham, buttered toast, and coffee with cream.

Another way to breach the endothelial barrier is by mechanical means, such as by a wayward catheter during an angiographic examination. The constant twisting and stretching of the coronary arteries during the normal pumping action of the heart can also injure endothelial linings. Endothelial injury may also be the result of shear-stress in areas where blood flow is not laminar, or it may be related to mechanical stretching of the arterial wall owing to hypertension *(55)*. In addition, increased endothelial permeability may be brought on by angiotension II-mediated hypertension.

Whatever the means, the damage will usually heal normally under conditions of ideal cholesterol levels. However, in the presence of cholesterol levels exceeding 5.2 mmol/L (200 mg%), atherosclerotic plaques may proliferate. Some kind of injury to the endothelial lining appears to be a prerequisite for atherosclerotic plaque formation. This endothelial damage can be brought about by a lowering of oxygen in the blood as a result of smoking, a high-fat diet in the context of elevated serum cholesterol, by dietary cholesterol itself, and by mechanical means *(56,57)* (Fig. 11).

Regardless of the mode of damage, it is usually only in the presence of excess cholesterol in the blood (and aggravated by hypertension) that plaque forms and grows. Elevated serum cholesterol levels, then, appear essential as

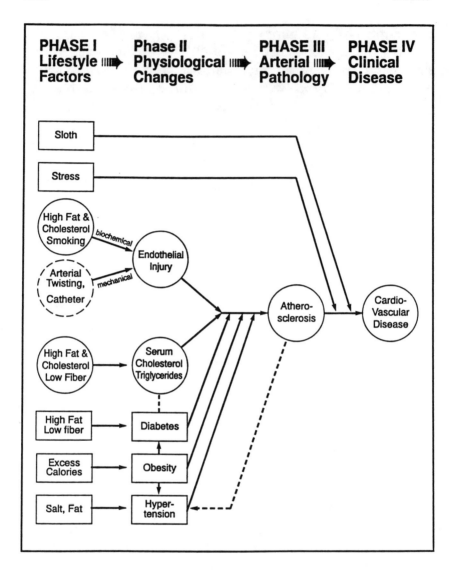

Fig. 11. Diet lifestyle-heart model (simplified): Cardiovascular disease and risk factors.

prerequisite for plaque-building, whereas serum cholesterol levels below 4.1 mmol/L (160 mg%) appear essential for the prevention and regression of atherosclerotic processes.

4. Hyperlipidemia Interventions

With the etiological preeminence of hyperlipidemia in CHD, drug and diet approaches have been utilized to lower total and LDL-cholesterol with the hope of primary and secondary prevention.

4.1. Effect of Drugs on Lipid Levels

The Coronary Drug Project, an ambitious 6-yr secondary prevention trial involving 8341 patients in 53 clinical centers, evaluated in a double-blind fashion the safety and efficacy of estrogen, thyroxine, clofibrate, and niacin, then the most popular hypolipidemic drugs. Although these drugs were successful in reducing serum cholesterol levels by 5 to 10%, none of them improved life expectancy.* On the contrary, they proved more harmful than helpful. They produced unpleasant and distressing side effects and *excessive* mortality *(59,60)*. Following the completion of the Coronary Drug Project in 1975, an editorial in the medical journal *Lancet* stated,

> The results are depressing for doctors who hoped that a daily drug dose would improve prognosis. The results are also discouraging for the pharmaceutical industry which has done so much to encourage clinical studies of cholesterol-lowering drugs *(61)*.

Whereas the Coronary Drug Project dealt with secondary prevention, the 5-yr WHO Clofibrate Trial evaluated clofibrate as a primary prevention in 15,745 hypercholesterolemic men in Europe. Even though clofibrate lowered the serum cholesterol levels by 9% and the incidence of nonfatal myocardial infarction by 20%, it had no effect on CHD mortality. On the contrary, mortality from all causes was significantly higher in the clofibrate-treated men because of diseases of the liver, gallbladder, and intestines *(62)*. An editorial sum-

---

*A 15-yr followup on niacin-taking patients suggests an apparent decrease in mortality *(58)*.

marized the results, "The conclusions must be that clofibrate can no longer be recommended for general use" *(63)*.

And the US National Institutes of Health (NIH) issued this directive:

> The widespread use of hypolipidemic drugs to treat and prevent coronary heart disease should be deferred until distinct benefits are demonstrated and significant toxicity can be excluded *(64)*.

In 1975, the NIH initiated the Lipid Research Clinics *Coronary Primary Prevention Trial* (LRC-CPPT) in an attempt to produce conclusive evidence that cholesterol-lowering was indeed beneficial. More than 480,000 asymptomatic men between the ages of 35–59 with serum cholesterol levels of ≥6.86 mmol/L (≥265 mg%) were screened to yield 3806 who eventually entered the multicentered, randomized double-blind study.

After receiving a mild cholesterol-reducing diet, which reduced the serum cholesterol levels by 3–5%, the participants were randomly allocated to receive the bile-acid sequestrant resin cholestyramine (Questran) or a placebo for an average followup of 7.4 yr. The cholestyramine group achieved 8.5 and 13% greater reductions in total and LDL-cholesterol levels than those obtained in the placebo group. The lipid changes were accompanied by a 19% reduction in nonfatal myocardial infarctions, a 24% drop in definite CHD deaths, and a 7% nonsignificant fall in all-cause mortality. In addition, the incidence rates for new positive stress tests, angina, and coronary bypass surgery were reduced by 25, 20, and 21%, respectively, in the cholestyramine group *(65,66)*.

Because of difficulties in adherence related to the bulk, texture and side effects of the drug* most participants fell

---

*Thirty percent of patients tolerate cholestyramine poorly at a dose of 12 g/d. This proportion may go up to 50% when the dose is increased to 16 g/d *(67)*.

short of the treatment goal of 24 g/d *(66)*. Those able to take the full dose, however, had much greater reductions in total and LDL-cholesterol (25 and 35%, respectively). This, in turn, was accompanied by a fall of 49% in the incidence of CHD, which highlighted the intimate dose–response relationship between drop in lipids and the drop in CHD risk (Fig. 12).

The LRC-CPPT confirmed the lipid hypothesis. This monumental study clearly showed that the lowering of serum cholesterol also lowers the risk of CHD. Although the benefits were accomplished by the use of a hypolipidemic drug, the researchers pointed out that similar results would be expected if the lipid levels were lowered by dietary intervention.

Similar to the LRC-CPPT study, the *Helsinki Heart Study* explored the efficacy of another hypolipidemic drug in reducing the risk of CHD. Some 4000 asymptomatic middle-aged men were randomly allocated to receive either 1200 mg of gemfibrozil (a fibric acid derivative) or a placebo. During the 5-yr study, gemfibrozil increased the level of HDL-cholesterol by 11%, whereas it decreased levels of total cholesterol, LDL-cholesterol, and triglycerides by 10, 11, and 35% as compared with the placebo. These lipid changes were accompanied by a reduction of 34% in the rates of fatal and nonfatal myocardial infarction and cardiac death as principal end products *(68)*. Although coronary mortality was significantly reduced, the all-cause mortality was not significantly altered by the treatment *(69)*.

The results of the Helsinki Heart Study strongly reinforced the findings of the LRC-CPPT. Together with other evidence, they suggest a rule of thumb: Each 1% reduction in serum cholesterol results in a 2–3% reduction in the incidence of CHD *(70–72)*.

The greater effect observed in the Helsinki Heart Study, where a 10% drop in total or LDL-cholesterol resulted in a 35% drop in cumulative endpoint rate, is possibly related

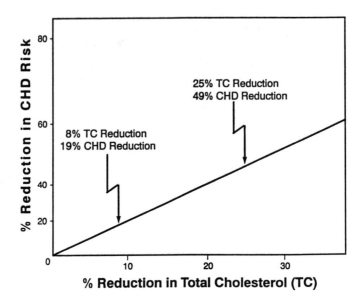

Fig. 12. LRC-CPPT. Relation of reduction in total cholesterol (TC) to CHD.

to the observed moderate increase in HDL-cholesterol and perhaps to the marked fall in triglycerides.

4.2. Effect of Diet on Lipid Levels

Much of the basic evidence of how diet composition influences human serum cholesterol levels derives from hundreds of meticulously controlled feeding experiments carried out by pioneers such as Ancel Keys, Mark Hegsted, William Connor, and Fred Mattson. By changing the amount and type of fat consumed and by manipulating the amount of cholesterol eaten, they were able to predictably raise and lower serum cholesterol levels. Their now-famous mathematical formulas can predict for groups of persons the mean changes in serum cholesterol produced by known changes in these dietary constituents (73–78).

For instance, by reducing his usual daily *cholesterol* intake of 600 mg by 500 mg (equivalent of two egg yolks), a man, on average, would decrease his serum cholesterol by an estimated 0.4–0.5 mmol/L (15–20 mg%). Assuming an initial serum cholesterol level of 5.7 mmol/L (220 mg%), this change in dietary cholesterol alone would lower the serum cholesterol by 8%, equivalent to a 20% lower risk of CHD. If he also cut his customary daily *saturated fat* intake of 50 g by 35 g, he would reduce his serum cholesterol by another 0.8 mmol/L (30 mg%). This change in saturated fat intake alone would lower his serum cholesterol 14% and his risk of CHD by an estimated 35%. The combined changes in dietary cholesterol and saturated fat intake in this example would, on average, lower his serum cholesterol level 1.16–1.3 mmol/L (45–50 mg%), or 22%. This would be equivalent to a substantial 50% drop in coronary risk *(79)*.

Although the impact of changes in saturated fat intake on serum cholesterol has been considered to be of greater quantitative importance than the effect of dietary cholesterol, the latter's contribution to hyperlipidemia is substantial and should not be devalued, especially in view of its *curvilinear* function (Fig. 13).

Cholesterol added to a very low cholesterol diet causes a much steeper rise in the serum cholesterol than when added to a diet that already contains large amounts of cholesterol. This *saturation* phenomenon is so pronounced that one egg (250 mg of cholesterol), for instance, added to a cholesterol-free diet may actually raise the serum cholesterol within a week by more than 0.5 mmol/L (20 mg%) *(80–82)*. When added, however, to an already cholesterol-rich diet (containing 600 mg, as typically found in Western diets), it may not make any significant contribution at all *(83,84)*. Divergent claims and contradictory reports on the effect of certain cholesterol-rich foods on serum cholesterol levels relate largely to this curvilinear effect, with a steep rise in effect up to 200–300 mg of cholesterol/d and a flat curve thereafter.

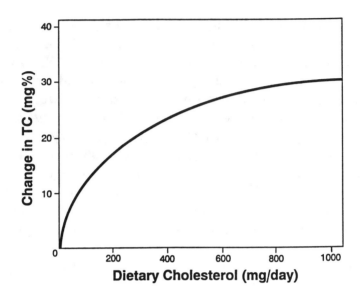

Fig. 13. Effect of dietary cholesterol on serum total cholesterol (TC).

As far back as 1960, Ancel Keys *(85)* observed that men in Naples, Italy, had lower serum cholesterol levels than could be fully explained by dietary lipid differences. Extending his suspicion that certain fibers, with their cholesterol-binding effect, may have a hypolipidemic effect, DeGroot *(86)*, shortly thereafter, conducted oatmeal feeding experiments that yielded a 10% drop in serum cholesterol and gave rise to a new wave of diet research. A recent meta-analysis suggests that *soluble dietary fiber* may lower serum cholesterol levels by about 5% *(87)*.

On a very low-fat diet, the liver and other parts of the body synthesize at least 500 mg of cholesterol/d, which is more than enough to meet the needs of the various functions of the body. On a Western high-fat diet, however, saturated fats can actually more than double the amount of *endogenous* cholesterol produced by increasing HMG-CoA reductase activity in the liver. Such a diet also provides in excess of 400 mg of *exogenou*s cholesterol/d. This dietary cholesterol not only contributes to higher serum cholesterol levels, but also acts as an independent risk factor in promoting atherogenesis *(56,57,79,88)*. It is this combined amount of cholesterol from endogenous and exogenous sources, then, that is the main determinant of elevated serum cholesterol levels commonly found in Western society.

Dietary cholesterol is found exclusively in animal products in varying amounts. In general, the higher the fat content of an animal product, the higher its cholesterol content (Fig. 14).

Table 3 displays the sources of dietary cholesterol in the United States. Notice that egg yolk accounts for 35% of the dietary cholesterol. The percentage is actually higher than that, since baked goods (8% of the total) get most of their cholesterol from eggs.

In addition to reducing the exogenous cholesterol, it is perhaps even more important to lower the total fat intake, and especially the saturated fats, which have such a powerful effect on endogenous cholesterol production.

Although the substitution of *polyunsaturated fats* for saturated ones in the diet can reduce total and LDL-cholesterol levels, no strong epidemiologic evidence exists that diets high in polyunsaturates actually will prevent CHD *(73,75,89,90)*. No large populations have ever consumed large quantities of these polyunsaturated fatty acids with proven benefit or safety. The lack of epidemiologic data supporting the safety of polyunsaturated fats has led many investigators to recommend that high intakes of polyunsaturates be avoided *(88)*. These fats contribute prominently to overweight, may predispose to gallstones, diabetes, breast and colon cancer, and they may be prominently involved in atherogenesis *(91–94)*. In addition, in laboratory animals, polyunsaturates can depress the immune system and promote the development of tumors *(95,96)*.

Around 36–40% of the calories eaten in North America today come from fat. Half of this amount comes from vis-

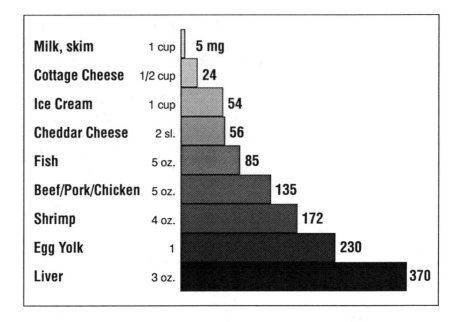

Fig. 14. Cholesterol content (in mg) in selected foods.

Table 3
Sources of Dietary Cholesterol (US)[a]

Food	Percentage of total cholesterol in US diet
Meats, poultry, fish	35%
Egg yolk	35%
Dairy products	16%
Commercially baked goods	8%
Table and cooking fats	6%
	100%

[a]Source: *National Dairy Council.*

ible or extracted fats and oils. The other half is not as readily seen, because most of these fats are well hidden, particularly in red meats and dairy products (Table 4). Red meat is actually the *single largest source of fat* in the North

Table 4
Fat Content of Selected Animal Products[a]

Animal product	Fat content (percentage of total calories)
Steak, choice	70–85%
Lamb, prime	70–85%
Ham	70–80%
Bacon	60–90%
Hot dog	80–85%
Whole milk	50%
Cheeses	65–75%
Cream cheese	90%

[a]Percent of energy.

American diet (Table 5). And, even more disturbing, most of it is saturated. Red meat is also—with egg yolks—*the single largest source of cholesterol*, and it is devoid of fiber.

Let us now examine the question of what lipid changes can be expected if people make substantial dietary changes.

Pekka Puska and his lifestyle intervention team *(97)* enrolled 54 middle-aged volunteers (average age 45.5 yr) in North Karelia, Finland, into an intensive 6-wk educational primary prevention program with the goal of lowering high serum cholesterol levels by adopting a more Mediterranean type diet. The participants and their wives received intensive dietary counseling and strategically important food items were provided free of charge. Milk, cheese, sausages, and fatty meats were replaced by skim milk, low-fat cheeses and sausages, and by lean meat, fish, and poultry. The use of vegetables and fruit was strongly encouraged. Five nutritionists visited the families at least twice a week during this 6-wk intervention period to supervise adherence to the experimental diet before the families switched back to their original diet. The changes in

Table 5
Sources of Dietary Fat (US)[a]

Animal product	Percentage of total fat in US diet
Meats, poultry, and fish	33%
Salad oils, shortening, lard	31%
Dairy products	17%
Margarine and butter	10%
Other	9%
	100%

[a]Source: *National Dairy Council.*

diet composition and in serum lipids during the three study periods are shown in Table 6.

The use of a diet substantially reduced in saturated fats, modest in cholesterol intake, and augmented with vegetables and fruits lowered the blood lipids consistently in every participant: The average reduction for total, LDL- and HDL-cholesterol and triglycerides was 23, 26, 19, and 23%, respectively. This study demonstrates that the high serum cholesterol and LDL-cholesterol levels in the North Karelia population are not so much under genetic control, but instead are largely related to dietary factors.

For more than 15 years, the *Pritikin Longevity Center* in Santa Monica, CA, has advocated a therapeutic diet where fats are largely replaced by complex carbohydrates (70%) as found in foods-as-grown, such as whole grains, legumes, tubers, and fresh fruits. Protein (15%) is primarily derived from vegetable sources, with small amounts of skim milk and very limited amounts of fat-free cheese, fish, or poultry. Consequently, the daily cholesterol intake is negligible (25 mg), the dietary fiber content is high (40 g), and the intake of total fat and saturated fat is very low,

Table 6
Diet and Lipid Changes[a]

	Baseline period	Intervention (6 wk)	Switch-back period
Diet changes			
Total fat	126 g	65 g	117 g
percentage of enery	39%	24%	36%
Saturated fat	71 g	21 g	65 g
percentage of enery	22%	9%	20%
P/S ratio	.15	1.2	.16
Cholesterol	537 mg	302 mg	542 mg
Lipid changes			
Total chol. (m%)	263	201	259
percentage of change		−23%	+22%
LDL-cholesterol (mg%)	185	137	183
percentage of change		−26%	+25%
HDL-cholesterol (mg%)	54	44	55
percentage of change		−19%	+20%
TC/HDL ratio	4.87	4.57	4.71
percentage of change		−6%	+3%
Triglycerides (mg%)	124	96	99
percentage of change		−23%	+20%

[a]Changes in diet and mean lipid levels in 30 men at the end of the study periods. (To convert values for cholesterol and triglycerides into mmol/L, multiply by 0.02586 and 0.01129, respectively.)

accounting for less than 10 and 2%, respectively, of daily energy with a P/S ratio of 2.4. At the same time, aerobic conditioning, and especially walking, is progressively promoted.

The subjects of a recent report were 4587 adults (average age 56 yr), who attended the 25-day residential lifestyle modification program between 1977–1988 (24,98). The majority of the participants entered with diagnoses of CHD, hypertension, diabetes, or gout. Most entered with multiple diagnoses, but some entered the program primarily

Table 7
Serum Lipid Changes[a]

Lipids		N	Before	After	Change
Total cholesterol	(mmol/L)	2685	5.99	4.53	−24%
	(mg%)		232	175	
LDL-cholesterol	(mmol/L)	1669	3.90	2.92	−25%
	(mg%)		151	113	
HDL-cholesterol	(mmol/L)	2044	1.03	.91	−12%
	(mg%)		40	35	
TC/HDL ratio		2044	6.26	5.30	−15%
Triglycerides	(mmol/L)	2685	2.51	1.56	−38%
	(mg%)		222	138	

[a]Changes in mean serum lipid levels in 2685 men attending a three-week residential lifestyle modification program: The Pritikin Longevity Center Study (98).

for preventive reasons. Aside from a careful medical workup and systematic clinical monitoring, the participants and family members received detailed instruction on the nature of their cardiovascular disease and its risk factors, and the roles that diet and exercise play in reversing, managing, and preventing them. The intensive 60-h educational curriculum includes how to understand medicines, purchase and prepare foods, and how to develop lifestyle management skills and behavioral techniques, all designed to help patients adhere to their new lifestyle once they return home.

During the 3-wk intervention, the 2685 men reduced their total serum cholesterol level by an average of 24% (Table 7). These reductions were accomplished without hypolipidemic drugs. The true changes as mediated primarily by the change in diet are actually even larger, since 5% of the patients were on hypolipidemic drugs when they entered the program that obviously depressed their lipid baseline values. These drugs, however, were in almost all cases discontinued during the first week of the intervention program.

The reduction in cholesterol values was proportional to their initial level. Participants with the highest initial cholesterol levels had the greatest reduction (Fig. 15).

Those, for instance, who started the intervention with initial cholesterol values between 4.1–4.7 mmol/L (160 and 180 mg%), achieved a 15% reduction, whereas those who started in the 6.2–6.7 mmol/L (240–260 mg%) category achieved a 26% reduction. The higher the initial cholesterol-mediated risk, the greater the risk reduction for CHD. Interestingly, the decreases in various cholesterol fractions were not significantly influenced by the age of the participants.

5. Effect of Dietary Intervention on CHD Incidence

5.1. Secondary Prevention Trials

As evidence from epidemiological and clinicopathological studies began to emerge linking diets high in fat and cholesterol to hyperlipidemia and atherosclerosis, Lester Morrison *(99,100)* began his dietary intervention study in Los Angeles.

Beginning in 1946, he enrolled 100 consecutive patients (85 men; mean age 61 yr) who had suffered from a coronary event within the previous 6 mo and alternatingly assigned them to control or experimental diets. Those on the control diet adhered to their customary daily diet of 600–1000 mg of cholesterol and about 120 g fat. Those assigned to the experimental diet consumed less than 25 g of fat (15% of total calories) and between 50 and 70 mg of cholesterol/d; this hardly allowed for any animal products. With these dietary changes, their weights adjusted well (20 pounds in 3 yr), their serum cholesterol dropped 29% (from 8.5 to 5.7 mmol/L, equivalent to 312 and 220 mg%), and at the end of the third year of observation, 43 of the 50 patients were still alive. In contrast, those who adhered to their customary precoronary diet did

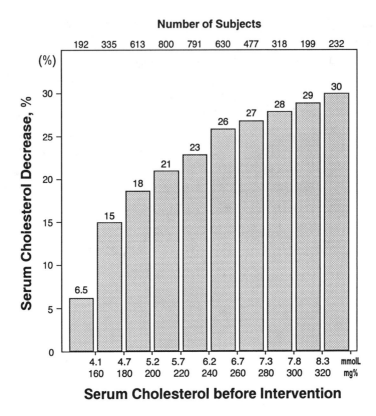

Fig. 15. Total serum cholesterol decreases (in percent) according to initial cholesterol levels in 4587 men and women during a 3-wk residential lifestyle intervention program.

not drop extra pounds or cholesterol, and after 3 yr, only 35 of these 50 patients were still alive. At the beginning of the twelfth year of intensive observation, all 50 of the control patients had died (100% mortality), mostly from recurrent forms of atherosclerosis, whereas 19 of the 50 patients treated with the diet were still alive (62% mortality).

Although formal matching between the two groups was not attempted, the method of assignment and statistical testing should have kept any baseline differences between the two groups to a minimum.

Morrison's original paper, published in 1951, stimulated researchers on both sides of the Atlantic to conduct dietary intervention studies that became increasingly more sophisticated and thus better equipped to understand the diet–hyperlipidemia–heart disease relationship. Unfortunately, all of the clinical trials involving individuals with already clinically apparent coronary artery disease (secondary prevention), fell short of the expected positive outcomes *(101–105)* (Table 8).

These dietary secondary prevention trials* largely ignored the therapeutic strategy of substantially reducing total fat and cholesterol intake, the very changes that appeared so promising from Morrison's experiment. Instead, these trials focused on increasing the P/S ratio by substituting *polyunsaturated* fats (like corn and safflower oils) for *saturated* ones (like animal fats and coconut oils), which, in feeding experiments, had been found beneficial in lowering serum cholesterol levels.

The differences in mean serum cholesterol levels between the experimental and control groups amounted to 6–14%, but there were no significant differences in the clinical outcome variables, such as infarction rates or coronary deaths, except for the Oslo Diet-Heart Study, which after 11 yr recorded some encouraging but inconclusive results *(105)* (*see* Table 8).

5.2. Primary Prevention Trials

When clinical trials were conducted shifting more to the *primary* prevention of CHD, the forthcoming results did not alleviate some of the controversy as to the precise

---

*The London Research Group study was an exception in that it lowered the total fat intake of the experimental diet. The researchers, however, accepted patients into the study immediately after their infarction, which is a time when hemodynamic and arrhythmic consequences of the recent infarct are often the overriding determinants of subsequent cardiac events.

Table 8
Dietary Trials[a]

Ref.	Study	1°/2°	Number of patients Diet	Control	Years	Dietary focus	TC Diff.	Attack rates Diet	Control	Significance, p	Syndrome
99	Morrison, 1951	2°	50	50	3	15% fat 60 mg C	−29%	14%	30%	.06	All deaths
100	Morrison, 1960	2°	28	52	12		−29%	62%	100%	.001	All deaths
101	Rose, 1965	2°	132	132	2	P/S	−9%	42%	29%	NS	SD/MI
102	London Res., 1965	2°	200	200	3	20% fat	−8%	23%	26%	NS	MI
103	MRC, 1968	2°	100	100	4	P/S	−14%	20%	20%	NS	MI
104	Bierenbaum, 1970	2°			5	P/S	−6%	18%	25%	NS	MI
105	Leren, 1970	2°	206	206	11	P/S	−14%	16%	28%	.001	Fatal reinfarction
								38%	46%	NS	CHD deaths
106	LA Vets, 1969	1°/2°	424	422	8	P/S	−13%	11%	17%	.02	Fatal CVD
								41%	42%	NS	All deaths
107	Finnish Diet Study, 1979	1°	3000	2700	6	P/S	−15%	3.0[b]	6.1[b]	.01	Coronary deaths
								4.2[b]	12.7[b]	.001	ECG changes
										NS	All deaths
108	Oslo Trial, 1981	1°	604	628	5	28% fat <300 mg C	−10%	19	36	.03	Coronary events
								22	39	.04	CVD events
								16	24	NS	All deaths
109	Oslo Trial, 1990				8		−8%	4	12	.02	Sudden coronary deaths
								19	31	.06	All deaths

[a]Controlled clinical dietary trials (1951–1981): Primary (1°) and secondary (2°) CHD prevention.
[b]Per 100 person-years
NS = no statistical significance; SD = sudden death; MI = myocardial infarction; CVD = cardiovascular disease; CHD = coronary heart disease; C = cholesterol; TC = total cholesterol.

benefits of a diet that emphasized the replacement of saturated fat with polyunsaturated fat but did not alter the high fat intake. Dietary intervention trials conducted in Los Angeles and Finland are classic examples of this approach (*see* Table 8).

The *Los Angeles Veterans Administration Study* involved 846 men with a mean age of 65 yr at entry. Since many of these men already had evidence of clinical atherosclerotic disease, this study was a mixed primary and secondary prevention trial. The men, residents of the Veterans Administration facility, were randomly assigned to the experimental or control group with excellent matching of the important risk factors, and were fed in two different cafeteria lines. Both diets contained about 40% fat, but the experimental diet contained much less saturated but much more polyunsaturated fat and a modestly reduced amount of cholesterol. The 8-yr study demonstrated lower serum cholesterol levels (13%) and a significant reduction in cardiovascular disease. The difference in mortality from all causes between the two groups, however, was not significant *(106)*.

The *Finnish Study* involved a sizable number of patients in two mental hospitals and used a crossover design. The intervention diet, whose milk and butter had been replaced with soybean emulsion and soft margarine in order to provide a more favorable P/S ratio, was fed in one hospital for the first 6 yr and then in the second hospital for the second 6 yr. Once again, the fat modified diet lowered the serum cholesterol (15%) and the number of coronary deaths, but the difference in mortality from all causes was not significantly different *(107)*.

The *Oslo Diet and Anti-Smoking Trial* involved 1232 nonhypertensive men aged 40–49 yr who were coronary prone because of high serum cholesterol levels (>7.5 mmol/ L, or >290 mg) and smoking *(108)*. Counseling for the 604 men randomly assigned to the intervention group involved advice to quit smoking and to reduce serum cholesterol levels by eating a diet considerably lower in total fat (by reducing substantially saturated fat) and cholesterol. It was the first primary prevention trial that broke with the pattern of increasing polyunsaturated fats (Table 9).

By the end of the 5-yr trial period, serum cholesterol levels for the intervention group had come down by an average of 13% (only 3% in the control group), and 25% had quit smoking (17% in the control group). At the same time, the incidence of coronary (fatal and nonfatal myocardial infarctions and sudden deaths) and cardiovascular events was significantly lower (45% [$p < .03$] and 41% [$p < .04$], respectively) in the intervention than in the control group. Multivariate analysis indicated that these improvements correlated exclusively with the serum cholesterol reduction and not with the reduction in smoking.

The sudden CHD death rate came down significantly ($p < .05$), but the differences in death rates from all CHD, from all cardiovascular disease (–45%), and from all causes (–33%) did not reach levels of statistical significance despite their substantial changes. A followup, 3 years after the trial's completion, however, showed a continuation of these favorable trends (109). The sudden CHD death rate difference had further increased ($p < .02$) and the all-cause mortality difference had now grown to 40%, approaching marginal statistical significance ($p = .055$) (Table 8).

The Oslo Trial demonstrated convincingly:

1. It is possible to alter dietary habits in high risk middle-aged men with modest effort;
2. Dietary changes that involve marked reduction in fat, and especially in saturated fat and cholesterol, are effective in reducing the level of serum cholesterol; and
3. Serum cholesterol reduction reduces the rate of subsequent CHD events and affects all-cause mortality favorably.

Table 9
Diet Changes[a]

Diet composition	I	C
Total calories	2248	2331
Protein (percentage of calories)	18.8	15.5
Total fat (%)	27.9	44.1
Saturated fat (%)	8.2	18.3
Polyunsaturated fat (%)	8.3	7.1
Cholesterol (mg/day)	209	527

[a]Diet composition of intervention (I) and control (C) groups after four years of trial: The Oslo Trial *(108)*.

5.3. Multiple Risk Factor Intervention Trial

Less than one year after the publication of the results of the 5-yr Oslo Trial, which clearly confirmed Morrison's idea that dietary intervention via substantial fat reduction can reduce coronary and cardiovascular incidence rates, results emerged from the $150 million NIH-sponsored *Multiple Risk Factor Intervention Trial* (MRFIT). The results were both ambiguous and disappointing, if not disturbing *(110)*.

The trial involved 22 clinical centers and 12,866 men, aged 35–57 yr, at high risk for CHD. It was the ultimate effort to ascertain whether simultaneous lowering of elevated serum cholesterol levels, hypertension, and cigaret smoking would also lower coronary death rates.

The goals were to reduce:

1. Serum cholesterol by 10% through a diet, where the emphasis, once again, was not on total fat reduction, but on P/S ratio changes (polyunsaturated fats 10%, saturated fats 10%, total fat 35%) and on limiting cholesterol intake to <300 mg/d;

2. Diastolic blood pressure by 10%, primarily through medication; and
3. Smoking by 10%.

Those, randomly allocated to the Special Intervention group (SI) received group education and therapeutic sessions regarding diet, blood pressure control, and smoking cessation, whereas those assigned to the Usual Care group (UC) were sent back to their own physician to receive their usual care.

After 6 yr the following risk factor changes emerged:

1. Serum cholesterol levels had dropped only 7% in the SI vs 5% in the UC group—a net change of 2% (!);
2. Diastolic blood pressures had dropped 10.5% vs 7.3%— a net change of only 3.2%; and
3. 46% of the smokers in the SI vs 29% in the UC group had quit smoking—a net change of 17%.

When the clinical endpoints in both groups were compared, no significant differences in the rates for CHD, cardiovascular disease, and all-cause mortality were found. Detailed analyses have suggested that the expected trial outcome was largely blunted by:

1. The unanticipated improvements in serum cholesterol and diastolic blood pressure in the UC group;
2. The lower than anticipated reduction in serum cholesterol in the SI group;
3. The cholesterol-elevating effect of diuretic drugs prescribed for hypertension control; and
4. Men with abnormal resting electrocardiograms at baseline and placed on high-dose diuretic treatment showed a 65% *greater* CHD mortality than those in the UC group.

The results of a recent report based on 10.5 yr of followup on MRFIT, although suggesting favorable differences in CHD and all-cause mortality rates for the SI group, fell short in reaching statistical significance *(111)*.

5.4. Summary

The failure of cholesterol reduction to reduce mortality from CHD and from all causes in most human intervention trials has prompted considerable debate regarding the validity of the lipid hypothesis and the probable benefits of cholesterol reduction as a national health objective *(112–119)*. Some have questioned the lipid hypothesis and the preferential citations of supportive trials in medical publications, but others have pointed out that most of these intervention trials had not been designed to address the issue of mortality inasmuch as they were primarily concerned with coronary event rates *(25,79,120)*. In their opinion, total mortality may not be the best indicator of the value of serum cholesterol lowering since it leaves out the issue of the quality of life, which, they claim, should improve with cholesterol reduction *(121)*.

Others, however, looking not only at the favorable results by Morrison, Pritikin, and the Oslo Trial, but also at the total epidemiological and experimental evidence, have suggested that it is not the lipid hypothesis that has failed but the dietary interventions, since they have been by far too modest and not rigorous enough to provide clinical efficacy *(24,42,122,123)*. Some even fear that the inability of achieving lower serum cholesterol levels and lower CHD incidence by diet (which basically has aimed only at a different P/S ratio without *substantially* lowering the total fat and cholesterol intake) may foster wide use of hypolipidemic drugs, which in turn, may transform asymptomatic people into patients.

To buttress their contention, they refer to prospective studies of Californian Seventh-Day Adventists, whose preference for nonsmoking on the one hand and for vegetables and fruits on the other, has apparently provided them with considerable protection against the major causes of death, such as CHD and cancer, and given them 7 yr of extra life *(124–126)*.

They also point to a recent cohort study involving 1904 vegetarians conducted at the German Research Center in Heidelberg *(127)*. After 11 yr of followup, mortality from all causes among men was reduced by more than one half compared with the general population (the Standardized Mortality Ratio [SMR] was 0.44). The lowest mortality was found for cardiovascular diseases (SMR = 0.39), particularly for CHD. CHD mortality among strict vegetarian men was reduced to about one third (SMR = 0.27) of that expected. Although one third of this advantage obviously relates to abstaining from smoking, the other half within this cohort is probably attributable to a diet that is very low in animal fats, markedly lower in total fat, and high in dietary fiber *(128)*.

6. Reversing CHD

The evidence linking elevated serum cholesterol as the atherogenic stimulus to CHD is overwhelming. Epidemiologic, clinical, and postmortem studies have established that high serum cholesterol levels are causally related to atherosclerosis and increased risk of CHD *(121)*. Moreover, in individuals with genetic forms of hypercholesterolemia, CHD commonly occurs at a young age even in the absence of other risk factors.

Notwithstanding this, sophisticated clinical trials designed to evaluate the efficacy of lipid-lowering interventions to reduce coronary morbidity and mortality, have not produced consistent data concerning clear-cut benefits, particularly in the area of all-cause mortality. Of course, the design and execution of such trials is difficult and very costly. This relates partly to the statistical need for a large enough sample size that, however, conflicts with the requirement for close supervision to ensure good adherence to the intervention regimen. Feeding experiments with animals, on the other hand, have made it possible to assess the progression and potential regression of atherosclerotic lesions more directly, in less time, and with less money.

6.1. Regression in Animals

Many animal species, including monkeys and baboons, consistently and predictably develop atherosclerosis when fed diets high in fats and/or cholesterol that largely determine serum cholesterol levels *(28,57,129–131)*. In 1963, Carl Taylor and colleagues *(28)* became the first team to feed a monkey to death. After being fed typically prepared American food for 30 mo, the rhesus monkey died of a massive myocardial infarction. The autopsy found myocardial damage and coronary atherosclerosis similar to that found in human hearts and coronary arteries after a serious myocardial infarction.

Seven years later, the first report came out that coronary atheromatosis could be reversed *(30)*. Since then, many studies with many different animal species have reported that even severe and established lesions shrink, undergo extensive anatomical and biochemical changes, and atherosclerotic stenoses begin to enlarge their lumina, often within weeks, once the atherogenic stimulus has been removed *(31,32,132–140)*. These remodeling processes have been called "atherosclerosis regression."

Regression is mainly characterized by plaque shrinkage, arrest of cell proliferation, decreased numbers of cells and necrotic foci, and lessened concentrations of cholesterol esters, free cholesterol, and elastin *(141)*. At the NATO-sponsored Advanced Study Institute on Regression of Atherosclerotic Lesions, Malinow *(142)* described an animal model of the arterial wall in which regression occurs when the outward flow exceeds the inward flow in transport and transformation of atheromatous materials.

6.2. Regression in Humans

In humans, anatomical evidence for regression remained largely circumstantial until the development of computer-assisted angiography (Coronary Angiography Analyses System, CAAS), a system that permits accurate delineation of the contours of user-selected arterial segments by means of automated edge-detection algorithms (143).

Aschoff was probably the first investigator to infer that regression of human atherosclerosis might be possible. In 1924 he wrote, "With the increasing duration of WWI, the atheromatous spots on the aorta were observed less frequently" (144). His findings were confirmed by other pathologists (145). Moreover, coronary morbidity and mortality rates dropped drastically during the "lean years" of World War II in European countries (7–11).

Although cholesterol-induced atherosclerotic plaques regress predictably in nonhuman primates once hypercholesterolemia as the key atherogenic stimulant has been reduced to less than 5.2 mmol/L (<200 mg%), many investigators and clinicians insisted that similar reversal would not occur in humans. Possible reasons for this are that atheromatous lesions have been present for a prolonged period, are calcified, are fixed, or are believed to be irreversible. Before too long, however, the evidence to the contrary started coming in.

6.2.1. Human Case Studies

In 1970, DePalma and colleagues described a diabetic, aged 57 yr, who reduced his serum cholesterol from 8.4 to 4.9 mmol/L (327 to 190 mg%), quit smoking, and began an exercise program. Serial angiograms showed the disappearance of a popliteal obstructive lesion after 9 mo (146).

In 1976 Basta reported the case of a woman with hypertension related to a 90% blockage of the right renal artery, who reduced her serum cholesterol from 8.6 to 4.1 mmol/L (334 to 160 mg%). Repeat angiography 3 yr later showed complete regression of the plaque and normalization of blood pressure (147).

Three years later, Crawford performed repeated femoral angiograms in two postinfarction patients and prescribed progressive exercise, diet modification, and weight reduction. After 2 yr, femoral regression was documented with a recording microdensitometer *(148)*.

Roth and Kostuk described a man, aged 46 yr, who was scheduled for bypass surgery to relieve his exertional angina owing to an 80% stenosis of the left anterior descending coronary artery. The patient, however, elected a nonsurgical approach. With self-imposed restrictions on meat, eggs, and dairy products, and with increased exercise, his serum cholesterol and triglyceride levels dropped 25 and 55%, respectively. One year later, angiography showed a greatly regressed lesion (now only 30%) accompanied by increased myocardial perfusion as documented by thallium scanning *(149)*.

In 1983 Buchwald and colleagues described serum cholesterol changes (from 20 to 14 mmol/L [757 to 540 mg%]) as a result of a partial ileal bypass operation in a man, whose coronary lesions subsequently regressed from 70 to 20% stenosis in 10 yr *(150)*.

Two years later, the *New England Journal of Medicine* reported the autopsy findings of Nathan Pritikin, aged 69 yr, who 27 years previously had been diagnosed with clear evidence of coronary artery disease but now had pliable, soft coronary arteries without any evidence of atherosclerotic plaques or arterial narrowing *(151)*. Soon after the CHD diagnosis, Pritikin had broken with his rich diet and instead had adopted a diet centered on plant foods (10% fat, 70% complex carbohydrates, 25 mg of cholesterol, and 40 g fiber/d), which is commonly eaten in most countries of the world where CHD and Western cancers are virtually absent. As a result, his serum cholesterol levels dropped from 7.2 to 3.1 mmol/L (280 to 120 mg%), where they stayed for the rest of his life. In the coroner's words, "The absence of developed atherosclerosis and the complete absence of its effects in a 69-yr-old man are remarkable" *(151)*.

6.2.2. Prospective Regression Studies

Since regression of atherosclerosis may occur spontaneously in the absence of lifestyle changes or drug treatment (possibly because of lysis of a thrombus, or arteriographic limitations), isolated case studies may be suggestive but their value is obviously limited *(152,153)*. More significant evidence comes from prospective studies of groups of patients (*see* Table 10).

Three patients among 31 reported by Ost and Stenson *(154)*, treated with nicotinic acid, showed improvement in femoral angiograms after 3 yr. Buchwald and colleagues *(155)*, in 1974, reported regression in 3 of 24 severely hypercholesterolemic patients, who underwent partial intestinal bypass operations. Their plaques clearly decreased or disappeared, and a totally occluded left anterior descending coronary artery became patent. The pioneer study of Barndt and Blankenhorn *(156)* at the University of Southern California constitutes an important landmark in regression research. A group of 25 middle-aged hypercholesterolemic patients with severe femoral occlusive disease were given a fat-modified, low-sodium diet plus hypolipidemic medication. Changes in femoral atherosclerosis were assessed with sensitive methods and with the aid of a computer-controlled image dissector. Contrasting serial angiograms were taken with a specific goal of achieving comparability. Regression of femoral atherosclerosis was demonstrated in nine patients (36%) after 13 mo (Fig. 16). The strongest predictors for regression were significant reductions in serum cholesterol, triglycerides and blood pressure *(18,156)*. Patients with regression had a drop of 21% in serum cholesterol (from 8.4 to 6.4 mmol/L [311 to 246 mg%]), whereas those with progression had a drop of only 6% (from 8.7 to 8.2 mmol/L [335 to 316 mg%]).

Based on serial angiographic examinations, Kuo and colleagues *(157)* reported that with successful management of hypercholesterolemia they were able to stabilize atherosclerotic lesions in middle-aged patients for periods as long

Table 10
Prospective Arteriographic Regression Studies

Ref.	Study	F/C	Time, mo	N	Progression, %	Static, %	Regression, %
154	Ost, 1967	F	42	31	55%	35%	10%
155	Buchwald, 1974	C	24	22	23%	63%	14%
156	Barndt,1977	F	13	25	52%	12%	36%
157	Kuo, 1979	C	42	12	33%	67%	—
158	Rafflenbeul, 1979	C	12	25	44%	36%	20%
159	Erikson,1983	F	18	8	12%	12%	76%

F = femoral or C = coronary arteries studied.

as 7.5 yr. Their "therapeutic diet" (40% fat with a favorable P/S ratio) was unable to alter the high serum cholesterol levels, but these levels dropped 35% [from 10.7 to 7.0 mmol/L (413 to 270 mg%)] when colestipol, a bile sequestrant drug, was added.

In the same year, Rafflenbeul and colleagues (158) reported on 25 patients whom they treated medically for unstable angina. After 1 yr, using a vernier device to measure coronary artery obstructions visualized on enlarged films, they found stenosis regression (from an average of 82 to 55%) in five patients. For 18 mo Brown and his team (152) at the University of Washington treated medically 47 patients, who had angiographically-proven and symptomatic coronary artery disease. They ascertained a total of 628 lesions and established the degree of severity with computerized image analysis. Objective results showed regression in 19 patients. This was significantly associated with reduced serum cholesterol levels. In a study by Erikson and colleagues (159) eight hyperlipidemic patients showed an average drop of 45% in LDL-cholesterol in response to a low cholesterol diet plus niacin and a bile sequestrant. Six of these patients showed significant plaque reduction within 18 mo as estimated with a computer microdensitometer.

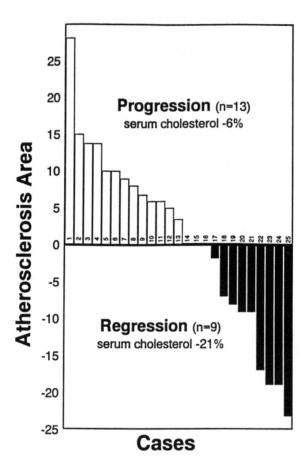

Fig. 16. Magnitude and direction of change in femoral atherosclerosis. Each patient is represented by a bar.

6.2.3. Controlled Clinical Regression Trials

The first study in which symptomatic patients were randomly allocated to intensive treatment (IT) or usual medical care (UC) took place at London's St. Thomas Hospital and was conducted by Duffield and colleagues *(160)* (*see* Table 11). Of the 24 patients with advanced femoral atherosclerosis and hyperlipidemia, 12 were treated with lipid-lowering drugs that reduced LDL-cholesterol levels

Table 11
Controlled Clinical Arteriographic Regression Trials

Ref.	Study	F/C[a]	Time yr	Number of patients — Treatment	Control	Progression %	Static %	Regression %	Change TC[a]
160	Duffield, 1983	F	1.5	12		7%	60%	33%	-25%
					12	17%	68%	15%	-3%
161	Nikkila, 1984	C	7	28		68%	32%	0%	-18%
					13	92%	8%	0%	-1%
162	NHLBI II, 1984	C	5	59		32%	63%	6%	-32%[b]
					57	49%	42%	7%	-11%[b]
163	CLAS I, 1987	F/C	2	80		10%		16%	-26%
					82	22%		3%	-4%
164	CLAS II, 1989	F/C	4	56				18%	-25%
					47			6%	-4%

[a]F = femoral or C = coronary arteries studied; TC = total cholesterol.
[b]LDL-cholesterol.

by 28%. The others served as controls with an LDL-cholesterol drop of 1%. After 18 mo, the number of femoral arterial segments that showed atherosclerotic progression was reduced from 17% in the UC group to 7% in the IT group. This randomized trial convincingly confirmed evidence from cross-sectional studies that the rate of atherosclerotic progression in hyperlipidemic patients is significantly related to ambient LDL-cholesterol concentrations.

Nikkila's group *(161)* at the University of Helsinki reported findings of 7-yr intervention study, where 28 coronary patients were treated with a fat-modified diet (30% fat) and medications to keep their serum cholesterol levels below 6.0 mmol/L (232 mg%), whereas a control group's cholesterol levels remained unchanged at around 7.8 mmol/L (300 mg%). Angiograms showed that the coronary lesions had progressed in all but one of the 13 surviving controls after 5 yr. By contrast, in the IT group the lesions had not progressed in 9 of the 28 patients after 7 yr. Nikkila also reported an almost fivefold coronary mortality differential in favor of the treated group as well as a substantial decrease in anginal pain.

The *National Heart, Lung, and Blood Institute Type II Study* (NHLBI II), a randomized, placebo-controlled, and double-blind study, enrolled 116 hypercholesterolemic coronary patients *(162)*. All patients were initially put on a fat-modified diet that lowered their LDL-cholesterol levels by 6%. They were then randomized into a drug treatment (IT) or placebo groups (UC). Those in the IT group received 24 g of cholestyramine for 5 yr. By the end of the study period, LDL-cholesterol levels had fallen another 5% in the UC group and an additional 26% in the IT group. Posttreatment angiograms indicated that coronary lesions had progressed in 49% of the UC group vs 32% of the IT group and suggested that cholestyramine retarded the progression of coronary atherosclerosis.

The *Cholesterol-Lowering Atherosclerosis Study* (CLAS), directed by David Blankenhorn and colleagues

(163), involved 162 nonsmoking men aged 40–59 yr who had previously had bypass surgery. This was a randomized placebo-controlled angiographic trial designed to test the combined effect of diet and drug therapy. The UC group received a fat-modified diet (250 mg of cholesterol, 26% total fat) and a placebo, whereas the IT group received 30 g of colestipol plus 3–12 g of niacin. In addition, the IT group was instructed to consume less than 25 mg of cholesterol/d and to keep saturated fat and total fat intake at levels of 5 and 22% of total calories, respectively.

During the 2 yr of treatment, total cholesterol and LDL-cholesterol levels in the IT group came down 26 and 43%, respectively, and HDL-cholesterol levels increased 37%. The cholesterol-lowering effect of the modified fat diet in the control group was only 4%. The substantial lipid reductions in the IT group resulted in a significant decrease in the average number of lesions per subject that progressed. They also lowered the percentage of subjects with new lesion formation in native coronary arteries. In addition, the percentage of patients with new lesions or any adverse changes in bypass grafts was significantly reduced. Deterioration of overall coronary status was much less in the IT group than in the UC group. Atherosclerotic regression occurred in 16% of the IT group vs 3% in the UC group.

The results of a 2-yr extension on 103 CLAS patients showed that lipid changes were well maintained and that the therapy effect seen at 2 yr was still strongly evident at 4 yr *(164)*. Perceptible regression of coronary lesions was seen in 18% of subjects in the IT group and in 6% of subjects in the UC group.

CLAS demonstrated that aggressive lowering of LDL-cholesterol with concomitant increase in HDL-cholesterol levels produced significant benefits to both native coronary arteries and various bypass grafts. It demonstrated the direct effect of lipid-lowering therapy on the process of atherogenesis. It offered the final piece of evidence, a logical, mechanical explanation for benefits from cholesterol-lowering therapy.

6.2.4. Dietary Intervention Regression Trials

Most of the dietary intervention trials of the 1960s and 1970s produced only marginal benefits, at best, on the course of CHD, coronary mortality, and mortality from all causes. Even though hyperlipidemia steadily emerged as the key member of the atherogenic team, traditional diet therapy, with its focus on only modestly decreasing cholesterol intake and increasing the polyunsaturated fats without altering the total fat intake, was generally unable to sufficiently affect the course of atherosclerosis.

It was at that time that some clinicians and researchers, such as William Connor, T. Colin Campbell, and William Castelli became increasingly vocal in advocating an alternative American diet, namely a diet considerably lower in fats, oil, cholesterol, sugars, and salt, and higher in fiber.

This was also the time when Nathan Pritikin, after a thorough literature review, decided to change his diet and eat himself out of his coronary ischemia and write his first book *Live Longer Now*. When it was published in 1974, he was flooded with inquiries from heart disease patients who had seen some of the best clinicians in some of the best treatment centers but whose underlying atherosclerotic processes rarely improved with medical or surgical therapy. In responding to this obvious need, Nathan Pritikin in 1976 opened his Longevity Center where he unapologetically applied his comprehensive lifestyle modification program aimed at improving physiological function in those affected by cardiovascular disease. The 25-d residential educational program emphasizes a radical restructuring of the American way of life. It was designed to simultaneously reduce recognized coronary risk factors, increase factors protective against CHD, alleviate symptoms and affect, over time, atherosclerotic processes. This new lifestyle revolves around a diet of foods-as-grown, resembling in its composition the diets consumed among peasant communities in Africa, Central America, and Asia, among whom CHD is very rare.

More than 35 clinical papers published in peer reviewed journals have documented significant clinical improvements that often take place within weeks. For instance, of the type II diabetics on insulin therapy, 50% are discharged with normal plasma glucose levels and no longer in need of insulin therapy or hypoglycemic drugs. Furthermore, 83% of those on antihypertensive medication leave the Center no longer requiring medication. Moreover, 65% of angina patients admitted with anti-anginal medication for documented coronary artery disease are discharged having little or no pain on exertion, so that they no longer require any medication. The effect of such a therapeutic diet on blood lipids and blood viscosity has been well documented *(24,98,123,165)*.

If such a more natural diet of foods-as-grown can possibly affect clinical symptomatology in such a short time by way of lowering blood lipids and viscosity, and with that improve the oxygen carrying capacity of the blood, what effect would this "therapeutic diet" have on the process of atherosclerosis? Would such a diet possibly alter the course of occlusive processes? Three clinical dietary intervention trials investigated this question.

The *Leiden Intervention Trial* enrolled 39 patients with stable angina pectoris in whom coronary arteriography had shown at least one vessel with 50% obstruction before intervention *(166)* (*see* Table 12). The intervention consisted of a 2-yr vegetarian diet that contained very little cholesterol (60 mg/d) and was very low in saturated fat (6.6%). The dietary changes were associated with a significant lowering of serum cholesterol (10%), body weight, and blood pressure. At initial angiography 307 lesions were

Table 12
Dietary Intervention Regression Trials[a]

Ref.	Study	Time mo	Diet focus	Number of patients IT	UC	Change TC	Progression %	Static %	Regression %
166	Leiden Trial, 1985	24	30% fat veget.	39		−10%	54%	46%	0%
167	Lifestyle Trial, 1990	12	7% fat veget.	28		−24%	18%	0%	82%
			30% fat		20	−5%	53%	5%	42%
171	STARS Trial, 1992	39	27% fat	26		−15%	15%	47%	38%
			Diet and drug	24		−23%	12%	55%	33%
			Usual care		24	−4%	46%	50%	4%

[a]The effect of diet and lifestyle changes on coronary atherosclerosis.
IT = intensive treatment, or UC = usual (medical) care; TC = total cholesterol.

found and quantified by computer-assisted analyses using the CAAS system. The mean percentage stenosis of the lesions increased from 44.1 to 48.6%. Of the 39 patients, 18 (46%), showed no progression of their atherosclerotic lesions. In those with lesions that progressed, mean serum cholesterol levels were 6.5 mmol/L (250 mg%) and TC/HDL ratio was 7.1, as against 5.8 mmol/L (224 mg%) and 5.7, respectively, in those whose lesions did not progress. The results emphasize the role of elevated total and LDL-cholesterol and diminished HDL-cholesterol in influencing atherosclerosis.

The *Lifestyle Heart Trial* was designed as a prospective, randomized, controlled trial to determine whether comprehensive lifestyle changes affect coronary atherosclerosis after one year *(167)*. Forty-eight patients with angiographically documented coronary artery disease were randomly assigned to an experimental group (IT), or to a usual care (UC) control group. The 28 IT patients were prescribed a lifestyle program that included a very low fat/ vegetarian diet, moderate aerobic exercise, stress management training, stopping smoking, and group support. The diet included fruits, vegetables, grains, and soybean products without caloric restriction. No animal products were allowed except egg whites and 1 cup/d of nonfat milk or yogurt. The diet contained under 10% of calories as fat, 70–75% predominantly as unrefined complex carbohydrates and 15–20% as protein. Cholesterol intake was limited to 5 mg/d and salt was restricted for hypertensive patients. Caffeine was eliminated, and alcohol was limited to no more than 2 drinks/d. The diet was nutritionally adequate and met the recommended daily allowance for all nutrients except vitamin B_{12}, which was supplemented. It resembled a diet that is commonly eaten in societies where Western foods have not yet arrived.

The UC patients were not asked to make any major lifestyle changes although they were free to do so. Progression or regression of coronary artery lesions was assessed in both groups by quantitative coronary angiography and positron emission tomography at baseline and after one year. More than 195 coronary artery lesions were analyzed. The average percentage diameter stenosis regressed from 40.0 to 37.8% in the IT group yet progressed from 42.7 to 46.1% in the UC group. When only lesions greater than 50% stenosed were analyzed, the average percentage diameter stenosis regressed from 61.1 to 55.8% in the IT group and progressed from 61.7 to 64.4% in the UC group. Overall, 82% of IT patients had an average change toward regression.

The regression results were accompanied by marked decreases in serum cholesterol and LDL-cholesterol levels (24 and 37%, respectively, in the IT group). These falls occurred even though patients had already reduced their fat consumption to 31.5% of calories and cholesterol intake to 230 mg/d on average before baseline testing. Patients assigned to UC and following the "prudent diet" of the American Heart Association reduced their total cholesterol and LDL-cholesterol by 5 and 6%, respectively.

IT group patients reported a 90% reduction in the *frequency* of angina, a 40% reduction in the *duration* of angina, and a 28% reduction in the *severity* of angina. In contrast, the UC group patients reported a 165% rise in frequency, a 95% rise in duration, and a 39% rise in severity of angina. The investigators noted that the functional status improvement occurred in only one month. This suggests that the improvement in angina may have preceded the regression of coronary atherosclerosis, perhaps by changing platelet-endothelial interactions, vasomotor tone, or blood viscosity. Adherence analysis found that those who made the greatest lifestyle changes showed the biggest improvement. They also found that severely stenosed lesions showed the greatest regression. Although the opposite had been expected, the latter finding is of vital importance since severely stenosed lesions are the most important clinically.

This clinical trial showed that a heterogeneous group of patients with CHD can be motivated to make comprehensive lifestyle changes. It also documented that adherence to and acceptability of a low-fat, vegetarian diet among cardiac patients is quite high *(168)*. Moreover, adopting such a diet would be very economical; it would save an average family of four $2100 a year when compared to a regular American diet, and even more when compared to the "prudent diet" of the American Heart Association *(169)* (Table 13).

Referring to the "prudent diet" used by the control group and presently advocated for the treatment and prevention of heart disease, Dean Ornish, the principal investigator of the study said *(170)*:

> These moderate dietary recommendations do not appear to go far enough to effectively influence the progression of CHD. People with clinically demonstrated disease need to go beyond the present dietary recommendations. We're finding out that staying on a moderate, 30% fat diet does not reverse the disease. It may actually make it worse.

Two years after the Lifestyle Heart Trial convincingly proved that CHD could be reversed without drugs or surgery through major lifestyle changes, including a very low-fat vegetarian diet, stress management, daily exercise, and smoking cessation, the *St. Thomas Atheroma Regression Study* (STARS) was published *(171)*. It became the first randomized, controlled trial to show that unifactorial dietary intervention was able to reverse coronary atherosclerosis. While the Lifestyle Heart Trial utilized multiple interventions, the STARS trial showed clear evidence that *diet alone* was capable of opening narrowed coronary arteries by reducing serum cholesterol levels.

In this 3-yr study, 90 middle-aged men with established CHD (serum cholesterols above 6.2 mmol/L, 50%

Table 13
Foodbasket Prices[a]

Diet	Cost	Excess cost
10% fat vegetarian	$5824	
37% fat US diet	$7924	$2100 (+36%)
30% fat AHA diet	$8312	$2488 (+43%)

[a]Cost comparison of foodbaskets for family with two grade school children for a year. Source: *Suzanne McNutt, George Washington University Lipid Research Clinic.*

with previous heart attacks and 75% with angina at entry) were randomly allocated to three treatment regimens:

1. *Usual Care (UC).*
2. *Dietary Intervention.* Fat was reduced from 42% to 27%, cholesterol from 650 to 140 mg/d, and soluble fiber was markedly increased.
3. *Diet and Drug.* Diet plus 16 g/d of cholestyramine.

At the end of 39 mo, the following results were observed.

Blood Cholesterol Levels. The cholesterol levels remained unchanged in the UC group, but the Dietary Intervention group had drops of 14% for total cholesterol, 16% for LDL-cholesterol, and 20% for triglycerides. In the Diet and Drugs group the total cholesterol and LDL-cholesterol levels dropped 25 and 35%, respectively, but the triglyceride levels remained the same.

Angiographic Findings. Sophisticated angiographic determinations of the internal diameter of the coronary lumen showed that the greatest improvements in arterial regression were seen in those with the most narrowing at entry (Table 14), a finding that confirms the observations made in the Lifestyle Heart Trial. Dietary change alone retarded overall progression and increased overall regression of coronary artery disease. Diet plus cholestyramine

Table 14
Effect of Dietary Intervention on Atherosclerosis[a]

	Usual care	Diet	Diet and drugs
Artery narrowing, *progressed* in percentage of patients	46%	15%	12%
Artery narrowing, *reversed* in percentage of patients	4%	38%	33%

[a]Results from the St. Thomas Atherosclerosis Regression Study (STARS).

was additionally associated with a net increase in coronary lumen diameter.

Clinical Events. Fourteen patients had cardiovascular events as indicated in Table 15. Angina symptoms markedly decreased within weeks in both the *Diet* and *Diet and Drugs* groups in contrast to the UC patients. This was also seen in the use of anti-anginal drugs.

The findings from this study support the use of a lipid-lowering diet and, if necessary, of an appropriate drug treatment in men with CHD. Barry Lewis, coinvestigator and Professor Emeritus at the University of London, said,

> Narrowing of the coronary arteries by atherosclerosis is caused largely by a diet that is too high in fat and cholesterol. Our study has shown conclusively that healthy eating can partly reverse this disease process, which can be demonstrated within three years. This approach for most coronary heart disease patients is not only very cost-effective, but it has also no surgical pain, no surgical mortality and no dependence on drugs. It is clearly emerging as a treatment of choice.

Clinicians had previously agreed that a change in diet may reduce chances of developing heart disease, but the

Table 15
Cardiovascular Events in the STARS Study[a]

	Usual care	Diet	Diet and drugs
Deaths	3	1	0
Heart attacks	2	1	1
Coronary surgery	4	1	0
Stroke	1	0	0

[a]Cardiac events in 39 months according to intervention modality (171).

British STARS study proved that dietary changes can actually *reverse* the process of atherosclerosis.

In view of the enormous costs and various side effects in treating hyperlipidemia pharmacologically, it would seem appropriate to reserve the drug approach for cases in which dietary intervention is not adequate—most often seen in genetic-lipid disorders—or in hyporesponders (5–10% of the population) (Table 16).

Optimal dietary interventions, that are educationally based and involve the whole family so as to facilitate the understanding of the rationale for lifestyle change and skill acquisition, can produce lipid changes similar to those achieved with drugs.

The simplest form of cholesterol intervention is clearly a very low-fat, very low-cholesterol, high-fiber diet. Such a diet would rely more on foods-as-grown. It would be centered on unrefined plant foods, high in fiber, nutrition, and volume, and lower in energy density. For most individuals this diet is the only intervention necessary to lower serum triglyceride and certain cholesterol levels, and to ameliorate other coronary risk factors such as overweight, hypertension, and diabetes (type II). It is also the diet of choice to favorably affect the course of atherosclerosis and its clini-

Table 16
Cholesterol-Lowering Drugs[a]

Drug	Daily dose, g	Effects on serum lipids				Cost per month
		TC	LDL	HDL	TG	
Colestipol	30	−18%	−24%	+6%	+10%	$200
Cholestyramine	24	−18%	−24%	+6%	+10%	$130
Nicotinic acid	3	−20%	−25%	+10%	−25%	$200
Gemfibrozil	1.2	−11%	−12%	+15%	−40%	$75
Lovastatin	0.08	−34%	−42%	+10%	−25%	$265

[a]A comparison of effects on serum lipids and current costs. Maximal recommended daily doses of hypolipidemic drugs presently available in the United States (172).

cal manifestations. This approach is also both cost effective and ecologically sound.

6.2.5. Summary

Professor David Blankenhorn, a pioneer in regression studies, summarized the evidence very well in his article *Atherosclerosis Regression in Humans (173)*:

Evidence that human atherosclerosis is a reversible condition has been accumulating for more than half a century. The first indications were from starved populations and patients with wasting diseases. More direct evidence became available when serial angiograms were used to follow the course of patients treated to reduce atherosclerotic risk factors. The earliest controlled clinical trials, which employed angiography with relatively small numbers of subjects and a moderate lowering of blood lipid levels, demonstrated that lesion progression could be reduced. Later trials with more subjects and more aggressive reduction of blood lipid levels have demonstrated regression.

Since then convincing evidence has emerged that a restructured diet can measurably reduce progression and induce regression within 1–3 yr and provide relief of angina within weeks.

7. Outlook

At present, US policy is based on US Dietary Goals (Table 17). These guidelines, issued by the United States Select Committee on Nutrition and Human Needs after a detailed study and years of testimony from the leading scientists in the world, have been reinforced by recent dietary guidelines from the National Cancer Institute for the prevention of cancer. Ernst Wynder, editor of *Preventive Medicine*, has estimated that 50% of all major cancers in the Western world, such as those of the colon, prostate, uterus, and breast, relate to over-nutrition, especially to the excesses of fats and cholesterol. This concern about diet and cancer was amplified by the US Surgeon General's report, *Health Promotion and Disease Prevention*, and by his more recent report, *Nutrition and Health*. The Surgeon General stated that the high consumption of animal protein and the excessive intake of fat, both from animal and vegetable sources, may be directly linked to colon and breast cancer.

The importance of the US Dietary Goals does not reside so much in its content as in the fact that it was the first major bipartisan effort delineating the dietary relationship to chronic killer diseases and making it publicly known. Although the advocated changes were moderate, they elicited strong reactions. Consumer advocates hailed the dietary goals as a major breakthrough toward improving health, whereas the food lobbies and the American Medical Association condemned them as being prematurely conceived. It was very clear: This historic report aimed at improving national health would usher in the turning away

Table 17
A Comparison of Diets
for the Prevention and Reversibility
of Cardiovascular Disease

	US Diet	US Dietary Goals	The Optimal Diet
Fats and oils[a]	38%	30%	<20%
Protein[a]	12%	12%	10%
Complex carbohydrates[a]	28%	42%	60%
Simple carbohydrates[a]	22%	16%	10%
Cholesterol[b]	500 mg	300 mg	<100 mg
Salt[b]	15 g	10 g	5 g
Fiber[b]	10 g	20 g	40 g

[a]Total calories.
[b]Per day.

from the rich American diet and turning towards a somewhat simpler diet. Major political and economic changes and alignments were in the making.

The Dietary Goals were reasonable and sound as a *first* step toward health improvement. They recommended only a modest reduction of total fat from 40 to 30%, but the authors suggested that further reductions may be desirable:

> There's increasing research that suggests that someday a dietary intake of 20% to 25% might be recommended, and even less for those people who already have heart disease. The basic research is strongly corroborated by studies of populations throughout the world who live quite well on a diet containing as little as 10% of total calories from fat.

Indeed, halfhearted changes in diet are not enough to turn the epidemic around. If we want to win the battle against the epidemic of Western lifestyle diseases, then we must adopt a simpler, more natural dietary lifestyle, that allows us to eat more, to lose weight, and to have better and more buoy-

Avoid visible fats and oils
Avoid fatty meats, cooking and salad oils, sauces, dressings, and shortening. Use margarine and nuts very sparingly. Avoid frying; sauté instead with a little water in nonstick pans.

 Avoid sugars
Limit sugar, honey, molasses, syrups, pies, cakes, pastries, candy, cookies, soft drinks, and sugar-rich desserts, like pudding and ice cream. Save these foods for special occasions.

 Severely restrict cholesterol
Strictly limit meat, sausages, egg yolks, and liver. Limit dairy products, if desired, to low-fat cheeses and nonfat milk products. If you use fish and poultry, use them sparingly.

 Severely limit salt
Don't salt food during cooking or at the table. Limit highly salted products like pickles, crackers, soy sauce, salted popcorn, nuts, chips, and pretzels.

 Avoid alcohol
Avoid alcohol in all forms as well as caffeinated beverages such as coffee, colas, and black tea.

 Freely eat whole grains
Freely use brown rice, millet, barley, corn, wheat, and rye. Also eat freely of whole grain products, such as breads, pastas, shredded wheat, and tortillas.

 Freely use tubers and legumes
Freely use all kinds of white potatoes, sweet potatoes, and yams (without high-fat toppings). Enjoy lentils, peas, and beans of every kind.

 Freely use fruits and vegetables
Eat several fresh, whole fruits every day. Limit fruits canned in syrup and fiber-poor fruit juices. Eat a variety of vegetables daily. Enjoy fresh salads with low-calorie, low-salt dressings.

 Drink plenty of water
Drink six to eight glasses of water a day. Vary the routine with a twist of lemon and occasional herb teas.

 Eat a hearty breakfast
Enjoy hot, multigrain cereals, fresh fruit, and whole wheat toast.

Fig. 17. Dietary guidelines for the prevention and reversibility of atherosclerosis-related diseases. Summary: Eat freely a wide variety of foods-as-grown, simply prepared with sparing use of fats and oils, sugars, and salt. Use refined products and animal products sparingly.

ant health. This is summarized in Fig. 17 and Table 17. In essence, what is needed is a more comprehensive dietary policy, educationally based, and nationally promoted, through public health agencies and the media.

Research with therapeutic nutrition has clearly demonstrated a *unitary* dietary principle in dealing with Western killer diseases: There is not a special diet for the treatment of heart disease, another diet for overweight, another for diabetes, and yet another for hypertension and high cholesterol levels. We now know that disease-specific diets are not necessary. There is one optimal diet, rich in whole grain cereals, legumes, tubers, vegetables, and fruits, low in fats, sugars, and salt, that will not only prevent a whole spectrum of Western diseases but will contribute to their reversibility. The rich Western diet emerges as the primary, essential, necessary cause of the current epidemic of premature atherosclerotic disease followed by cigaret smoking as an important secondary or complementary cause. On the other hand, a simpler diet of foods-as-grown emerges more and more as the primary, essential lifestyle agent to prevent and reverse these Western diseases.

References

1. Kuller L, Lilienfeld A, Fisher R. Epidemiological Study of sudden and unexpected deaths due to arteriosclerotic heart disease. *Circulation* 1966; **34**:1056–1068.
2. Enos WF, Holmes RH, Beyer J. Coronary disease among US soldiers killed in action in Korea. *JAMA* 1953; **152**: 1090–1093.
3. Strong JP. Coronary atherosclerosis in soldiers. *JAMA* 1986; **256**:2863–2866.
4. World Health Organization. World health statistics manual 1991. Geneva: WHO, 1992; 25,26.
5. Anitschkow N. Ueber experimentelle Cholesterin Steatose and ihre Bedeutung fuer die Entstehung einiger pathologischer Prozesse. *Zbl Path* 1913; **26**:1-8.
6. DeLanger CD. Cholesterol metabolism in racial pathology. *Geneesk Tydschr Nederl Indie* 1916; **56**:1–34.
7. Schettler G. Cardiovascular diseases during and after World War II: a comparison of the Federal Republic of Germany with other European countries. *Prev Med* 1979; **8**:581–590.
8. Malmros H. The relation of nutrition to health. *Acta Med Scand* 1950; **246(Suppl)**:128–150.
9. Vartiainen I, Kanerva K. Arteriosclerosis in wartime. *Ann Intern Med (Finland)* 1947; **36**:748–758.

10. Strom A. Mortality from circulatory diseases in Norway 1940–1945. *Lancet* 1951; **260**:126–129.
11. Castelli, WP. 1986 (personal communication).
12. Keys A, Taylor HL, Blackburn H, Brozek J, Anderson JT, Simonson E. Coronary heart disease among Minnesota business and professional men followed 15 years. *Circulation* 1963; **28**: 381–395.
13. The Framingham Study. Kannel WB, Gordon T, eds. Washington, DC: USD HEW Public Health Service, National Institutes of Health, 1973; #74-618.
14. Kannel WB, Castelli WP, Gordon T. Cholesterol in the prediction of atherosclerotic disease; new perspective on the Framingham study. *Ann Intern Med* 1979; **90**:85–91.
15. Seidel D, Cremer P, Nagel D. Significance of risk factors in the prediction of atherosclerosis. *Atherosclerosis Rev* 1991; **23**:243–250.
16. Gohlke H, Gohlke-Baerwolf C, Stuerzenhofecker P, Goernandt L, Thilo A, Haakshorst W, Roskamm H. Myocardial infarction at young age: correlation of angiographic findings with risk factors and history in 619 patients. *Circulation* 1980; **62**:III–39.
17. Gohlke H, Stuerzenhofecker P, Goernandt L, Haakshorst W, Roskamm H. Progression und Regression der koronaren Herzerkrankung im chronischen Infarktstadium bei Patienten unter 40 Jahren. *Schweiz Med Wochenschr* 1980 (Nov. 8); **110**:1663–1665.
18. Barndt R. Serial angiographic correlations of progression and regression of atherosclerosis with risk factor levels. *Arteriosclerosis* 1981; **1**:1–8.
19. Truett J, Cornfield J, Kannel W. A multivariate analysis of the risk of coronary heart disease in Framingham. *J Chron Dis* 1967; **20**:511–524.
20. *Coronary Risk Handbook.* Estimating risk of coronary heart disease in daily practice. New York: American Heart Association, 1973.
21. Cooper T. The scientific foundation for the prevention of coronary heart disease. *Am J Cardiol* 1981; **47**:720–724.
22. Byington R, Dyer AR, Garside D, Liu K, Moss D, Stamler J, Tsong Y. Recent trends of major coronary risk factors and CHD mortality in the US. In: Hawlik RJ, Feinleib M, eds. *Proceedings of the Conference on the Decline in CHD Mortality.* USD HEW Public Health Service, National Institutes of Health 1979; #79-1610.

23. Stamler J, Wentworth D, Neaton JD. Is the relationship between serum cholesterol and risk of premature death from coronary heart disease continuous and graded? The MRFIT Study. *JAMA* 1986; **256**:2823–2828.
24. Diehl H, Mannerberg D. Regression of hypertension, hyperlipidaemia, angina and coronary heart disease. In: Trowell HC, Burkitt DP, eds. *Western Diseases: Their Emergence and Prevention*. London: Edward Arnold, 1981; 392–410.
25. Stamler J. Lifestyles, major risk factors, proof and public policy. *Circulation* 1978; **58**:3–18.
26. Stamler J. The marked decline in coronary heart disease mortality rates in the United States, 1968–1981. Summary of findings and possible explanations. *Cardiology* 1985; **72**:5–22.
27. Ignatowski A. Ueber die Wirkung des tierischen Eiweisses auf die Aorta und die parenchymatosen Organe der Kaninchen. *Virchows Arch* 1909; **198**:248–259.
28. Taylor CB, Patton DE, Cox GE. Atherosclerosis in rhesus monkeys. *Arch Pathol* 1963; **76**:404–412.
29. Wissler RW. 1990 (personal communication).
30. Armstrong ML, Warner ED, Connor WE. Regression of coronary atheromatosis in rhesus monkeys. *Circ Res* 1970; **27**: 59–67.
31. Armstrong ML, Megan MB. Lipid depletion in atheromatous coronary arteries in rhesus monkeys after regression diets. *Circ Res* 1972; **30**:675–680.
32. Connor, WE. *Regression of Atherosclerotic Plaques*. Presentation at University of California, San Francisco, Dental Alumni Convention, Jan 10, 1992.
33. McGill HC. The geographic pathology of atherosclerosis. *Lab Invest* 1968; **18**:463–478.
34. Keys A. Coronary heart disease in seven countries. *Circulation* 1970; **41(Suppl. 1)**:1–196.
35. Keys A, Kimura N, Kusukawa A. Lessons from serum cholesterol studies in Japan, Hawaii and Los Angeles. *Ann Intern Med* 1958; **48**:83–94.
36. Keys A. Lessons to be learned from serum cholesterol level. *Ann Intern Med* 1976; **61**:421–430.
37. Welch CC, Proudfit WL, Sones FM, Shirey EK, Sheldon WC, Razavi M. Cinecoronary arteriography in young men. *Circulation* 1970; **42**:647–653.
38. Page IH, Berrettoni Jr, Butkus A, Sones FM. Prediction of coronary heart disease based on clinical suspicion, age, total cholesterol and triglycerides. *Circulation* 1970; **42**:625–640.

38a. Roberts, WG. Atherosclerotic risk factors—Are there ten or is there only one? *Am J Cardiol* **64:**552–554.

39. Gotto A. Hearing before the Select Committee on Nutrition and Human Needs of the US Senate. Part I. *CVD-Diet Related to Killer Diseases*. Washington, DC: US Government Printing Office, 1977; 325–326.

40. Castelli WP. *Reversing Heart Disease*. Med World News 1979 (Sept. 3).

41. Consensus Conference. Lowering blood cholesterol to prevent heart disease. *JAMA* 1985; **253:**2080–2086.

42. Pritikin N. Optimal dietary recommendations: a public health responsibility. *Prev Med* 1982; **11:**733–739.

43. Blackburn H. Diet and atherosclerosis: epidemiologic evidence and public health implications. *Prev Med* 1983; **12:**2–10.

44. Castelli WP, Wilson PWF, Levy D, Anderson K. Serum lipids and risk of coronary artery disease. *Atherosclerosis Rev* 1990; **21:**7–19.

45. Ross R, Glomset JA. The pathogenesis of atherosclerosis, Parts I and II. *N Engl J Med* 1976; **295:**369–375, 420–426.

46. Ross R. Lipoproteins, endothelial injury and atherosclerosis. *Cardiovascular Res Rep* 1982; **3:**1026–1033.

47. Majno G. The diet/atherosclerosis connection: new insights. *J Cardiovasc Med* 1984 (Jan.); 21–28.

48. Constantinides P, Robinson M. Ultrastructural injury of arterial endothelium. *Arch Pathol* 1969; **88:**99–116.

49. Davies MJ, Woolf N, Katz DR. The role of endothelial denudation injury, plaque fissuring, and thrombosis in the progression of human atherosclerosis. *Atherosclerosis Rev* 1991; **23:**105–113.

50. Astrup P. Carbon monoxide, smoking and cardiovascular disease. *Circulation* 1973; **48:**1167,1168.

51. Wells RE. Rheology of blood microvasculature. *N Engl J Med* 1964; **270:**832–839.

52. Swank RD, Nakamura H. Oxygen availability in brain tissues after lipid meals. *Am J Physiol* 1960; **198:**217–220.

53. Kuo PT, Joyner CR. Angina pectoris induced by fat ingestion in patients with coronary artery disease. *JAMA* 1955; **158:**1008–1012.

54. Williams AV. Increased cell agglutination following ingestion of fat, a factor contributing to cardiac ischemia, coronary insufficiency and anginal pain. *Angiology* 1957; **8:**29–36.

55. Gresham GA. Is atheroma a reversible lesion? *Atherosclerosis* 1976; **23:**379–391.

56. Stamler J. Towards cardiovascular health. *Am Med* 1989; **21**:141–155.
57. Armstrong ML, Megan MB, Warner ED. Intimal thickening in normocholesterolemic rhesus monkeys fed low supplements of dietary cholesterol. *Circ Res* 1974; **34**:447–454.
58. Canner PL, Berger KG, Wenger NK. Fifteen-year mortality in Coronary Drug Project patients: long-term benefit of niacin. *J Am Coll Cardiol* 1986; **6**:1245–1255.
59. Coronary Drug Project Research Group. Factors influencing long-term prognosis after recovery from myocardial infarction. Three-year findings of the Coronary Drug Project. *J Chron Dis* 1974; **27**:267–279.
60. Coronary Drug Project Research Group. Clofibrate and niacin in coronary heart disease. *JAMA* 1975; **231**:360–381.
61. Lipid-lowering drugs after myocardial infarction. *Lancet* 1975; **i**:501,502.
62. Committee of Principal Investigators. WHO Cooperative trial on primary prevention of ischemic heart disease using clofibrate to lower serum cholesterol: mortality follow-up. *Lancet* 1980; **ii**:379–384.
63. The clofibrate dilemma. *Br Med J* 1978; **2**:1585,1586.
64. Rifkind BM, Levy RI. Testing the lipid hypothesis. *Arch Surg* 1978; **113**:80–83.
65. Lipid Research Clinics Program. The Lipid Research Clinics Coronary Primary Prevention Trial results. I. Reduction in incidence of coronary heart disease. *JAMA* 1984; **251**:351–364.
66. Lipid Research Clinics Program. The Lipid Research Clinics Coronary Primary Prevention Trial results. II. The relationship of reduction in incidence of coronary heart disease to cholesterol lowering. *JAMA* 1984; **251**:365–374.
67. Tikkanen MJ, Nikkila EA. Current treatment of elevated serum cholesterol. *Circulation* 1987; **76**:529–533.
68. Frick MH, Elo MO, Haapa K, et al. Helsinki Heart Study: primary prevention trial with gemfibrozil in middle-aged men with dyslipidemia. *N Engl J Med* 1987; **317**:1237–1245.
69. Manninen V, Elo MO, Frick MH, et al. Lipid alterations and decline in the incidence of coronary heart disease in the Helsinki Heart Study. *JAMA* 1988, **260**:641–651.
70. Tyroler, HA. Review of lipid-lowering clinical trials in relation to observational epidemiologic studies. *Circulation* 1987; **76**: 515–522.
71. National Research Council, Committee on Diet and Health: *Diet and Health Implications for Reducing Chronic Disease Risk*. Washington, DC: National Academic Press, 1989.

72. Holme I. An analysis of randomized trials evaluating the effect of cholesterol reduction on total mortality and CHD incidence. *Circulation* 1990; **82:**1916–1924.

73. Keys A, Anderson JR, Grande F. Serum cholesterol response to changes in the diet. *Metabolism* 1965; **14:** 747–787.

74. Anderson JR, Jacobs DR, Foster N, Keys A. Scoring systems for evaluation of dietary pattern effect on serum cholesterol. *Prev Med* 1979; **8:**525–530.

75. Hegsted DM, McGandy RB, Myers ML, Stare FJ. Quantitative effects of dietary fat on serum cholesterol in man. *Am J Clin Nutr* 1965; **17:** 281–295.

76. Hegsted DM. Serum cholesterol response to dietary cholesterol: a re-evaluation. *Am J Clin Nutr* 1986; **44:**299–305.

77. Connor WE, Stone DB, Hodges RE. The interrelated effects of dietary cholesterol and fat on human serum lipid levels. *J Clin Invest* 1964; **143:**1691–1696.

78. Mattson FH, Erickson BA, Klingman AM. Effect of dietary cholesterol on serum cholesterol. *Am J Clin Nutr* 1972; **25:**589–594.

79. Stamler J, Shekelle R. Dietary cholesterol and human coronary heart disease: the epidemiologic evidence. *Arch Pathol Lab Med* 1988; **112:**1032–1040.

80. Beveridge JMR. The response of man to dietary cholesterol. *J Nutr* 1960; **71:** 61–65.

81. Sacks FM. Effect of ingestion of meat on plasma cholesterol of vegetarians. *JAMA* 1981; **246:**640–644.

82. Sacks FM. Ingestion of egg raises plasma LDL-cholesterol in free-living subjects. *Lancet* 1984; **i:**647–649.

83. Alfin-Slater R. Plasma cholesterol and triglyceride levels in men with added eggs in the diet. *Nutr Rep Intl* 1976; **14:** 249–259.

84. Flynn MA. Effect of dietary egg on human serum cholesterol and triglycerides. *Am J Clin Nutr* 1979; **32:**1051–1057.

85. Keys A, Anderson JT, Grande F. Diet-type (fats constant) and blood lipids in man. *J Nutr* 1960; **70:**257–269.

86. DeGroot AP, Luyken R, Pikaar NR. Cholesterol-lowering effect of rolled oats. *Lancet* 1963; **ii:**303–307.

87. Ripsin CM, Keenan JM, Jacobs DR, et al. Oat products and lipid lowering-a meta-analysis. *JAMA* 1992; **267:**3317–3325.

88. Grundy SM. Dietary therapy for different forms of hyperlipoproteinemia. *Circulation* 1987; **3:**523–528.

89. Ahrens EH, Hirsch J, Stoffel W, Peterson ML, Farquar JW. The effect on human serum lipid of a dietary fat, highly unsaturated, but poor in essential fatty acids. *Lancet* 1959; **i:**115–119.

90. Vega LG, Groszek E, Wolf R, Grundy SM. Influence of polyunsaturated fats on plasma lipoprotein composition and apolipoprotein. *J Lipid Res* 1982; **23**:811–818.

91. Bennion LJ, Grundy SM. Risk factors for the development of cholelithiasis in man. *N Engl J Med* 1978; **299**:1221–1227.

92. Broitman S. Polyunsaturated fat, cholesterol and tumorigenesis. *Cancer* 1977; **40**:2455–2459.

93. Felber JP, Vannotti A. Effects of fat infusion on glucose tolerance and insulin plasma levels. *Med Exp* 1964; **10**:1536–1541.

94. Cullen C. Intravascular aggregation and adhesiveness of the blood elements associated with alimentary lipemia and injections of large molecular substances. *Circulation* 1954; **9**:335–339.

95. Bansal BR, Rhoads JE, Bansal SC. Effect of diet on colon carcinogenesis and the immune systems in rats treated with 1,2-dimethylhydrazine. *Cancer Res* 1978; **38**:3293–3298.

96. Carroll K. Experimental evidence of dietary factors in hormone-dependent cancer. *Cancer Res* 1975; **35**:3374–3379.

97. Ehnholm C, Huttunen JK, Pietinen P, et al. Effect of diet on serum lipoproteins in a population with a high risk of coronary heart disease. *N Engl J Med* 1982; **307**:850–855.

98. Barnard RJ. Effects of lifestyle modification on serum lipids. *Arch Intern Med* 1991; **151**:1389–1394.

99. Morrison LM. Reduction of mortality rate in coronary atherosclerosis by low-cholesterol, low-fat diet. *Am Heart J* 1951; **42**:538–545.

100. Morrison LM. Diet in coronary atherosclerosis. *JAMA* 1960; **173**:104–108.

101. Rose GA, Thompson WB, Williams RT. Corn oil in treatment of ischemic heart disease. *Br Med J* 1965; **i**:1531–1533.

102. London Research Group. Low-fat diet in myocardial infarction—a controlled trial. *Lancet* 1965; **ii**:501–504.

103. Medical Research Council, Research Committee. Controlled trial of soybean oil in myocardial infarction. *Lancet* 1968; **ii**:693–700.

104. Bierenbaum ML, Fleischman AI, Green DP. The 5-year experience of modified fat diets on younger men with CHD. *Circulation* 1970; **42**:943–949.

105. Leren P. The Oslo Diet-Heart Study. Eleven-year report. *Circulation* 1970; **42**:935–942.

106. Dayton S, Pearce ML, Hashimoto S, Dixon Wj, Tomiyasn M. A Controlled trial of a diet high in unsaturated fat in preventing complications of atherosclerosis. *Circulation* 1969; **40(Suppl. 2)**:1–63.

107. Turpeinen O, Karvonen MJ, Pekkarinen M. Dietary prevention of CHD—the Finnish mental hospital study. *Int J Epidemiol* 1979; **8**:99–118.
108. Hjermann I, Byre KV, Holme I, Leren P. Effect of diet and smoking intervention on the incidence of CHD. Report from the Oslo Study Group of a randomized trial in healthy men. *Lancet* 1981; **ii**:1303–1310.
109. Hjermann I. The Oslo Study: Some trial results. *Atherosclerosis Rev* 1990; **21**:103–108.
110. MRFIT Research Group. MRFIT—Risk factor changes and mortality results. *JAMA* 1982, **248**:1465–1477.
111. MRFIT Research Group. Mortality rates after 10.5 years for participants in the MRFIT. *JAMA* 1990; **263**:1795–1801.
112. McNamara DJ. Diet and hyperlipidemia—A justifiable debate. *Arch Intern Med* 1982; **142**: 112–114.
113. Ahrens EH. The diet–heart question in 1985: has it really been settled? *Lancet* 1985; **i**:1085–1087.
114. Kolata G. Heart panel's conclusions questioned. *Science* 1985; **227**:40–41.
115. Taylor WC, Pass TM, Shepard DS, Kamaroff AL. Cholesterol reduction and life expectancy; a model incorporating multiple risk factors. *Ann Int Med* 1987; **106**: 605–614.
116. Oliver MF. Reducing cholesterol does not reduce mortality. *J Am Coll Cardiol* 1988; **12**: 814–817.
117. Muldoon MF, Manuck SB, Matthews KA. Lowering cholesterol concentrations and mortality: a quantitative review of primary prevention trials. *Br Med J* 1990; **301**: 309–314.
118. Oliver MF. Doubts about preventing CHD. Multiple interventions in middle-aged men may do more harm than good. *Br Med J* 1992; **304**:393,394.
119. Ravnskov U. Cholesterol lowering trials in CHD: frequency of citation and outcome. *Br Med J* 1992; **305**:15–19.
120. Fraser GF. *Preventive Cardiology*. New York: Oxford University Press, 1986.
121. Taskforce on Cholesterol Issues. The Cholesterol Facts—the summary of the evidence relating dietary facts, serum cholesterol, and CHD. A joint statement by the American Heart Association and the National Heart, Lung and Blood Institute. *Circulation* 1990; **81**:1721–1733.
122. Ramsey LE, Yee WW, Jackson PR. Dietary reduction of serum cholesterol concentrations: time to think again. *Br Med J* 1991, **303**:953–957.

123. Robinson MH. *Nutrition and Physical Fitness in Public Health: the Pritikin Program.* Hearing before the Committee on Labor and Human Resources, US Senate. Washington, DC: US Government Printing Office, 1986 (SHRG 99-513); 166–215.

124. Phillips RL, Kuzma JW, Beeson WL, Lotz T. Influence of selection versus lifestyle on risk of fatal cancer and cardiovascular disease among Seventh-day Adventists. *Am J Epidemiol* 1980; **112:**296–314.

125. Phillips, RL, Garfinkel L, Kuzma JW, Beeson WL, Lotz T, Brin B. Mortality among California Seventh-day Adventists for selected cancer sites. *J Natl Cancer Inst* 1980; **65:** 1097–1107.

126. Kahn HA, Phillips RL. Association between reported diet and all-cause mortality: Twenty-one year follow-up on 27,530 adult Seventh-day Adventists. *Am J Epidemiol* 1984; **119:**775–787.

127. Chang-Claude J, Frentzel-Beyme R, Eilber U. Mortality pattern of German vegetarians after 11 years of follow-up. *Epidemiology* 1992; **3:**395–401.

128. McNichael AJ. Vegetarians and longevity—imagining a wider reference population. *Epidemiology* 1992; **3:**389–391.

129. Katz LN, Stamler J. *Experimental Atherosclerosis.* Springfield, IL: Charles Thomas, 1953.

130. Wissler RW, Hughes RH, Frazier LE, Getz GS, Turner DF. Aortic lesions and blood lipids in rhesus monkeys fed "table prepared" human diets. *Circulation* 1965; **32(Suppl. 2):**220.

131. Clarkson TB, Lehner ND, Wagner WD, St. Clair RW, Bond MG, Bullock BC. A study of atherosclerosis regression in Macaca mulatta. I. Design of experiment and lesion induction. *Exp Mol Pathol* 1979; **30:**360–385.

132. Tucker CF, Catsulis C, Strong JP, Eggen DA. Regression of early cholesterol-induced aortic lesions in rhesus monkeys. *Am J Pathol* 1971; **65:**493–514.

133. Stary HC. Progression and regression of experimental atherosclerosis in rhesus monkeys. In: Goldsmith E, Morr-Hankowsky J, eds. *Medical Primatology.* Basel: Karger, 1972; 356,357.

134. Vesselinovitch D, Wissler RW. Reversal of advanced atherosclerosis in rhesus monkeys. *Atherosclerosis* 1976, **23:**155–176.

135. Bond MG, Bullock BC, Clarkson TB, Lehner ND. The effect of plasma cholesterol concentrations on regression of primate atherosclerosis. *Am J Pathol* 1976; **82:**69a.

136. Armstrong ML. Evidence of regression of atherosclerosis in primates and man. *Postgrad Med J* 1976; **52**:456–461.
137. Wissler RW, Vesselinovitch D. Regression of atherosclerosis in experimental animals and man. *Mod Concepts Cardiovasc Dis* 1977; **46**:27–32.
138. Wagner WD, St. Clair RW, Clarkson TB, Connor JR. A study of atherosclerosis regression in Macaca mulatta. III. Chemical changes in arteries from animals with atherosclerosis induced for 19 months and regressed for 48 months at plasma cholesterol concentrations of 300 or 200 mg%. *Am J Pathol* 1980; **100**:633–650.
139. Clarkson TB, Bond MG, Bullock BC, Marzetta CA. A study of atherosclerosis regression in Macaca mulatta. IV. Changes in coronary arteries from animals with atherosclerosis induced for 19 months and then regressed for 24 or 48 months at plasma cholesterol concentrations of 300 or 200 mg%. *Exp Mol Pathol* 1981; **34**:345–368.
140. Clarkson TB, Bond MG, Bullock BC, McLaughlin KJ, Sawyer JK. A study of atherosclerosis regression in Macaca mulatta. V. Changes in abdominal aorta and carotid and coronary arteries from animals with atherosclerosis induced for 38 months and then regressed for 24 or 48 months at plasma cholesterol concentrations of 300 or 200 mg%. *Exp Mol Pathol* 1984; **41**:96–118.
141. Malinow MR. Atherosclerosis progression, regression, and resolution. *Am Heart J* 1984; **108**:1523–1537.
142. Malinow MR, Blaton V. Regression of atherosclerotic lesions (Meeting Summary). *Arteriosclerosis* 1984; **4**:292–295.
143. Reiber JHC, Gerbrands JJ, Booman F. Objective characterization of coronary obstructions from monoplane cineangiograms and three-dimensional reconstruction of an arterial segment from two orthogonal views. In: Schwartz MD, ed. *Applications of Computers in Medicine. Circulation* 1985; **71**:280–288.
144. Aschoff L. *Lectures in Pathology.* New York: Hoeber Medical Division, Harper & Row, 1924.
145. Breitske H. Zur Entstehung der Atherosklerose. *Virchows Arch* 1928; **267**:625–631.
146. DePalma RG, Hubay CA, Insull W, Robinson AV, Hartman PH. Progression and regression of experimental atherosclerosis. *Surg Gynec Obstet* 1970; **131**:633–647.
147. Basta LL. Regression of atherosclerotic stenosing lesions of the renal arteries and spontaneous cures of systemic hypertension through control of hyperlipidemia. *Am J Med* 1976; **61**:420–423.

148. Crawford DW, Sanmarco ME, Blankenhorn DH. Spatial reconstruction of human femoral atheromas showing regression. *Am J Med* 1979; **66**:784-789.

149. Roth D, Kostuk WJ. Non-invasive and invasive demonstration of spontaneous regression of coronary artery disease. *Circulation* 1980; **62**:888–896.

150. Buchwald H, Moore RB, Rucker RD, Amplatz K, Castenada WR, Francoz RA, Pasternak AC, Varco RL. POSCH Arteriography Review Panel. Clinical angiographic regression of atherosclerosis after partial ileal bypass. *Atherosclerosis* 1983; **16**:117–124.

151. Hubbard JD, Inkiles S, Barnard RJ. Nathan Pritikin's Heart. *N Engl J Med* 1985; **313**:52,53.

152. Brown BG, Bolson EL, Dodge HT. Arteriographic assessment of coronary atherosclerosis: review of current methods, their limitations, and clinical applications. *Arteriosclerosis* 1982; **2**:2–15.

153. Bruschke AVG, Wijers TS, Kolsters W, Landmann J. The anatomic evolution of coronary artery disease demonstrated by coronary arteriography in 256 nonoperated patients. *Circulation* 1981; **63**:527–534.

154. Ost RC, Stenson S. Regression of peripheral atherosclerosis during therapy with high doses of nicotinic acid. *Scand J Clin Lab Invest* 1967; **99(Suppl.)**:241–245.

155. Buchwald H, Moore RB, Varco RL. Surgical treatment of hyperlipidemia. *Circulation* 1974; **49(Suppl. 1)**:1–12.

156. Barndt R, Blankenhorn DH, Crawford DW, Brooks SH. Regression and progression of early femoral atherosclerosis in treated hyperlipoproteinemic patients. *Ann Int Med* 1977; **86**:139–146.

157. Kuo PT, Hayase K, Kostis JB, Moreyra AE. Use of combined diet and colestipol in long-term (7–7-$\frac{1}{2}$ years) treatment of patients with hyperlipoproteinemia. *Circulation* 1979; **59**:199–207.

158. Rafflenbeul W, Smith LR, Rogers WJ, Mantle JA, Rackley CE, Russell RO. Quantitative coronary arteriography. Coronary anatomy of patients with unstable angina pectoris reexamined one year after optimal medical therapy. *Am J Cardiol* 1979; **49**:699–706.

159. Erikson U, Helmius G, Hemmingsson A, Ruhn G, Olsson AG. Measurement of atherosclerosis by arteriography and microdensitometry: model and clinical investigation. In: Schettler FG, Gotto AM, Middelhoff F, Habenicht AJ, Jurutka KR, eds. *Atherosclerosis.* Proceedings of the Sixth International Symposium on Atherosclerosis. Berlin: Springer Verlag, 1983.

160. Duffield RG, Lewis B, Miller NE, Jamieson CW, Brunt JN, Colchester AC. Treatment of hyperlipidaemia retards progres-

sion of symptomatic femoral atherosclerosis. A randomised controlled trial. *Lancet* 1983; **ii**:639–642.

161. Nikkila EA, Viikinkoski P, Valle M, Frick MH. Prevention of progression of coronary atherosclerosis by treatment of hyperlipidaemia: a 7-year prospective angiographic study. *Br Med J* 1984; **289**:220–223.

162. Brensike JF, Levy RI, Kelsey SF, et al. Effects of therapy with cholestyramine on progression of coronary arteriosclerosis: Results of the NHLBI Type II Coronary Intervention Study. *Circulation* 1984; **69**:313–324.

163. Blankenhorn D, Nessim S, Johnson R, Sanmarco M, Azen A, Cashin-Hemphill L. Beneficial effects of combined colestipol-niacin therapy on coronary atherosclerosis and coronary venous bypass grafts. *JAMA* 1987; **257**:3233–3240.

164. Cashin-Hemphill L, Sanmarco ME, Blankenhorn DH. Augmented beneficial effects of colestipol niacin therapy at four years in the CLAS trial [Abstract]. *Circulation* 1989; **80**:II–381.

165. Dintenfass L. Effect of a low-fat, low-protein diet on blood viscosity factors in patients with cardiovascular disorders. *Med J Australia* 1982; **1**:543,544.

166. Arntzenius AC, Kromhout D, Barth JD, et al. Diet, lipoproteins and the progression of coronary atherosclerosis. *N Engl J Med* 1985; **312**:805–811.

167. Ornish D, Brown SE, Scherwitz LW, et al. Can lifestyle changes reverse coronary heart disease? The Lifestyle Heart Trial. *Lancet* 1990; **ii**:129–133.

168. Barnard ND, Scherwitz LW, Ornish D. Adherence and acceptability of a low-fat, vegetarian diet among patients with cardiac disease. *J Cardiopulmonary Rehab* 1992; **12**:423–431.

169. Diehl HA. Does it really cost more to eat healthy? *Lifeline Health Lett* 1992; **7(6)**:10,11.

170. Ornish D. Can lifestyle changes reverse coronary atherosclerosis? *Hosp Pract* 1991 (May 15); 123–132.

171. Watts GF, Lewis B, Brunt JNH, et al. Effects on coronary artery disease of lipid-lowering diet, or diet plus cholestyramine, in the St. Thomas' Atherosclerosis Regression Study (STARS). *Lancet* 1992; **ii**:563–569.

172. Roberts WC. Lipid-lowering therapy after an atherosclerotic event. *Am J Cardiol* 1989; **64**:693–695.

173. Blankenhorn, DH. Atherosclerosis regression in humans. *Atherosclerosis Rev* 1990; **21**:151–157.

The Reversibility of Obesity, Diabetes, Hyperlipidemia, and Coronary Heart Disease

James W. Anderson
and Abayomi O. Akanji

1. Introduction

The most important cause of death in Western countries is cardiovascular disease. It accounts for about 36% of all deaths in the United States; in 1988, total health spending on prevention and treatment of heart disease was about $84 billion *(1,2)*. The situation is likely similar in the other developed northern European countries. Coronary heart disease (CHD) is easily the most frequent and most important cardiovascular disease in these countries in the second half of the twentieth century, with the eradication of infectious disease and consequent reduction in the prevalence of rheumatic and syphilitic heart disease, and with improved technology for the earlier detection and management of congenital heart disease. This high prevalence of CHD is directly attributable to lifestyle and dietary factors, most of which are modifiable *(3)*. The lifestyle-related risk factors are cigaret smoking, physical inactivity, and poor stress management; the diet-related factors are hyper-

From: *Western Diseases: Their Dietary Prevention and Reversibility*
Edited by: N. J. Temple and D. P. Burkitt Copyright ©1994 Humana Press, Totowa, NJ

lipidemia, hypertension, diabetes, and obesity (3). These factors have independent and additive effects not only in the genesis of CHD but also in the pathogenesis of the other factors. As an example, obesity can precipitate noninsulin-dependent diabetes mellitus (NIDDM) in the individual so predisposed; also the diabetic state can worsen atherogenic lipid profiles in individuals already genetically prone to hyperlipidemia. Indeed, with better recognition and management of some of these risk factors, the prevalence of CHD in some countries, including the United States, has declined in recent years (4). Probably also significant in this respect is the increasing role of a single factor, increased dietary fiber intake, in the prevention and control of all the diet-related risk factors.

In parts of the United States and western Europe, obesity prevalence rates approach 50% of the adult population (5). Additionally, many epidemiologic studies indicate NIDDM prevalence rates of 2–5% in the United States and in western Europe, which is responsible for about 2% of deaths in the United States (6,7). In certain populations in the South Pacific and the Pima Indians of the Arizona desert, NIDDM prevalence rates approach 50% of the adult population (7,8). Also, alarmingly, at least 50% of adult Americans have elevated plasma cholesterol levels (cholesterol >5.16 mmol/L) (9,10), although there is evidence that this high prevalence of hypercholesterolemia has declined significantly in the past 20 years (11). These statistical data therefore prove that populations in North America and northern Europe are presently in an epidemic of potent risk factors for CHD. Even more disturbing is an increasing percentage of children with high blood lipid levels (12), which suggests that the epidemic will persist for a long time unless drastic changes in dietary mores and lifestyle patterns are effected.

Indeed, the prevalence rates of some of these disorders, notably, hyperlipidemia and CHD, have declined in recent years, although they are still relatively high *(13)*. This is consequent on improved dietary awareness and adoption of healthy eating and living habits, including physical exercise. In our estimation the most important dietary change has been the adoption of diets high in fiber content. This is based on the hypothesis of Trowell *(14)*, proposed in 1975, that these Western diseases—diabetes, hyperlipidemia, and CHD—are essentially fiber deficiency disorders.

We have followed up on Trowell's suggestion that high fiber diets are most effective when they also contain a high carbohydrate content. These HCF diets have subsequently been shown to be effective in controlling most of the adverse risk factors for CHD, including diabetes, obesity, and elevated atherogenic blood lipid levels. Indeed, most health practitioners and professional medical organizations, such as the American Heart Association, American Diabetic Association, British Diabetic Association, Canadian Diabetic Association, and the European Association for the Study of Diabetes *(15–19)*, now routinely recommend diets high in complex carbohydrates (and fiber) and low in saturated fat. These dietary guidelines apply equally to the healthy population, and this has dramatically improved compliance in the high-risk population. There is increasing evidence that these measures, coupled with reduction in unhealthy habits such as smoking, and adoption of more physical exercise, result in improved physical health and amelioration of diabetes and hyperlipidemia as well as reduction in body weight. All these changes should, in the long term, reduce the prevalence of CHD.

This chapter reviews the effectiveness of these diets. It should be stated that the concept of regression is considered in its widest perspective. For instance, there are fairly objective measures for identifying changes in body weight in the obese, symptoms of angina in CHD, and reduction of plaque size in coronary atheromatous lesions. Regression in hyperlipidemia refers principally to a reduction in blood levels of the atherogenic lipids such as LDL-cholesterol and triglycerides and, to a lesser extent, an increase in HDL levels. With diabetes, which is essentially incurable, regression would indicate attainment and maintenance of normoglycemia as well as prevention or amelioration of the long-term complications without excessive drug intake.

2. Obesity

Quite recently, obesity has acquired increased recognition as a morbid state (5). It is well known to be associated with such conditions as degenerative joint disease, increased susceptibility to trauma, heart disease including hypertension, respiratory disorders including the Pickwickian syndrome, noninsulin-dependent diabetes mellitus, certain cancers, and obstetric complications. Indeed, the Framingham study showed that obesity is an important predictor of cardiovascular risk in men in whom there was at least a twofold increase in CHD risk with body weight over 130% ideal (20). The degree of obesity and pattern of excess body fat distribution can also indicate the extent of this increased susceptibility to cardiovascular disease caused by obesity, more so as abdominal (upper body) obesity has consistently been shown to be more dangerous than upper thigh (lower body) obesity (21,22).

The important metabolic abnormalities in obesity include increases in serum total and LDL-cholesterol, triglycerides and reduced HDL-cholesterol levels with increased liability to insulin resistance and NIDDM (23). Blood pressure also tends to be elevated (24). The main modalities of treatment have always been hypocaloric (weight-reducing) diets that are fiber-supplemented. These have been shown to reduce body weights and blood pressure as well as total and LDL-cholesterol levels, although HDL-cholesterol levels are also reduced (25). Studies with these high fiber, high carbohydrate diets containing 6–20 kcal/kg as energy, 70% energy as carbohydrate, 18% protein, 12% fat, and 30–40 g fiber/1000 kcal in obese diabetic subjects, showed weight loss to be directly correlated with severity of energy restriction, reaching up to about 1 kg/wk in those with marked energy restriction, a pattern similar to that described in individuals on very low calorie liquid diets (Fig. 1) *(26)*. Glycemic control was improved in the subjects, and there was a reduction in insulin requirements and blood cholesterol and triglyceride levels. Weight loss was maintained in the subjects for as long as 42 mo, whereas the improvement in glycemic and lipidemic control persisted even on weight maintenance diets containing 55% energy as carbohydrate, 20% protein, 25% fat, and 30–40 g fiber/1000 kcal *(26)*.

These hypocaloric diets supplemented with dietary fiber are also effective in nondiabetic subjects. A recent study in overweight, nondiabetic men and women reproduced the observations made in the diabetic subjects above, and, furthermore, indicated an extra benefit of exercise on dietary control in improving the blood lipid parameters, particularly the HDL-cholesterol levels *(27)*.

Various studies have indicated that soluble fiber supplements such as guar gum and oat bran are more effective in weight reduction than the insoluble fiber in wheat bran and methylcellulose *(28)*. Many supplements are available as over-the-counter laxative or bulking agents *(26)*.

3. Diabetes Mellitis

Diabetes mellitus is most prevalent among populations with low fiber intakes *(14,29)*. This and other observations led Trowell *(14)* to postulate that diabetes was a fiber deficiency disorder. Since the early work in 1976 of Jenkins et al. *(30)* and our group *(31)*, many workers have reported on the effectiveness of fiber in the management of diabetes and its complications. There have been numerous recent

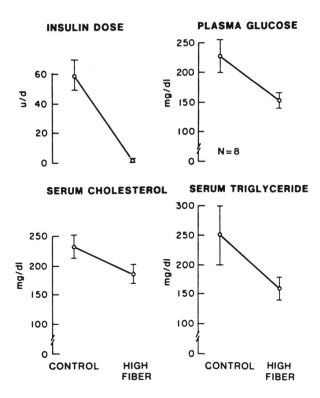

Fig. 1. Response of eight obese diabetic subjects to high-fiber, weight-reducing diets providing approx 6–8 kcal/kg. Values on control weight-maintaining and high-fiber, weight-reducing diets are presented.

reviews on the subject *(32–36)*. The usefulness of fiber in the prevention, amelioration, or retardation of the development of diabetes and its complications is, however, not yet established. It has been suggested that high fiber diets for subjects with diabetes may reduce:

1. The frequency of reactive hypoglycemia in 'chemical diabetics' *(37)*;
2. The vulnerability to atherothrombotic complications from their favorable effects on blood rheology and hemostatic variables *(38,39)*;

3. The aggregate risk factors for atherosclerosis *(13,40)*; and
4. The frequency of pain and discomfort associated with intermittent claudication *(41)*.

Other lifestyle changes also contribute to the development of the diabetic state. Certainly, changes in the consumption of other food components, especially saturated fat, often accompany increases in dietary fiber intake. Indeed, Swinburn and others *(42)* in a recent study among the Pima Indians in Phoenix, AZ, adduced evidence to the effect that the change in the traditional Pima Indian diet from one high in carbohydrate and fiber but low in fat content to a modern diet containing less carbohydrate and fiber but significantly more fat (up to 50% of total energy), within the past century, may have significantly contributed to the high prevalence of diabetes (about 50% of the adult population over 35-yr-old have NIDDM). A recent questionnaire study on the patterns of physical activity and other personal characteristics in relation to the development of NIDDM in 5990 male alumni of the University of Pennsylvania provides further proof of the importance of lifestyle patterns in the development of diabetes *(43)*. That study, which extended from 1962–1976, concluded that leisure time physical activity inversely correlated with the development of NIDDM, an association that persisted even when the data were adjusted for obesity, hypertension, and family history of diabetes. This suggests that NIDDM is associated with a sedentary lifestyle and may be prevented by physical exercise.

3.1. Clinical Studies

We recently reviewed the English language literature for clinical studies on diabetic subjects using either fiber supplements or diets generous in high fiber foods (herein termed high fiber diets). Although our detailed report is published elsewhere *(44)*, we shall summarize the essential findings here. The overall consensus is that intake of

high fiber, high carbohydrate, low saturated fat diets provide many benefits for diabetic patients. These diets lower
blood glucose concentrations, reduce postprandial insulin
levels, lower the requirements for antidiabetic drugs, and
decrease blood lipid concentrations. For lean individuals
with type I diabetes, these diets can reduce insulin requirements by 25–50% and improve glycemic control *(45)*. For
lean individuals with type II diabetes these diets can lower
insulin needs by 50% or more and often eliminate the need
for insulin *(45)*. For obese persons with diabetes, as indicated above, these highly satiating diets promote weight
loss and can provide reasonable glycemic control without
specific antidiabetic medication. These diets also lower
blood glycohemoglobin levels, plasma cholesterol levels
(20–30%), triglycerides (approx 15%), and blood pressure
(average of 10%) in patients with diabetes. However, these
effects are not consistently seen, and are still controversial *(46,47)*.

3.1.1. Fiber Supplemented Meals

We reviewed 35 publications published between 1976–
1990, which described the responses of diabetic subjects
to glucose loads or meals without or with fiber supplements
(44). The commonly used supplements were guar (13 studies), mixed high fiber (8 studies), pectin (6 studies), wheat
bran (5 studies), beans (4 studies), and cellulose/hemicel-
lulose (2 studies). Other useful supplements included cot-
tonseed fiber, soy fiber, apple powder, Mexican nopal
leaves, Indian fenugreek seeds, Japanese glucomannan,
and xanthan gum *(44)*. Water-soluble fibers such as guar,
pectin, and psyllium extract have greater effects on the glycemic response than do water-insoluble fibers such as cellulose and wheat bran. The glucose loads or meals supplemented
with these soluble fibers were followed by lower glycemic
responses in 30 of the studies and unchanged in 5; in no
study was the glycemic response worsened. Usually, insu-

lin and C-peptide levels were reduced, suggesting some improvement in insulin sensitivity. These findings suggest a probable beneficial effect of fiber supplements on long-term glycemic control.

3.1.2. Fiber-Supplemented Diets

We also reviewed 54 publications on the response of diabetic subjects (IDDM in 29 studies, NIDDM in 36 studies) to fiber-supplemented diets over 10–365 d *(44)*. Most investigators used guar (36 studies) or another fiber source rich in soluble fiber such as psyllium (2 studies), or apple fiber (2 studies); wheat bran was used in 9 studies. To improve palatability, and hence acceptability, these fiber supplements, especially guar, were incorporated into everyday foods such as bread, biscuits, and chocolate bars. Glycemic profile was reduced in 36 of the studies, unchanged in 12, whereas 10 did not comment on the glycemic status. Most of the studies suggested that fiber supplements lower blood levels of glycohemoglobin, plasma cholesterol (especially LDL-cholesterol), and triglyceride (variably) and reduce requirements for insulin or oral hypoglycemic agents. Changes in HDL-cholesterol levels and body weight change were inconsistent. The controlled studies of Uusitupa et al. *(48)* in NIDDM and Vaaler et al. *(49)* in IDDM demonstrate the effects of guar supplementation on long-term (3-month) glycemic control and blood cholesterol levels. Some studies suggested that wheat bran supplements also improve glycemic control and reduce the requirements for insulin and/or oral hypoglycemic agents *(49,50)*.

3.1.3. High Fiber Diets

We also reviewed 53 publications on the responses of diabetic subjects (22 studies in IDDM, 36 studies in NIDDM) to high fiber diets developed from high fiber foods rather than fiber supplements *(46)*. Most of the investigators used high carbohydrate, high fiber (HCF) diets, although some tested high fiber (HF) diets that were similar in carbohydrate content to the control diets. Our initial studies with HCF diets documented that they improve glycemic control, lower insulin requirements, and reduce blood lipid levels (31,51). These studies have been extended (52–55) and have also been confirmed by many other groups (44). It is now generally accepted that diabetic subjects derive distinct advantages from both HCF and HF diets, the former probably being more beneficial. A particularly useful study is that reported by O'Dea et al. (56; see following chapter by O'Dea), which investigated the effects of varying proportions of carbohydrate, fiber, and fat on metabolic control in NIDDM and concluded that high carbohydrate diets for diabetic patients should select carbohydrates that are unrefined and high in fiber.

When subjects with diabetes consume diets high in carbohydrate but deficient in fiber, glycemic control may become worse and plasma triglyceride levels increase with a reduction in HDL cholesterol levels (57,58). This metabolic profile is definitely atherogenic and therefore undesirable. It is thus essential that high carbohydrate diets are also fiber rich. If a subject is unable or unwilling to tolerate marked increases in fiber intake, an alternative diet containing less carbohydrate (about 40–50% total calories) and more unsaturated fat (about 30–40%, rich in monounsaturated fatty acids) could be offered. A high monounsaturated fat diet has been shown to reduce plasma triglyceride and increase HDL-cholesterol levels, and thus to reduce the potentially significant cardiovascular risk from hypertriglyceridemia recently identified with diets high in carbohydrate but deficient in fiber (59,60). The long-term effects of these monounsaturated fat diets, however, are still unknown (61), although they have been consumed in Mediterranean countries for centuries and are associated with low rates of cardiovascular mortality (40,60).

3.2. Use of Dietary Fiber
in Distinct Diabetes Groups

Various reports confirm that dietary fiber confers distinct advantages to both insulin dependent (IDDM) and noninsulin-dependent (NIDDM) subjects with diabetes, improving their glycemic and lipidemic control, irrespective of body weight (lean or obese), state of diabetic control or degree of compliance with treatment *(44)*. Special groups such as children *(62,63)*, pregnant women *(64,65)*, and geriatric subjects *(66,67)* also benefit from fiber supplementation. Dietary fiber confers additional benefit in the management of diabetic patients who are hypertensive *(68–70)*, have chronic renal failure *(71,72)*, or have hepatic encephalopathy from liver cirrhosis *(73)*.

4. Hyperlipidemia

There is a strong relationship between the blood cholesterol level and CHD risk; this important link is conferred by the LDL fraction *(74)*. On the other hand, the blood HDL-cholesterol is inversely related to CHD risk, probably because of the role of HDL in the transport of cholesterol from peripheral tissues to the liver *(75)*. An independent link between CHD and triglyceride levels is controversial *(76,77)*, although the Framingham study *(78)* and many other workers *(79,80)* have presented evidence for such a relationship. More recently, the focus of CHD risk prevention has shifted to apolipoprotein measurements. Apolipoprotein B found in LDL is positively related to CHD risk, whereas apolipoprotein A found predominantly in HDL inversely relates to risk. Apo E (in VLDL) has also been shown to be higher in patients with myocardial infarction than in controls *(81)*. It is thus obvious that elevated blood levels of LDL, triglycerides, and their respective apolipoproteins are atherogenic, whereas increased HDL and apo-

A levels reduce atherogenic risk. Measures to reduce the prevalence of CHD should therefore aim to appropriately change the levels of these respective blood lipids and apolipoproteins.

4.1. Dietary Fiber and Plasma Cholesterol Levels

Nutrition management is the treatment of choice for most individuals with hypercholesterolemia *(13)*. Intensive dietary education and advice on prudent diets containing increased amounts of dietary fiber result in a significant reduction in blood cholesterol levels, often exceeding 20% *(13)*. In long-term studies, we have demonstrated that water-soluble fiber from oat and bean products incorporated into diets low in fat and cholesterol cause a 26% or so reduction in serum cholesterol level in hypercholesterolemic men after 24 wk and 23% after 99 wk (Fig. 2) *(82)*. In the same time period, HDL-cholesterol levels increased over baseline by about 10%. Many other shorter-term studies reached essentially the same conclusions, using several different soluble fiber sources including guar gum, psyllium, and pectin *(9,44)*. Less consistent results were obtained when mixed fiber sources or insoluble fiber such as wheat bran or cellulose were used *(83)*. Generally, however, plasma total and LDL-cholesterol levels are reduced and HDL-cholesterol is unchanged or increased when soluble fiber is added to the diet.

4.2. Dietary Fiber and Plasma Triglyceride Levels

Dietary effects on the blood triglyceride levels are more controversial. We have consistently shown that high carbohydrate (about 70% energy) and high fiber diets (about 35–40 g/1000 kcal energy) favorably affect fasting and postprandial plasma triglyceride levels in patients with dia-

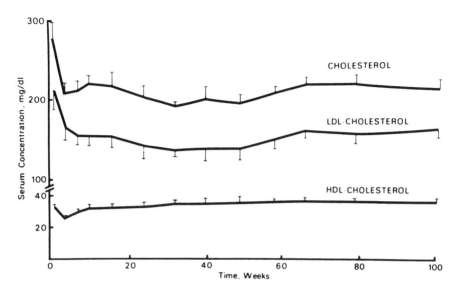

Fig. 2. Short-term and long-term response of ten hypercholes-terolemic men to high carbohydrate, high fiber (HCF) diets over 100 weeks. The initial 4 wk were on a metabolic research ward and the remaining 96 wk were as free-living subjects.

betes (Fig. 3) *(84)*; low fiber, high carbohydrate diets however may have different effects *(57,58)*. HCF diets may produce a short-term increase in plasma triglycerides in normotriglyceridemic individuals, but these diets usually lower values in hypertriglyceridemic subjects (Fig. 3) *(13)*. A reduction in plasma triglyceride levels of 10% was observed in NIDDM after 8 wk on a high fiber (46 g/d) diet in comparison to changes on a low fiber (14 g/d) diet *(85)*. This effect may not however be demonstrable if only insoluble fiber is used *(86)*. The Cholesterol Prevention Evaluation Program diet, which is low in saturated fat and cholesterol, moderate in polyunsaturated fat and high in carbohydrate, also demonstrated reductions in plasma triglyceride levels of 40–60% and total cholesterol levels of

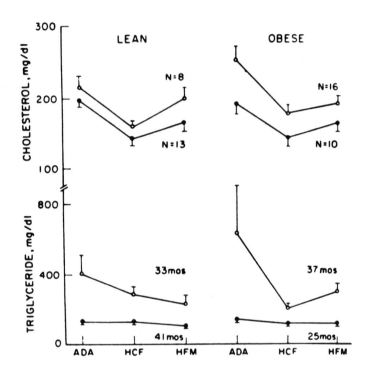

Fig. 3. Fasting serum lipid values for lean and obese diabetic subjects on American Diabetes Association (ADA), high-carbohydrate, high-fiber (HCF), and high-fiber maintenance (HFM) diets. Subjects were initially treated with HCF diets for 3 wk on a metabolic research ward and then followed for the indicated number of months on HFM diets. Number (N) of subjects in each group are indicated. The closed circles identify subjects whose initial fasting serum triglycerides were normal, whereas the open circles identify subjects who initially had hypertriglyceridemia.

30–40% over 2 yr in 119 men *(87)*. A similar study in hyper-triglyceridemic men but over 4 wk showed essentially the same effect on blood lipids *(88)*. The variable effects of high carbohydrate and fiber diets on plasma triglyceride levels are also seen with fiber supplements such as guar *(13)*.

In general, however, it is probably safe to advise high fiber diets in the management of hypertriglyceridemia. If the diet is high in carbohydrate but deficient in fiber, blood triglyceride levels could rise even higher and constitute a potent risk factor for CHD *(57,58)*. The suggestions highlighted above for diabetic individuals unable to tolerate a high fiber diet certainly also apply to nondiabetic subjects with hyperlipidemia. Some of the carbohydrate content could be replaced by fat, predominantly monounsaturated, similar to the Mediterranean diet *(40,60)* with the goal of reducing plasma triglyceride and increasing HDL levels *(59,60)*.

Another treatment that has received much attention in recent years is the incorporation of fish and marine oils as adjuncts in the dietary management of hyperlipidemia *(89,90)*. This development arose from epidemiologic studies in Greenland Eskimos that showed marine oils to be rich in n-3 fatty acids and to consistently lower plasma triglyceride levels and increase HDL levels even if only slightly; changes in LDL-cholesterol are variable and may even increase if the saturated fat intake is unchanged. Plasma apolipoprotein levels after dietary fish oil diets, change in concert with their associated lipoprotein cholesterol levels *(89,90)*.

5. Coronary Heart Disease

5.1. Epidemiology

Coronary heart disease mortality has declined significantly in the United States in recent years—up to 50% between the mid-1960s and 1985 *(4,74)*. This is likely to be caused by a combination of factors, including improvements in hypertension awareness and control, treatment of cardiovascular disease, and lifestyle management *(13)*.

The major lifestyle changes have been cessation of smoking (an independent risk factor for CHD) *(13)* and dietary modifications. In the same 20-yr period, per capita butter consumption decreased 33%, egg consumption decreased 12%, and animal fat and oil consumption decreased 39%. Simultaneously, there were increases in per capita consumption of vegetable oils (58%) and fish (23%). These changes, accompanied by improved media awareness of the need for increased dietary fiber intake, must have contributed to the decrease in mean cholesterol levels of the US adult population in recent years *(11,91)*.

Reducing the blood cholesterol level will clearly reduce the prevalence of CHD. A study of 14,919 American men enrolled in the Physicians' Health Study showed that the levels of total cholesterol and apolipoprotein B 100 were significantly associated with increased risk of myocardial infarction, whereas HDL, HDL2, and HDL3 levels were associated with a substantially decreased risk of myocardial infarction *(95)*. Even in a Chinese population with a low cholesterol concentration, Chen et al. *(96)* showed, in a prospective observational study based on 8–13 yr followup of 9021 men and women aged 35–64 yr at baseline, a direct relationship between serum cholesterol and death from CHD. The Chicago Western Electric Study over 19 yr showed that a drastically reduced cholesterol intake reduced the risk of death from any cause, including CHD, by 37% *(97)*.

It is thus clear that modifying the diet can favorably influence the levels of blood lipids, which in turn should result in reduction in prevalence rates of coronary atherosclerosis. Diets high in complex carbohydrate, vegetable protein, fish, and polyunsaturated fat are associated with a reduced risk for CHD *(13)*; conversely, diets high in total fat, cholesterol, saturated fat, and simple sugars are associated with increased CHD risk *(13,42,92)*. The intake of saturated fat correlates more closely with the prevalence of CHD than any other dietary variable *(93)*.

This contention is supported by the findings from the Ireland-Boston Diet–Heart study, which involved 1001 middle-aged Irish and Irish-American men in 3 cohorts and clearly showed that the mortality from CHD was higher in individuals taking increased amounts of saturated fat and cholesterol and reduced amounts of polyunsaturated fat and vegetable fiber (94).

It is likely that the addition of fiber to the diet will confer protection from CHD independent of its effect in reducing the cholesterol concentration. Liu et al. (98) in analyzing food consumption patterns from 20 economically advanced countries showed that fiber intake inversely correlated with CHD mortality rate. Khaw and Barrett-Connor (99) observed that increasing the daily fiber intake by 6 g resulted in a 25% reduction in CHD mortality. This observation is supported by other epidemiological studies, including those in vegetarians (100), in the Baltimore Longitudinal Study (101), in London bank and transport workers (102), in Dutch men (103), and in the Oslo study (104). Other population studies in Framingham (105), Puerto Rico (105,106), and Honolulu (92,105) also show that increasing the intake of dietary carbohydrate (though not specifically fiber) may independently and favorably modify the lipid risk factors for CHD.

It is therefore clear that a low saturated fat, high carbohydrate, high fiber diet can protect against CHD. We have prescribed this diet for the past 15 years. It also forms the basis of diets recommended by various diabetic associations (15–18), although the American Heart Association recommendation did not include a specific statement on the need to increase dietary fiber intake (19).

5.2. Regression of Established Lesions

Recent studies indicate that a reduction (17–37%) in CHD events is associated with an 8–10% reduction in plasma cholesterol levels by pharmacological methods (13). Different large-scale epidemiological diet–heart studies show that dietary change has a marked effect on CHD morbidity and mortality (13). These data offer additional proof of the effectiveness of hypolipidemic treatment (diet and/or drugs) in reducing CHD risk. Many of the drugs available are expensive and associated with potentially serious side effects (13). Dietary treatment alone, using high carbohydrate, high fiber, low fat diets, can reduce blood lipid levels to the extent seen with the currently available drugs and should be the first line of treatment for hyperlipidemic disorders. Dietary treatment, needless to say, is far cheaper and safer than drug treatment.

Apart from a clinical endpoint assessment of regression of coronary atherosclerotic lesions, angiographic methods are currently available to directly quantitate plaque size in coronary and other peripheral arteries such as the femoral (13). These methods are of low risk, and the angiographic films generated can be objectively analyzed by computer.

The major clinical regression trials with coronary angiographic outcome were recently summarized (13). Most of the study patients were treated with drugs and diet. However, one study, the Leiden Intervention Trial, involved patients treated only by dietary means (100). In this study, 39 patients with stable angina (and at least one coronary vessel narrowed by half) were given a vegetarian diet for 2 yr. Their body weight, blood pressure, total cholesterol, and ratio of total cholesterol to HDL cholesterol were decreased. Coronary lesions did not progress in 46% of the subjects. Coronary lesion growth correlated only with total/HDL cholesterol ratio. There was no lesion growth in individuals with ratios <6.9. This important study indicates that dietary modification alone can result in regression (or at least cessation of progression) of coronary atherosclerotic lesions.

The HCF diets can reduce total cholesterol levels to the extent achievable with drugs and, with variable changes in HDL levels, could reproduce the observations from the Leiden trial *(100)*. Dietary fiber therefore holds much promise in the prevention of CHD and the regression of established atheromatous lesions. Much further work in this area appears justified *(13)*.

6. Safety of Fiber Preparations

The long-term efficacy and safety of high fiber diets in diabetic subjects has been exhaustively investigated *(107–110)*. The consensus is that there is no significant risk. Vitamin, mineral, and trace element levels are generally unaffected, as is absorption of simultaneously administered drugs *(111,112)*. A common problem is acceptability and palatability of the diets, especially in light of the abdominal discomfort experienced by many patients. This has been partially obviated by the use of low dose guar preparations *(113,114)*, impregnation of guar with fructose for children *(115)*, and use of natural high-fiber foods rather than fiber-enriched supplements *(45)*. The major potential problems with long-term fiber supplementation are hypertriglyceridemia, especially with HCF diets *(57,58)* as indicated above, and small bowel obstruction *(116)*, although the latter appears quite rare.

7. Mechanism of Effect of Dietary Fiber

7.1. Weight Reduction

High fiber diets aid in weight reduction by improving satiety, prolonging food ingestion and gastric emptying times, reducing the intestinal absorptive efficiency and transit time, and improving insulin sensitivity by as yet unclear mechanisms *(26,53)*. These various mechanisms interact to reduce postprandial glycemia and triglyceride synthesis and deposition in adipose tissue.

7.2. Glycemic Control

High fiber diets, especially those with a high carbohydrate content, probably exert their major effect on glycemic control by improving insulin sensitivity. This can be inferred from the reduced need for antidiabetic medication in subjects on these diets. Insulin clamp studies are not consistent; they report that insulin sensitivity in these diabetic subjects is variously unchanged *(117,118)* or improved *(119)*. However, some studies *(120–122)* reported increased insulin binding to monocytes and adipocytes in subjects on HF diets. It has also been reported in both normal subjects and diabetic patients *(123–125)* that fiber taken at a meal can improve glycemic response to subsequent meals. Other possible mechanisms of action of fiber may be via modulation of the secretion of gut hormones *(126–128)* or the intermediary metabolic effects of short-chain fatty acids—acetate, propionate, butyrate—derived from the colonic fermentation of fiber in the diet *(129,130)*.

7.3. Blood Lipids

Dietary fiber probably acts to reduce blood lipid levels via many mechanisms, including modification of bile acid absorption and metabolism, interference with lipid absorption and metabolism, influence on hormone secretion (particularly insulin and the enteral hormones that influence peripheral glucose and lipid utilization), and increase in the production of short-chain fatty acids (i.e., acetate, propionate, and butyrate) from colonic bacterial fermentation *(13)*. Some of these mechanisms also occur with the glycemic effects of dietary fiber.

8. Conclusions

High-fiber diets offer numerous advantages to the subject with diabetes, obesity, hyperlipidemia, or coronary heart disease. They provide improved glycemic control, reduce blood levels of the atherogenic lipids, and may reduce body weight. These beneficial effects are even increased when these diets contain more complex carbohydrates.

For those groups of subjects with definitely increased risk for CHD and also for a health-conscious general population, we routinely recommend a prudent weight-maintaining diet containing (34,45):

1. 55–60% of energy from carbohydrate (two-thirds derived from the complex forms);
2. 12–16% of total calories from protein (or daily intake of 0.8 g/kg desirable body weight). This amount should be reduced in individuals with kidney disease (131);
3. Less than 30% of total calories from fat, consistent with the American Heart Association recommendations (19), with less than 10% saturated fat and daily cholesterol intake of less than 200 mg; and
4. Dietary fiber of about 40 g/d (or 15–25 g/1000 kcal), to include soluble and insoluble fiber from commonly available foods.

Other features of the everyday diet of the diabetic and nondiabetic individual, especially with respect to total calorie intake, use of alternative sweeteners, intake of salt, and vitamin and mineral supplements as well as alcohol ingestion could be considered on the basis of individual need, general clinical state, and common sense.

Strict adherence to high carbohydrate, high fiber, low fat diets should delay the onset, slow the progression, and, in some cases effect regression of CHD (13,100,132,133). Innovative clinical studies (56,70,134,135) and well-designed epidemiologic studies (136) indicate that diets rich in complex carbohydrates and low in fat protect from development of obesity, hypertension, and diabetes.

Acknowledgment

This research was supported in part by a grant from the National Institutes of Health (HL-36552).

References

1. US Department of Health and Human Services. *The Surgeon General's Report on Nutrition and Health.* Washington, DC: US Government Printing Office, 1988; 1–12.
2. American Heart Association. *Heart Facts.* Dallas: American Heart Association, 1988; 1.
3. Atherosclerosis Study Group. Optimal resource; 1s for primary prevention of atherosclerotic diseases. *Circulation* 1984; **70:**155A.
4. Stamler J. The marked decline in coronary heart disease mortality rates in the United States, 1968-1981; summary of findings and possible explanations. *Cardiology* 1985; **72:**11–22.
5. Burton BT, Foster WR. Health implications of obesity: NIH consensus development conference. *J Am Diet Assoc* 1985; **85:**1117–1121.
6. National Diabetes Advisory Board *Sixth Annual Report.* Washington, DC: US Department of Health and Human Services. NIH Publication No. 84-1587, May 1984.
7. King H, Zimmet P. Trends in the prevalence and incidence of diabetes: non-insulin dependent diabetes mellitus. *Wld Hlth Statist Quart* 1988; **41:** 190–196.
8. Knowler WC, Bennett PH, Hamman RH, Miller M. Diabetes incidence and prevalence in Pima Indians: a 19-fold greater incidence than in Rochester, Minnesota. *Am J Epidemiol* 1978; **108:** 497–505.
9. Anderson JW, Tietyen-Clark J. Dietary fiber hyperlipidemia, hypertension and coronary heart disease. *Am J Gastroenterol* 1986; **81:** 907–919.
10. Kannel WB, Castelli WP, Gordon T. Cholesterol in the prediction of atherosclerotic disease. New perspectives based on the Framingham study. *Ann Intern Med* 1979; **90:**85–91.
11. Burke GL, Sprafka JM, Folsom AR, Hahn LP, Luepker RV, Blackburn H. Trends in serum cholesterol levels from 1980–1987. The Minnesota Heart Survey. *N Engl J Med* 1991; **324:**941–946.
12. Newman WP, Freedman DS, Voors AW, Gard PD, Srinivasan SR, Creasanta JL, Williamson GD, Webber LS, Berenson GS. Relation of serum lipoprotein levels and systolic blood pressure to early atherosclerosis: the Bogalusa Heart Study. *N Engl J Med* 1986; **314:** 138–144.
13. Anderson JW, Deakins DA, Floore TL, Smith BM, Whitis SE. Dietary fiber and coronary heart disease. *CRC Critical Rev Food Sci Nutr* 1990; **29:** 95–147.

14. Trowell HC. Dietary fiber hypothesis of the etiology of diabetes mellitus. *Diabetes* 1975; **24**:762–765.

15. American Diabetes Association. Policy Statement. Nutritional recommendations and principles for individuals with diabetes mellitus: 1986. *Diab Care* 1987; **10**:126–132.

16. The Nutrition Sub-committee of the British Diabetic Association's Medical Advisory Committee. Dietary recommendations for diabetics for the 1980s: a policy statement by the British Diabetic Association. *Hum Nutr: Appl Nutr* 1982; **36A**:378–382.

17. Canadian Diabetes Association. Position Statement. Guidelines for the nutritional management of diabetes mellitus in the 1990's. *Beta Release* 1989; **13**:8,9.

18. Diabetes and Nutrition Study Group of the EASD. Nutritional recommendations for individuals with diabetes mellitus. *Diab Nutr Metab* 1988; **1**:145–149.

19. American Heart Association. Position Statement. Dietary guidelines for healthy American adults. *Circulation* 1988; **77**:721A–724A.

20. Hubert HB, Feinleb M, Mcnamara P, Castelli WP. Obesity as an independent risk factor for cardiovascular disease: a 26-year follow-up of participants in the Framingham Heart Study. *Circulation* 1983; **67**:968–977.

21. Kalkhoff RK, Hartz AH, Rulley D, Kissebah AH, Kelber S. Relationship of body fat distribution to blood pressure, carbohydrate intolerance and plasma lipids in healthy obese women. *J Lab Clin Med* 1983; **102**:621–627.

22. Larsson B, Svardsudd K, Welin L, Wilhemsen L, Bjorntorp P, Tibblin G. Abdominal adipose tissue distribution, obesity and risk of cardiovascular disease and death: 13-year follow-up of participants in a study of men born in 1913. *Br Med J* 1984; **288**:1401–1404.

23. Gordon T, Castelli WP, Hjortland MC, Kannel WB, Dawber TR. Diabetes, blood lipids and the role of obesity in coronary heart disease risk for women: the Framingham study. *Ann Intern Med* 1977; **87**:393–397.

24. Kannel WB, Gordon T. Physiological and medical concomitants of obesity: the Framingham Study. In: Bray GA, ed. *Obesity in America*. Washington, DC: US Department of Health, Education and Welfare, Public Health Service, National Institutes of Health, NIH Publication No. 79-359, 1979; 125–163.

25. Anderson JW. High fiber, hypocaloric vs. very-low-calorie diet effects on blood pressure of obese men. *Diabetes* 1986; **35(Suppl. 1):**217A.

26. Anderson JW, Bryant CA. Dietary fiber: diabetes and obesity. *Am J Gastroenterol* 1986; **81:**898.

27. Wood PD, Stefanick ML, Williams PT, Haskell WL. The effects on plasma lipoproteins of a prudent weight-reducing diet, with or without exercise, in overweight men and women. *N Engl J Med* 1991; **325:**461–466.

28. Krotkiewski M. Use of fibres in different weight reduction programs. In: Bjorntorp P, Vahouny GV, Kritchevsky D, eds. *Dietary Fiber and Obesity*. New York: Alan R Liss, 1984; 85–109.

29. Anderson JW. The role of dietary carbohydrate and fiber in the control of diabetes. *Adv Intern Med* 1980; **26:**67–95.

30. Jenkins MA, Goff DV, Leeds AR, Alberti KGMM, Wolever TMS, Gassull MA, Hockaday TDR. Unabsorbable carbohydrates and diabetes: decreased postprandial hyperglycaemia. *Lancet* 1976; **ii:**172–174.

31. Kiehm TG, Anderson JW, Ward K. Beneficial effects of a high carbohydrate, high fiber diet on hyperglycemic diabetic men. *Am J Clin Nutr* 1976; **29:**895–899.

32. Wahlquist ML. Dietary fiber and carbohydrate metabolism. *Am J Clin Nutr* 1987; **45:**1232–1236.

33. Vinik AI, Jenkins MA. Dietary fiber in management of diabetes. *Diab Care* 1988; **11:**160–173.

34. Anderson JW, Geil PB. New perspectives in nutrition management of diabetes mellitus. *Am J Med* 1988; **85:**159–165.

35. Council on Scientific Affairs, American Medical Association. Dietary fiber and health. *JAMA* 1989; **262:**542–546.

36. Anderson JW. Treatment of Diabetes with High Fiber Diets. In: Spiller GA, ed. *CRC Handbook of Dietary Fiber in Human Nutrition*. Boca Raton, FL: CRC Press, 1984; 349–359.

37. Monnier LH, Colette C, Aquirre L, Orsetti A, Combeaue D. Restored synergistic entero-hormonal response after addition of dietary fiber to patients with impaired glucose tolerance and reactive hypoglycaemia. *Diab Metab* 1982; **8:**217–227.

38. Simpson HC, Mann JI, Chakrabarti R, Imeson JD, Stirling Y, Tozer M, Woolf L, Meade TW. Effect of high fibre diet on haemostatic variables in diabetes. *Br Med J* 1982; **284:**1608.

39. Koepp P, Hegewisch S. Effect of guar on plasma viscosity and related parameters in diabetic children. *Eur J Pediatr* 1981; **137:**31–33.

40. Rivellese A, Riccardi G, Giacco A, Postiglione A, Mastranzo P, Mattioli PL. Reduction of risk factors for atherosclerosis in diabetic patients treated with a high-fiber diet. *Prev Med* 1983; **12**:128–132.

41. Pacy PJ, Dodson PM, Taylor MP. The effect of a high fiber, low fat, low sodium diet on diabetics with intermittent claudication. *Br J Clin Pract* 1986; **46**:313–317.

42. Swinburn BA, Boyce VL, Bergman RN, Howard BV, Bogardus C. Deterioration in carbohydrate metabolism and lipoprotein changes induced by modern, high fat diet in Pima Indians and Caucasians. *J Clin Endocrinol Metab* 1991; **73**:156–165.

43. Helmrich SP, Ragland DR, Leung RW, Paffenberger RS. Physical activity and reduced occurrence of non-insulin-dependent diabetes mellitus. *N Engl J Med* 1991; **325**:147–152.

44. Anderson JW, Akanji AO. Treatment of diabetes with high fiber diets. In: Spiller GA, ed. *CRC Handbook of Dietary Fiber.* Boca Raton, FL: CRC Press, 1992; 443–470.

45. Anderson JW. *Nutrition Management of Metabolic Conditions.* Lexington, KY: HCF Diabetes Research Foundation, 1981.

46. Hockaday TDR. Controversies in therapeutics. I. Natural fiber useful as part of total dietary prescription. *Br Med J* 1990; **300**:1334–1336.

47. Tattersall R, Mansell P. Controversies in therapeutics. 2. Benefits of fibre itself are uncertain. *Br Med J* 1990; **300**:1336,1337.

48. Uusitupa M, Siitonen O, Savolainen K, Silvasti M, Penttila I, Parvianen M. Metabolic and nutritional effects of long-term use of guar gum in the treatment of noninsulin-dependent diabetes of poor metabolic control. *Am J Clin Nutr* 1989; **49**:345–351.

49. Vaaler S, Hanssen KF, Dahl-Jorgensen K, Frolic W, Aaseth J, Odegaard H, Aagenaes O. Diabetic control is improved by guar gum and wheat bran supplementation. *Diab Med* 1986; **3**: 230–233.

50. Nygren C, Berglund O, Hallmans G, Lithner F, Taljehahl IB. The effect of a high bran diet on diabetes in mice and humans. *Acta Endocrinol* 1980; **94(Suppl. 237)**:66.

51. Anderson JW, Ward K. Long-term effects of high carbohydrate, high fiber diets on glucose and lipid metabolism: a preliminary report on patients with diabetes. *Diab Care* 1978; **1**: 77–82.

52. Anderson JW, Ward K. High carbohydrate, high fiber diets for insulin-treated men with diabetes mellitus. *Am J Clin Nutr* 1979; **32**:2312–2321.

53. Anderson JW, Sieling B. High fiber diets for obese diabetic patients. *Obesity/Bariatric Med* 1980; **9:**109–113.
54. Anderson JW, Chen WL, Sieling B. Hypolipidemic effects of high-carbohydrate, high fiber diets. *Metabolism* 1980; **29:** 551–558.
55. Anderson JW, Zeigler JA, Deakins DA, Floore TL, Dillon DW, Wood CL, Oeltgen PR, Whitley RJ. Metabolic effects of high carbohydrate, high fiber diets for insulin-dependent diabetic individuals. *Am J Clin Nutr* 1991; **54:**936–943.
56. O'Dea K, Traianedes K, Ireland P, Niall M, Sadler J, Hopper J, De Luise M. The effects of diet differing in fat, carbohydrate, and fiber on carbohydrate and lipid metabolism in type II diabetes. *J Am Diet Assoc* 1989; **89:**1076–1086.
57. Coulston AM, Hollenbeck CB, Swislocki ALM, Chen Y-D, Reaven GM. Deleterious metabolic effects of high carbohydrate, sucrose containing diets in patients with non-insulin dependent diabetes mellitus. *Am J Med* 1987; **82:**213–220.
58. Coulston AM, Hollenbeck CB, Swislocki ALM, Reaven GM. Persistence of hypertriglyceridemic effect of low-fat high-carbohydrate diets in NIDDM patients. *Diab Care* 1989; **12:**94–101.
59. Garg A, Bonanome A, Grundy SM, Zhang Z-J, Unger RH. Comparison of a high-carbohydrate diet with a high monounsaturated fat diet in patients with noninsulin-dependent diabetes mellitus. *N Engl J Med* 1988; **319:**829–834.
60. Riccardi G, Rivellese AA. Dietary fiber and carbohydrate: effects on glucose and lipoprotein metabolism in diabetic patients. *Diab Care* 1991; **14:**115–125.
61. Anderson JW, Akanji AO. Dietary fiber—an overview. *Diab Care* 1991; **14:**1126–1131.
62. Kinmonth AL, Angus RM, Jenkins A, Smith MA, Baum JD. Whole foods and increased dietary fiber improve blood glucose control in diabetic children. *Arch Dis Child* 1982; **57:**187–194.
63. Lindsay AN, Hardy A, Jarrett L, Rallison ML. High carbohydrate, high fiber diet in children with type I diabetes. *Diab Care* 1984; **7:** 63–67.
64. Kuhl C, Molsted-Pederson L, Hornnes PJ. Guar gum and glycemic control of pregnant insulin-dependent diabetic patients. *Diab Care* 1983;**6:**152–154.
65. Ney D, Hollingsworth DR, Cousins L. Decreased insulin requirement and improved control of diabetes in pregnant women given a high carbohydrate, high fiber, low fat diet. *Diab Care* 1982; **5:** 529–533.

66. Kyllastinen M, Lahikainen T. Long-term dietary supplementation with a fiber product (guar gum) in elderly diabetics. *Curr Ther Res* 1982; **30**:872–879.

67. Sels JP, Flendrig JA, Postmes Th J. The influence of guar gum bread on the regulation of diabetes mellitus type II in elderly patients. *Br J Nutr* 1987; **57**:177–183.

68. Dodson PM, Pacy PJ, Beevers M, Bal P, Fletcher RF, Taylor KG. The effects of a high fiber, low fat and low sodium dietary regime on diabetic hypertensive patients of different ethnic groups. *Postgrad Med J* 1983; **59**:641–644.

69. Pacy PJ, Dodson PM, Kubicki AJ, Fletcher RF, Taylor KG. Comparison of the hypotensive and metabolic effects of bendrofluazide therapy and a high fiber, low fat, low sodium diet in diabetic subjects with mild hypertension. *J Hypertension* 1984; **2**:215–220.

70. Dodson PM, Pacy PJ, Bal P, Kubicki AJ, Fletcher RF, Taylor KG. A controlled trial of a high fiber, low fat and low sodium diet for mild hypertension in type 2 (non-insulin-dependent) diabetic patients. *Diabetologia* 1984; **27**:522–526.

71. Rivellese A, Parillo M, Giacco A, Marco FD, Riccardi G. A fiber-rich diet for the treatment of diabetic patients with chronic renal failure. *Diab Care* 1984; **8**:620,621.

72. Parillo M, Riccardi G, Pacioni D, Iovine C, Contaldo F, Carmela I, De Marco F, Perrotti N, Rivellese A. Metabolic consequences of feeding a high-carbohydrate, high-fiber diet to diabetic patients with chronic kidney failure. *Am J Clin Nutr* 1988; **48**:255–259.

73. Uribe M, Dibildox M, Malpica S, Guillermo E, Villallobos A, Nieto L, Vargas F, Ramos G. Beneficial effect of vegetable protein diet supplemented with psyllium plantago in patients with hepatic encephalopathy and diabetes mellitus. *Gastroenterology* 1985; **88**:901–907.

74. The Lipid Research Clinic Program. The Lipid Research Clinic Coronary Primary Prevention Trial results: reduction in incidence of coronary heart disease. *JAMA* 1984; **251**:351–364.

75. Castelli WP, Doyle JT, Gordon T, Hames CG, Hjortland MC, Hulley SB, Kagan A, Zukel WJ. HDL cholesterol and other lipids in coronary heart disease: the Cooperative Lipoprotein Phenotyping Study. *Circulation* 1977; **55**:767–772.

76. Hulley SB, Rosenman RH, Bawol RD, Brand RJ. Epidemiology as a guide to clinical decisions: the association between triglyceride and coronary heart disease. *N Engl J Med* 1980; **302**:1383–1389.

77. Wilhemsen L, Wedel H, Tibblin G. Multivariate analysis of risk factors for coronary heart disease. *Circulation* 1973; **48**: 950–998.

78. Gordon T, Castelli WP, Hjortland MC, Kannel WB, Dawber TR. High density lipoprotein as a protective factor against coronary heart disease: the Framingham study. *Am J Med* 1977; **62**:707–714.

79. Aberg H, Lithell H, Selinius I, Hedstrand H. Serum triglycerides are a risk factor for myocardial infarction but not for angina pectoris. *Atherosclerosis* 1985; **54**:89–97.

80. West KM, Ahuja MMS, Bennett PH, Czyzyk A, De Acosta OM, Fuller JH, Grab B, Grabauskas V, Jarrett RJ, Kosaka K, Keen H, Krolewski AS, Miki E, Schliack V, Teuscher A, Watkins PJ, Stober JA. The role of circulating glucose and triglyceride concentrations and their interactions with other 'risk factors' as determinants of arterial disease in nine diabetic population samples from the WHO multinational study. *Diab Care* 1983; **6**:361–369.

81. Mahley RW. Apolipoprotein E: cholesterol transport protein with expanding role in cell biology. *Science* 1988; **240**:622–630.

82. Anderson JW, Story L, Sieling B, Chen WL, Petro MS, Story J. Hypocholesterolemic effects of high fiber diets rich in water-soluble plant fibers: long term studies with oat bran and bean-supplemented diets for hypercholesterolemic men. *J Can Diet Assoc* 1984; **45**:140–149.

83. McIvor ME, Cummings CC, Van Duyn MA, Leo TA, Margolis S, Behall KM, Michnowski JE, Mendeloff AI. Long-term effects of guar gum on blood lipids. *Atherosclerosis* 1986; **60**:7–13.

84. Anderson JW, Chen W-JL. Plant fiber. Carbohydrate and lipid metabolism. *Am J Clin Nutr* 1979; **32**:346–363.

85. Hagander B, Asp N-G, Efendic S, Nilsson-Ehle P, Scherstein B. Dietary fiber decreases fasting blood glucose levels and plasma LDL concentration in noninsulin-dependent diabetes mellitus patients. *Am J Clin Nutr* 1988; **47**:852–858.

86. Hollenbeck CB, Coulston AM, Reaven GM. To what extent does increased dietary fiber improve glucose and lipid metabolism in patients with noninsulin-dependent diabetes mellitus (NIDDM). *Am J Clin Nutr* 1986; **43**:16–24.

87. Hall Y, Stamler J, Cohen DB, Mojonnier L, Epstein MB, Berkson DM, Whipple IT, Cathcings S. Effectiveness of a low saturated fat, low cholesterol, weight reducing diet for the control of hypertriglyceridemia. *Atherosclerosis* 1972; **16**: 389–403.

88. Sommariva D, Scotti L, Fasoli A. Low fat diet versus low carbohydrate diet in the treatment of type IV hyperproteinemia. *Atherosclerosis* 1978; **29**:43–51.

89. Kinsella JE. Effects of polyunsaturated fatty acids on factors related to cardiovascular disease. *Am J Cardiol* 1987; **60:** 23G–32G.

90. Harris WS. Fish oils and plasma lipid and lipoprotein metabolism in humans: a critical review. *J Lipid Res* 1989; **30:** 785–807.

91. National Center for Health Statistics—National Heart, Lung and Blood Institute Collaborative Lipid Group: Trends in serum cholesterol levels among US adults aged 20–74 years— Data from the National Health and Nutrition Examination Surveys, 1960–1980. *JAMA* 1987; **257:**937–942.

92. McGee DL, Reed DM, Yano K, Kagan A, Tillotson J. Ten year incidence of coronary heart disease in the Honolulu Heart Program. *Am J Epidemiol* 1984; **119:**667–676.

93. Hegsted DM, Ausman LM. Diet, alcohol and coronary heart disease. *J Nutr* 1988; **118:**1184–1189.

94. Kushi LH, Lew RA, Stare FJ, Ellison CR, Lozy ME, Bourke G, Daly L, Graham I, Hickey N, Mulcahy R, Kevaney J. Diet and 20-year mortality from coronary heart disease. The Ireland-Boston Diet–Heart Study. *N Engl J Med* 1985; **312:**811–818.

95. Stampfer MJ, Sacks FM, Salvini S, Willett WC, Hennekens CH. A prospective study of cholesterol, apolipoproteins and the risk of myocardial infarction. *N Engl J Med* 1991; **325:**373–381.

96. Chen Z, Peto R, Collins R, MacMahon S, Lu J, Li W. Serum cholesterol concentration and coronary heart disease in population with low cholesterol concentrations. *Br Med J* 1991; **303:**276–282.

97. Stamler JS, Shekelle R. Dietary cholesterol and human coronary heart disease. *Arch Pathol Lab Med* 1988; **112:**1032–1040.

98. Liu K, Stamler J, Trevisan M, Moss D. Dietary lipids, sugar, fiber, and mortality from coronary heart disease. *Arteriosclerosis* 1982; **2:**221–227.

99. Khaw KT, Barrett-Connor E. Dietary fiber and reduced ischemic heart disease mortality rates in men and women: a 12 year prospective study. *Am J Epidemiol* 1987; **126:** 1093–1102.

100. Arntzenius AC, Kromhout D, Barth JD, Reiber JHC, Bruschke AVG, Buis B, vanGent CM, Kempen-Voogd N, Strikwerda S, van der Velde EA. Diet, lipoproteins and the progression of coronary atherosclerosis. The Leiden Intervention Trial. *N Engl J Med* 1985; **312:**805–811.

101. Hallfrisch J, Tobin JD, Muller DC, Andres R. Fiber intake, age and other coronary risk factors in men of the Baltimore Longitudinal Study (1959–1975). *J Gerontol Med Sci* 1988; **43M:**64–68.

102. Morris JN, Marr JW, Clayton DG. Diet and heart: a postscript. *Br Med J* 1977; **2**:1307–1314.
103. Kromhout D, Bosschieter EB, Coulender CDL. Dietary fiber and 10-year mortality for coronary heart disease, cancer and all other causes. *Lancet* 1982; **ii**:518–522.
104. Hjermann I. Dietary prevention of coronary heart disease. *Biblthca Nutr Dieta* 1987; **40**:28–32.
105. Gordon T, Kagan A, Garcia-Palmieri M, Kannel WB, Zukel WJ, Tillotson J, Sorlie P, Hjortland M. Diet and its relation to coronary heart disease and death in three populations. *Circulation* 1981; **63**:500–515.
106. Garcia-Palmieri MR, Sorlie P, Tillotson J, Costas R, Rodriguez M. Relationship of dietary intake to subsequent coronary heart disease incidence: the Puerto Rico Heart Health Program. *Am J Clin Nutr* 1980; **33**:1818–1827.
107. Ray TK, Mansell KM, Knight LC, Malmud LS, Owen OE, Boden G. Long-term effects of dietary fiber on glucose tolerance and gastric emptying in noninsulin-dependent diabetic patients. *Am J Clin Nutr* 1983; **37**:376–381.
108. Van Duyn MAS, Leo TA, McIvor ME, Behall KM, Michnowski JE, Mendeloff AI. Nutritional risk of high-carbohydrate, guar gum dietary supplementation in non-insulin-dependent diabetes mellitus. *Diab Care* 1986; **9**:497–503.
109. Behall KM, Scholfield W, McIvor ME, Van Duyn MS, Leo TA, Michnowski JE, Cummings CC, Mendeloff AI. Effect of guar gum on mineral balances in NIDDM adults. *Diab Care* 1989; **12**:357–364.
110. Garg A, Bonanome A, Grundy SM, Unger RH, Breslau NA, Pak CYC. Effects of dietary carbohydrates on metabolism of calcium and other minerals in normal subjects and patients with non-insulin-dependent diabetes mellitus. *J Clin Endocrinol Metab* 1990; **70**:1007–1013.
111. Huupponen R, Karhuvaara S, Seppala P. Effect of guar gum on glipizide absorption in man. *Eur J Clin Pharmacol* 1985; **28**:717–719.
112. Huupponen R. The effect of guar gum on the acute metabolic response to glyburide. *Res Commun Chem Path Pharmacol* 1986; **54**:137–140.
113. Jones DB, Slaughter P, Jelfs R, Lousley S, Carter RD, Mann JI. Low dose guar improves diabetic control. *J R Soc Med* 1985; **78**:546–548.
114. Peterson DB, Ellis PR, Baylis JM, et al. Low dose guar in a novel food product: improved metabolic control in non-insulin-dependent diabetes. *Diab Med* 1987; **4**:111–115.

115. Paganus A, Maenpaa J, Akerblom HK, Stenman U-H, Knip M, Simell O. Beneficial effects of palatable guar and guar plus fructose diets in diabetic children. *Acta Paediatr Scand* 1987; **76**:76–81.

116. Miller DL, Miller PF, Dekker JJ. Small bowel obstruction from bran cereal. *JAMA* 1990; **263**:813–814.

117. Ebeling P, Yki-Jarvinen H, Aro A, Helve E, Sinisalo M, Koivisto VA. Glucose and lipid metabolism and insulin sensitivity in type I diabetes: the effect of guar gum. *Am J Clin Nutr* 1988; **48**:98–103.

118. Nestel PJ, Nolan C, Bazelmans J, Cook R. Effects of a high-starch diet with low or high fiber content on postabsorptive glucose utilization and glucose production in normal subjects. *Diab Care* 1984; **7**:207–210.

119. Fukagawa NK, Anderson JW, Hageman G, Young VR, Minaker KL. High carbohydrate, high fiber diets increase peripheral insulin sensitivity in healthy young and old adults. *Am J Clin Nutr* 1990; **52**:524–528.

120. Pedersen O, Hjollund E, Lindkov HO, Helms P, Sorensen NS, Ditzel J. Increased insulin receptor binding to monocytes from insulin dependent diabetic patients after a low fat, high starch, high fiber diet. *Diab Care* 1982; **5**:284–291.

121. Ward GM, Simpson RW, Simpson HCR, Naylor BA, Mann JI, Turner RC. Insulin receptor binding increased by high carbohydrate low fat diet in non-insulin-dependent diabetics. *Eur J Clin Invest* 1982; **12**:93–96.

122. Hjollund E, Pedersen O, Richelsen B, Beck-Nielsen H, Sorensen NS. Increased insulin binding to adipocytes and monocytes and increased insulin sensitivity of glucose transport and metabolism in adipocytes from non-insulin-dependent diabetics after a low-fat/high starch/high fiber diet. *Metabolism* 1983; **32**:1067–1075.

123. Nestler JE, Barlascini CO, Clore JN, Blackard WG. Absorption characteristic of breakfast determines insulin sensitivity and carbohydrate tolerance for lunch. *Diab Care* 1988; **11**:755–760.

124. Trinick TR, Laker MF, Johnston DG, Keir M, Buchanan KD, Alberti KGMM. Effect of guar on second-meal glucose tolerance in normal man. *Clin Sci* 1986; **71**:49–55.

125. Sundell IB, Hallmans G, Nilsson TK, Nygren C. Plasma glucose and insulin, urinary catecholamine and cortisol responses to test breakfasts with high or low fiber content: the importance of the previous diet. *Ann Nutr Metab* 1989; **33**:333–340.

126. Morgan LM, Goulder TJ, Tsiolakis D, Marks V, Alberti KGMM. The effect of unabsorbable carbohydrate on gut hormones. *Diabetologia* 1979; **17:**85–89.

127. Groop P-H, Groop L, Totterman KJ, Fyhrquist F. Relationship between changes in GIP concentrations and changes in insulin and C-peptide concentration after guar gum therapy. *Scand J Clin Lab Invest* 1986; **46:**505–510.

128. Sestoft L, Krarup T, Palmvig B, Meinertz H, Faergeman O. High carbohydrate, low fat diet: effect on lipid and carbohydrate metabolism, GIP and insulin secretion in diabetics. *Danish Med Bull* 1985; **32:** 64–69.

129. Anderson JW, Bridges SR. Short-chain fatty acid fermentation products of plant fiber affect glucose metabolism of isolated rat hepatocytes. *Proc Soc Exp Biol Med* 1984; **177:**372–376.

130. Akanji AO, Peterson DB, Humphreys S, Hockaday TDR. Change in plasma acetate levels in diabetic subjects on mixed high fiber diets. *Am J Gastroenterol* 1989; **84:**1365–1370.

131. Ciavarella A, Gianfranco DM, Stefoni S, Borgnino LC, Vannini P. Reduced albuminuria after dietary protein restriction in insulin-dependent diabetic patients with clinical nephropathy. *Diab Care* 1987; **10:**407–413.

132. Ornish D, Brown SE, Scherwitz LW, Billings JH, Armstrong WT, Ports TA, McLanahan SM, Kirkeeide RL, Brand RJ, Gould KL. Can lifestyle changes reverse coronary heart disease? *Lancet* 1990; **336:**129–133.

133. Watts GF, Lewis B, Brunt JNH, Lewis ES, Coltart DJ, Smith LDR, Mann JI, Swan AV. Effects on coronary artery disease of lipid-lowering diet, or diet plus cholestyramine, in The St. Thomas' Atherosclerosis Regression Study (STARS). *Lancet* 1992; **339:**563–569.

134. McMurry MP, Cerqueira MT, Connor SL, Connor WE. Changes in lipid and lipoprotein levels and body weight it Tarahumara Indians after consumption of an affluent diet. *N Engl J Med* 1991; **325:**1704–1708.

135. Shintani TT, Hughes CK, Beckham S, O'Connor HK. Obesity and cardiovascular risk intervention through the ad libitum feeding of traditional Hawaiian diet. *Am J Clin Nutr* 1991; **53:**1647S–1651S.

136. Marshall JA, Hamman RF, Baxter J. High-fat, low-carbohydrate diet and the etiology of non-insulin-dependent diabetes mellitus: The San Luis Valley diabetes study. *Am J Epidemiol* 1991; **134:**590–603.

The Therapeutic and Preventive Potential of the Hunter-Gatherer Lifestyle

Insights from Australian Aborigines

Kerin O'Dea

Prehistorians believe that Aborigines came to Australia from Southeast Asia at least 40,000–50,000 years ago *(1)*. Until European colonization of Australia just over 200 years ago, Aborigines lived as hunter-gatherers all over the continent under widely varying geographic and climatic conditions, ranging from the tropical coastal regions of the north (latitude 11–20° S), through the vast arid regions of the center (latitude 20–30° S), to the cool-temperate regions of the south (latitude 30–43° S). The more fertile coastal areas, both north and south, could sustain larger populations than the arid inland or desert areas. Each tribal group hunted and gathered food in a defined territory, which could be as vast as 100,000 sq km in the desert regions or as small as 500 sq km in fertile coastal country *(2)*.

1. Western Diseases in the Contemporary Australian Aboriginal Population

Most Aborigines in Australia today live a Westernized lifestyle, deriving their diet completely, or in large part, from Western foods, and leading sedentary, physically

From: *Western Diseases: Their Dietary Prevention and Reversibility*
Edited by: N. J. Temple and D. P. Burkitt Copyright ©1994 Humana Press, Totowa, NJ

inactive lives. Under these circumstances, they develop extremely high prevalence rates of obesity and noninsulin-dependent diabetes mellitus (NIDDM) *(3–10)* (Table 1). They share this vulnerability with other populations around the world that have been subjected to a similar rapid lifestyle change in the twentieth century: Examples include the Pima Indians and other native American Indians *(11)*, Nauruans and other Pacific Islanders *(12)*, the multiethnic groups in Mauritius *(13)*, and emigrant Indian populations *(14)*. Examples of such populations are now so numerous that it was recently suggested that Europeans may be the exception among the world's diverse population groups in being relatively resistant to the diabetogenic effects of the Western lifestyle *(15)*.

1.1. Noninsulin-Dependent Diabetes

Diabetes prevalence in Australian Aborigines is highest in the most Westernized communities *(3,4,8,9)* and lower in the least Westernized *(7,10)* or those with significant non-Aboriginal genetic admixture *(5)*. As illustrated in Fig. 1, Aborigines not only develop much higher prevalence rates of diabetes than Australians of European descent, but the average age of onset is also much younger. Cases of NIDDM have been diagnosed in Aboriginal teenagers and the prevalence reaches its maximum at about the age of 40—which is about 30 years earlier than in Europeans *(16)*. In the 20–50-year age group the prevalence of diabetes is more than ten times higher in Aborigines than in Europeans. As such, it represents an extremely serious public health problem.

1.2. Obesity

Obesity is very common in most Westernized Aboriginal communities *(6)*. Average body mass index (BMI) increases steadily over the 15–45-year age range in both

Table 1
Lifestyle-Related Chronic Diseases
in Aborigines After Westernization

Android obesity
Hypertension
Body weight and blood pressure increase with age
Noninsulin-dependent diabetes mellitus
Coronary heart disease
Elevated triglycerides, low HDL-cholesterol levels
Hyperinsulinemia
Insulin resistance

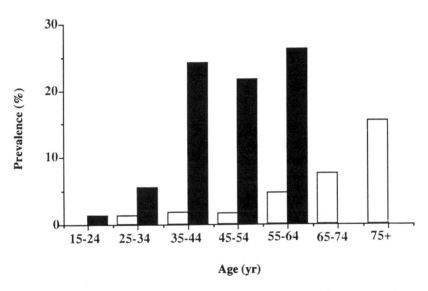

Fig. 1. Diabetes prevalence as a function of age in white Australians (open bars *n* = 3197, ≥25 yr [Busselton Survey]) *(16)* and Aborigines from six communities in northern and central Australia (black bars *n* = 1345, ≥15 yr). There were insufficient numbers in the Aboriginal cohort to include data from those aged ≥65 yr.

men and women, with more than 60% of those over 35 years of age being overweight or obese (BMI > 25 kg/m^2). Both men and women have an android pattern of fat distribution (i.e., fat deposited in the abdominal region rather than on the thighs and hips), with little difference between the sexes. For example, the range of waist-to-hip ratios of the adults in one community in northern Australia was 0.74–1.07 in the women and 0.79–1.10 in the men, and all overweight people had an android pattern of fat distribution (9). Furthermore, anthropometric measurements indicate that Australian Aborigines have a "linear" body build, with relatively long legs and short trunks, and narrow across the shoulders, chest, and hips. There is evidence that for a given BMI, Aboriginal women have significantly more body fat than European women (17), consistent with their linear, lighter build.

1.3. Hyperinsulinemia

Hyperinsulinemia (both in the fasting state and in response to oral glucose) occurs with high frequency in Australian Aborigines. As illustrated in Fig. 2 there is a strong association between insulin and glucose levels 2 h after 75 g oral glucose in the range 4–12 mmol/L glucose, consistent with increasing insulin resistance over this concentration range. Above a 2 h glucose of 12 mmol/L the 2 h insulin concentrations fell off sharply, consistent with pancreatic β-cell dysfunction once clear-cut diabetes had developed.

1.4. Syndrome X

Associated with the high prevalence of android obesity, abnormal glucose tolerance, and hyperinsulinemia are high frequencies of dyslipidemia (elevated VLDL and total triglycerides, and low HDL-cholesterol levels), hypertension, and coronary heart disease (3–9). These metabolic and physiologic characteristics may all be linked to insu-

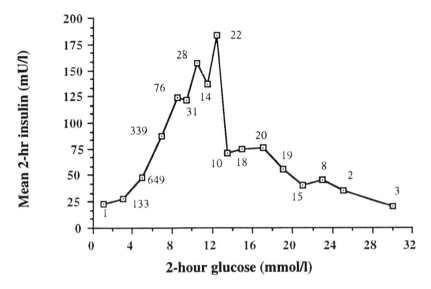

Fig. 2. The relationship between mean glucose and insulin concentrations 2 h after a 75 g oral glucose load in Aborigines from six communities in northern and central Australia.

lin resistance. We have observed that they frequently occur together in individual Aborigines—as has been observed in other populations *(18)*, and the cluster is now described as Syndrome X *(19)*.

2. Aborigines as Hunter-Gatherers

2.1. Body Weight and Blood Pressure

There is no evidence that Aborigines experienced any of these Western diseases when they lived as hunter-gatherers (Table 2). Although there is little reliable quantitative data on the health of Aborigines with little or no European contact, numerous early reports described them as lean and apparently physically fit. Most information on the health and traditional lifestyle of Australian Aborigines has come from the study of groups who continued living as

Table 2
Health of Australian Aborigines
as Hunter-Gatherers

Physically fit
Lean (body mass index < 20 kg/m²)
Blood pressure at the low end of the normal range
No age-related increase in BMI or blood pressure
Low fasting glucose
Low fasting cholesterol
No evidence of diabetes or coronary heart disease

hunter-gatherers well into the twentieth century *(20)*. These people lived in the most remote and, to many white Australians, inhospitable parts of the country in the center and north of the continent (Fig. 3). Data collected from such groups indicate that they were extremely lean (BMI in the range 14–20 kg/m²) and had blood pressure at the low end of the normal range (systolic b.p. usually below 110 and diastolic b.p. usually below 70 mm Hg), and neither body weight nor blood pressure increased with age (Table 2). These data are in striking contrast to those from Westernized Aborigines and Australians of European descent, but similar to data from other traditionally-living populations *(21)*, which indicate that increasing blood pressure is not an inevitable consequence of aging, but rather a consequence of Western lifestyle (possibly owing to the age-related increase in body weight).

2.2. Other Risk Factors for NIDDM and Coronary Heart Disease (CHD)

2.2.1. Studies in Traditionally Oriented Groups

There are no published data on glucose tolerance or lipid levels in Aborigines living traditionally as hunter-gatherers, but studies in the least Westernized groups who have continued to live a partially traditional lifestyle indi-

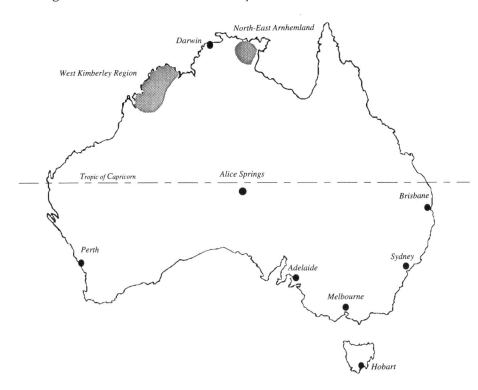

Fig. 3. Map of Australia showing the West Kimberley Region and North-East Arnhemland.

cate that they were also protected from diabetes and hyperlipidemia. We had the opportunity to examine this question recently (10) as part of a long-term study by White (22) on the relationship between lifestyle and health in small traditionally oriented groups in northeast Arnhem Land. By standard criteria, all adults were "underweight," with BMI in the range 13.4–19.3 kg/m². Despite this, they displayed no biochemical evidence of malnutrition: None were anemic and red cell folate levels were all within the normal range for European Australians (a unique observation in itself).

The mean fasting glucose levels (3.8 ± 0.4 mmol/L) and fasting cholesterol levels (3.9 ± 0.2 mmol/L) in the 18 adults were low relative to Westernized Aborigines and Europeans, and typical of nonwesternized populations elsewhere *(23)*. However, their fasting insulin levels (13 ± 4 mU/L) were similar to other more Westernized young Aboriginal men and higher than those of European men who had a higher mean BMI (about 21 kg/m²) *(7)*. These data suggest that, in view of their very low BMI (about 17 kg/m²) and low fasting glucose levels, the insulin levels of these traditionally oriented Aborigines were inappropriately elevated. Their fasting triglyceride concentrations (1.13 ± 0.09 mmol/L), although within the normal range for Europeans, were higher than would be expected in view of their extreme leanness, but consistent with insulin resistance. This indirect evidence for insulin resistance in these people despite their extreme leanness, regular physical activity (daily hunting and foraging), and traditionally oriented diet suggests that, in common with other Aboriginal communities all over Australia, they may become susceptible to obesity and NIDDM if they Westernize further.

2.2.2. Studies in Transitional Groups

We have made similar observations in another isolated Aboriginal community in northwestern Australia which, at the time of our survey (1979), had less exposure to Western diet and lifestyle than other Aboriginal communities in the region *(7)*. Mean fasting glucose levels were low (4.1 ± 0.2 mmol/L), as were the cholesterol levels (3.56 ± 0.11 mmol/L). HDL-cholesterol levels (measured in young men only) were also low in this population (0.85 ± 0.04 mmol/L). However, VLDL-triglyceride and total triglyceride levels were not low and the group had insulin levels which were similar to more Westernized Aboriginal groups and significantly higher than age- and BMI-matched white Australian men *(7)*. They also exhibited mild impairment of glucose tolerance despite the hyperinsulinemia, strongly suggestive of insulin resistance.

2.3. Nature vs Nurture

These studies allowed us to differentiate between those metabolic characteristics that are strongly influenced by Westernization (fasting glucose concentrations, total cholesterol, LDL-cholesterol, and HDL-cholesterol levels), and those that appear to be under genetic control and related to underlying insulin resistance (hyperinsulinemia, elevated triglycerides and VLDL-triglycerides, and mild impairment of glucose tolerance). This latter cluster of metabolic characteristics may indicate susceptibility to NIDDM in Aborigines—which was not expressed when they lived as hunter-gatherers under a set of lifestyle conditions that minimized insulin resistance.

It is clear that the hunter-gatherer lifestyle protected Aborigines from a range of Western diseases that occur in epidemic proportions once they make the transition to a Western lifestyle: obesity, NIDDM, hypertension, dyslipidemia, and CHD. We were interested in whether this lifestyle could also be therapeutic. Could it be used to treat these chronic degenerative conditions once they were established?

2.4. Impact of Temporary Reversion of Westernized Aborigines to Their Hunter-Gatherer Lifestyle

For some Aboriginal communities in remote areas of Australia, westernization has been a relatively recent process, occurring within the lifespan of people still alive today. Older people in these communities retain the knowledge and ability to survive as hunter-gatherers. Collaboration with such people has provided us with the opportunity to document the impact on the risk factors for chronic degenerative diseases such as NIDDM and CHD of a "reverse lifestyle change" or "dewesternization"—the temporary reversion of Westernized Aborigines (with all the associated contemporary health problems) to traditional hunter-gatherer lifestyle. This type of study has also provided information on the characteristics of the hunter-gatherer lifestyle itself: dietary composition, eating patterns, food preferences, and physical activity levels.

We have examined the impact of temporary reversion to traditional lifestyle on carbohydrate and lipid metabolism in nondiabetic Aborigines in two studies, one lasting 3 mo and the other for 2 wk (24,25). In both, there were significant improvements in glucose tolerance associated with marked reductions in hyperinsulinemia and fasting triglyceride levels. Although these effects were more pronounced in the longer study, it is important to note that they were clearly evident after a period as short as 2 wk. The longer study was also associated with significant weight loss. The metabolic changes accompanying temporary reversion to hunter-gatherer lifestyle in these two groups of nondiabetic Aborigines were consistent with improved insulin sensitivity. The lifestyle change involved at least three factors that are known to affect insulin sensitivity directly: increased physical activity, reduced energy intake (and weight loss over the longer term), and changes in dietary composition. The results of these studies in nondiabetic Aborigines provided a strong rationale for testing the therapeutic potential of the hunter-gatherer lifestyle in diabetic Aborigines. It was a reasonable proposition that the major abnormalities of carbohydrate and lipid metabolism of NIDDM that are related to insulin resistance should respond positively to the lifestyle change.

3. The Therapeutic Efficacy of the Hunter-Gatherer Lifestyle in the Treatment of NIDDM

This study was conducted in close collaboration with the Mowanjum Aboriginal community in the Kimberley region of northwest Australia (Fig. 3) in July and August of 1982 (the "dry season"). A survey of the community in 1977 had revealed that 17% of adults over the age of 20 years had diabetes *(4)* and there was a great deal of concern within the community about the particularly high prevalence of diabetes and heart disease among older people. For this reason, the community supported the testing of a group of diabetics before and after temporarily returning to the bush to live as hunter-gatherers. The community's own cattle station (to the northeast of their urban settlement) was used for the study and the community council was closely involved in all stages of the planning and organization of the study. Without this strong support, the study would have been impossible.

3.1. The Field Study

The field study was carried out at Pantijan, the Mowanjum community's cattle station and traditional country of many of the Aborigines now resident at Mowanjum. It is an extremely isolated location northeast of Derby, 1.5 day's journey by four-wheel drive vehicle or 1 h by light airplane. There was no access to storefoods or beverages from the time the subjects left Derby until they returned 7 wk later. The only food eaten after they left Derby was that which they hunted or gathered. The 7-wk period was spent as follows: en route to the cattle station, 10 d; at the coastal location, 2 wk; inland on a river, 3.5 wk.

3.1.1. The Subjects

Ten diabetic and four nondiabetic full-blood Aborigines participated in this study. There were equal numbers of men and women in both groups. They were aged between 47–62 years and weighed between 64–97 kg (equivalent to a BMI range of 22.6 to 32.8 kg/m^2). Five of the diabetics and three of the nondiabetics were moderate to heavy drinkers in the urban setting, but all had abstained from alcohol for at least 2 d prior to the baseline metabolic test. Of the 10 diabetics, only one was currently treated with oral hypoglycemic agents (tolbutamide) and none was on insulin. This subject's medication was withdrawn beginning on the morning of the baseline glucose tolerance test. The same subject was also on antihypertensive medication (atenolol, amiloride, and hydrochlorthiazide), which was withdrawn under close medical supervision.

3.1.2. Experimental Diet

During the 10-d trip from Derby to the coastal location, the diet was mixed and included locally killed beef, since supplies of bush foods were inadequate: meat (beef, kangaroo), freshwater fish and turtle, vegetables, and honey. It was estimated that beef comprised 75% of the energy intake during this 10-d period and the overall dietary composition was estimated to be: protein 50%, fat 40%, and carbohydrate 10%. No further beef was consumed once the group arrived at the coastal location.

During the 2-wk period spent on the coast, the diet was derived predominantly from seafood with supplements of birds and kangaroo. The lack of vegetable food in this area eventually precipitated the move inland to the abandoned site of the old homestead. The estimated dietary composition while on the coast was: protein 80%, fat 20%, and carbohydrate <5%.

At the inland location, which is on a river, the diet was much more varied: kangaroo, freshwater fish and shellfish, turtle, crocodile, birds, yams, figs, and bush honey. A detailed analysis of the food intake was conducted over a 2-wk period during this phase of the study (Table 3). All food was weighed before it was eaten and samples were collected and stored in liquid nitrogen before being flown back to Melbourne for analysis. Energy intake over this period averaged 1200 kcal/person/d. In terms of total dietary energy consumed over the 2-wk period, kangaroo accounted for 36%, freshwater bream 19%, and yams 28%. The remaining 17% was made up from wild honey, figs,

Table 3
Design of the Study and Composition of the Diet
During the 7-wk Reversion to Traditional Hunter-Gatherer Lifestyle

Phase of the study	Traveling	Coast	Inland
Main foods (as percentage of total calories)	75%—Beef 25%—Kangaroo, turtle, bream	80%—Fish 20%—Birds, kangaroo, crocodile	36%—Kangaroo 19%—Freshwater fish (bream) 28%—Yams 17%—Birds, crocodile, crustacea, figs, honey
Composition of the diet	Estimate only	Estimate only	Measured over 2-wk period
Carbohydrate	10%	<5%	33%
Protein	50%	80%	54%
Fat	40%	20%	13%
Energy (cal/pers/d)	1100–1300	1100–1300	1200

Week	0	1	2	3	4	5	6	7
	↑ Baseline metabolic test	↑ Bleeding time	↑ Bleeding time	↑ Bleeding time	↑ Bleeding time		↑ Bleeding time	↑ Final metabolic test

birds, turtle, crocodile, and crustacea. The dietary composition in terms of total energy was 54% protein, 13% fat, and 33% carbohydrate. Animal foods accounted for 64% of total energy.

3.1.3. Urban Diet

The main dietary components in the urban setting were flour, sugar, rice, carbonated drinks, alcoholic beverages (beer and port), powdered milk, cheap fatty meat, potatoes, onions, and variable contributions of other fresh fruit and vegetables. At the time of the study the composition of the diet was estimated to be: carbohydrate 50%, fat 40%, and protein 10%. There was considerable variation within the group depending on the contribution of alcohol to the diet. In the urban environment the nondrinkers were more diet conscious and tended to eat more fresh fruit and vegetables and whole meal bread.

3.1.4. Metabolic Tests

Immediately before the 7-wk experimental period, baseline studies were performed at the local Derby Regional Hospital after a 12-h overnight fast. A fasting blood sample was taken (for the measurement of glucose, insulin, lipoprotein lipids, and plasma fatty acid composition) before the 75 g oral glucose load was consumed. Blood samples were taken 0.5, 1, 2, and 3 h postprandially. At the conclusion of the 7-wk study, the subjects were flown back to Derby at dawn and the baseline measurements were repeated.

3.1.5. Measurements During the Field Study

Fasting blood glucose was measured regularly in the diabetic subjects using a battery-operated Ames Glucometer and Dextrostix. Blood pressure and body weight were monitored regularly in all subjects. An attempt was made to assess physical activity over a 2-wk period on a scale of 1–5: with 1 being equivalent to sleeping all day and 5 being equivalent to vigorous activity (e.g., hunting or digging for yams for at least 6 h). Ivy bleeding times were performed using a disposable Simplate II device under standardized conditions. Because of the high frequency of pain-killer ingestion by the participants in the urban setting, bleeding time measurements were not made before leaving town. No aspirin or related pain-killing preparations were ingested by the subjects after they left Derby. The first bleeding time measurements were made 10 days after leaving Derby—which should have been long enough for any effects of aspirin to have disappeared. The second measurement was made 3.5 wk into the study after 2 wk on a diet derived almost exclusively from seafood, and the final measurement was made 6.5 wk into the study after 3 wk on a mixed, traditional diet (Table 3).

Details of the study design, analytical methods, and results of this study have been published previously (26,27).

3.2. Results

3.2.1. Dietary Composition

The overall design of the study and composition of the diet over the 7-wk lifestyle change period is presented in Table 3. Animal foods comprised 90% of the energy intake during the first phase (traveling), essentially all of the energy intake on the coast, and 64% during the final (inland) phase of the study. When the Aborigines were eating wild animal foods exclusively (coast and inland, 5.5 wk) the diet was low in fat because of the low fat content of wild animals and fish (28,29). This would have been an important factor in their weight loss. During the one period when their energy intake was measured accurately (2 wk in the third phase of the study), it was found to average 1200 kcal/person/d. Since weight loss was relatively constant over the 3 phases of the study (Fig. 4), it was assumed

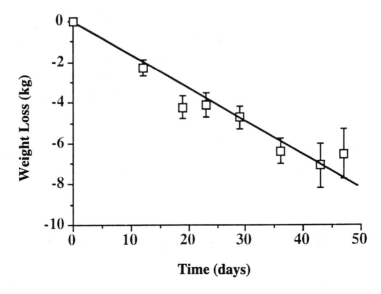

Fig. 4. Fall in body weight of the nine diabetic Aborigines over the 7-wk period of traditional lifestyle (mean ± SEM).

that the energy intake was also fairly constant (1100–1300 kcal/person/d) over the 7-wk period.

3.2.2. Weight Reduction

The mean change in body weight of the ten diabetic subjects is shown in Fig. 4. Initial and final body weights for all subjects are reported in Table 4. All subjects lost weight steadily over the 7-wk period: The mean total weight loss was 8.0 ± 0.9 kg (range 3.1–12.0 kg). The three leanest diabetic subjects (baseline BMI < 24 kg/m^2) lost the least weight (3.1–5.1 kg), whereas the remaining seven diabetic subjects (baseline BMI > 26 kg/m^2) lost between 7.8–12.0 kg. Weight loss was highly correlated with baseline BMI ($r = 0.819$, $p < 0.01$). Despite this impressive weight loss, most of the diabetic subjects were still overweight at the end of the study (BMI > 24 kg/m^2). Physical activity over the 2-wk period of assessment (weeks five and six) did not correlate with total weight loss.

Table 4
Summary of the Changes in Ten Diabetic Aborigines
in Response to 7-wk Reversion
to a Traditional Hunter-Gatherer Lifestyle[a]

	Before	After 7 wk	p Value
Body weight (kg)	81.9 ± 3.4	73.8 ± 2.8	<0.001
Body mass index (kg/m^2)	27.2 ± 1.1	24.5 ± 0.8	<0.001
Fasting glucose (mmol/L)	11.6 ± 1.2	6.6 ± 0.5	<0.001
2-h glucose (mmol/L)	18.5 ± 1.3	11.9 ± 0.9	<0.001
ΔAUC glucose (mmol/L/h)[b]	15.0 ± 1.2	11.7 ± 1.2	<0.005
Fasting insulin (mU/L)	23 ± 3	12 ± 1	<0.005
2-h insulin (mU/L)	49 ± 9	59 ± 11	n.s.
ΔAUC insulin (mU/L/h)[b]	61 ± 18	104 ± 21	<0.05

[a]Mean ± SEM.
[b]ΔAUC is the incremental area under the curve for 3 h after 75 g oral glucose. ΔAUC insulin is similarly defined.

3.2.3. Glucose Tolerance and Insulin Response

Fasting plasma glucose concentrations in the ten diabetic subjects fell markedly during the study (Table 4). Blood glucose measurements taken during the study on a Glucometer with Dextrostix indicated that fasting glucose levels fell very rapidly during the first 10 d of the study and plateaued thereafter (Fig. 5). There was also an improvement in glucose tolerance as illustrated by the significant reduction in the postprandial rise in glucose concentration in the glucose tolerance test (Fig. 6). However, the most important contributor to the improved glucose tolerance was the fall in fasting glucose level.

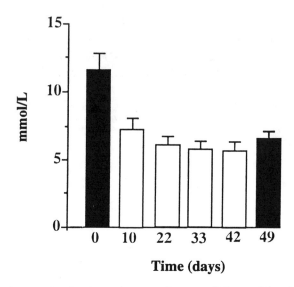

Fig. 5. Change in fasting plasma glucose (■), and fasting blood glucose (□) measured by Detrostix, during the 7-wk intervention period.

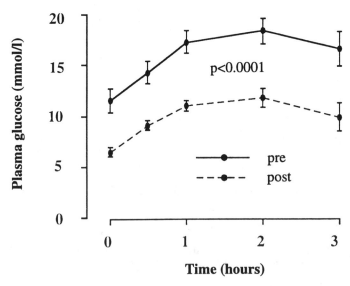

Fig. 6. Change in plasma glucose concentrations in ten diabetic Aborigines after 75 g oral glucose before and after 7-wk of traditional lifestyle (mean ± SEM).

Fasting plasma insulin concentrations also fell significantly in the diabetic subjects, and the insulin *response* to oral glucose *increased* as illustrated by the increase in the incremental area under the curve (Table 4).

3.2.4. Cardiovascular Disease Risk Factors

The diabetic Aborigines were extremely hypertriglyceridemic before the study. Their triglyceride levels were completely normalized by the end of the study (Table 5). Although there was a trend toward reduced cholesterol levels, it was not statistically significant.

Blood pressure also fell in response to the lifestyle change, with the fall in diastolic pressure being statistically significant (Table 5).

The baseline bleeding time measurements were made after 10 d away from aspirin and alcohol at the end of phase one where the diet was rich in beef and estimated to contain 40–50% energy from fat. Subsequent bleeding time measurements were made after 2 wk on the diet derived predominantly from tropical seafood (phase two), followed by 3 wk on a diet (phase three) containing a wider variety of traditional animal and vegetable foods (primarily kangaroo, freshwater fish, and yams). Two weeks after the Aborigines changed from the beef-based diet to the seafood diet, bleeding time rose in all subjects (4.1 ± 0.4 min to 5.3 ± 0.4 min), a mean increase of 29%. When the Aborigines switched to the mixed, traditional diet the mean bleeding time rose a further 15% (to 5.9 ± 0.4 min), although the effect was less consistent than the response to the seafood diet. There was a 44% increase in bleeding time overall, consistent with a reduced tendency to thrombosis (Table 5).

The increased bleeding time was associated with an unexpected change in the plasma fatty acid composition. There was a threefold increase in the proportion of arachidonic acid in the plasma fatty acids. There were also

Table 5
Changes in Cardiovascular Disease Risk Factors in Ten Diabetic
Aborigines in Response to 7-wk Reversion to a Traditional
Hunter-Gatherer Lifestyle[a]

	Before	After 7 wk	p Value
Fasting plasma triglycerides (mmol/L)	4.02 ± 0.46	1.15 ± 0.10	<0.001
Fasting plasma cholesterol (mmol/L)	5.65 ± 0.23	4.98 ± 0.34	n.s.
Blood pressure (mm Hg)			
Systolic	121 ± 5	114 ± 4	<0.08
Diastolic	80 ± 2	72 ± 2	<0.02
Bleeding time (min)	4.1 ± 0.4	5.9 ± 0.4	<0.01
Plasma fatty acids (percentage of total FA)			
Linoleic	14.3 ± 1.5	16.1 ± 2.1	n.s.
Arachidonic	3.3 ± 0.3	10.2 ± 0.1	<0.001
Eicospentenoic (EPA)	0.4 ± 0.1	0.9 ± 0.1	<0.02
Docosahexenoic (DHA)	0.8 ± 0.3	2.2 ± 0.4	<0.005

[a]Mean ± SEM.

increases in the proportions of the long chain n-3 polyunsaturated fatty acids (PUFA), EPA, and DHA (Table 5).

3.3. Summary of the Effects of Lifestyle Change

The most important finding in this study was the striking improvement in all of the metabolic abnormalities of diabetes in this group of Aborigines: Fasting glucose was greatly reduced, consistent with reduced hepatic glucose output; the impaired insulin secretory response was improved; and insulin sensitivity was improved as illustrated by the reduced fasting glucose in the face of lower insulin levels and the normalized triglyceride levels.

In addition, there were important reductions in a number of cardiovascular disease risk factors: abolition of the hypertriglyceridemia, reduction in blood pressure, and increased bleeding time (interpreted as indicating a reduced thrombosis tendency).

The hunter-gatherer lifestyle was effective in the treatment of the major chronic lifestyle-related health problems of contemporary Aborigines: obesity, NIDDM, hypertension, and CHD.

4. Which Aspects of this Lifestyle Were More Critical for the Treatment and Prevention of Western Disease?

4.1. Diet

All animals were potential food sources—mammals, birds, reptiles, insects, and marine species (mammals, reptiles, fish, crustacea, shellfish). Everything edible on an animal carcass was eaten, including muscle, fat depots, bone marrow, and internal organs (although usually not intestinal contents). One of the most striking characteristics of wild (or nondomesticated) animals is the low fat content of their carcases (28). In this they differ strikingly from domesticated meat animals such as cattle, sheep, pigs, and chickens, which have fat depots under the skin, within the abdomen, and between and within muscles (30). Wild animals do have discrete depots of fat within the abdomen (primarily surrounding the gonads, kidneys, and intestines), and these depots frequently increase in size at particular times of the year. However, they generally tend to be small and usually have to be shared among many people. The meat (muscle) is always lean irrespective of the season and does not "marble." Because of this, a high proportion of what little fat is present is structural fat (28)—part of the membranes of the muscle cells, and relatively rich in the long-chain highly polyunsaturated fatty acids PUFA (n-6 and n-3).

Although on a weight basis, muscle constituted the major edible portion of an animal carcass, fat depots and organs (including emptied intestines and stomach) were also consumed. The organ meats (particularly brain and liver) were rich sources of cholesterol and fat. However, the fat in these organ meats was relatively rich in the long-chain PUFA (n-6 and n-3) (28). Even the depot fat on some species was unusually rich in polyunsaturated fat (consistent with its "softness" relative to beef tallow or mutton fat). Such foods were highly valued.

Insects also provided significant contributions to the diet seasonally, both directly and indirectly. Honey, a product of wild bees and the honey ant, was an important dietary carbohydrate in season. Witchetty grubs are still a seasonal delicacy in many parts of the country. Interestingly, although they are rich in fat and have a nutty-buttery taste, they have a fat composition very similar to olive oil (28)! The bogong moth was so plentiful in the summer months in the mountains of southeastern Australia, that large gatherings of people could be supported while participating in ceremonies.

The detailed and thorough knowledge the Aborigines had of the ecology of their environment allowed them to take full advantage of a wide range of plants as sources of food over the year—tuberous roots, seeds, fruits, nuts, gums, and nectar. Dietary carbohydrate in the hunter-gatherer diet was derived from uncultivated plants (tuberous roots, fruits, berries, seeds, nuts, beans) and honey. Cereal grains, the dietary staples of humans since the development of agriculture, were not a major component of the traditional Aboriginal diet. Relative to many of their cultivated forms, wild plant foods are particularly rich in protein and vitamins *(31)*. Although many of the vegetable foods are not rich fat sources (tuberous roots, leafy vegetables, fruits, and berries), they nevertheless contain both n-6 and n-3 PUFA and the ratio of n-6/n-3 was often much lower than in most seeds. The wild vegetable foods are also rich in dietary fiber, low in sodium, but rich in potassium, magnesium, and calcium. In general, the wild vegetable foods were bulky, with high nutrient density but low energy density. The only carbohydrate source with high energy density was honey—a particularly favored food.

The carbohydrate in many of these traditional foods has been shown to be more slowly digested and absorbed than the carbohydrate in equivalent domesticated plant foods. It has been suggested that the lower postprandial glucose and insulin levels elicited by the ingestion of these slowly digested wild plant foods may have been a factor in helping to protect these populations from NIDDM *(32)*—although other factors, in particular their leanness, physical fitness, and low fat, high fiber diet, were undoubtedly also very important.

4.2. Physical Activity Levels of Aborigines as Hunter-Gatherers

Although it has been argued that hunter-gatherers spent less time on average each day ensuring their livelihood (3–5 h) than agriculturalists or the employed in societies such as our own, food procurement and preparation for Aboriginal hunter-gatherers were energy-intensive processes that could involve sustained physical activity: walking long distances, digging in rocky ground for tubers deep below the surface, digging for reptiles, eggs, honey ants, and witchetty grubs, chopping with a stone axe (honey, grubs, and so on), winnowing and grinding of seeds, digging pits for cooking large animals, gathering wood for fires (for cooking and warmth).

4.3. Eating Patterns of Aborigines
as Hunter-Gatherers

Food intake could vary enormously both on a day-to-day and a seasonal basis *(10,22)*. This has been described as a "feast-and-famine" pattern of food intake. The actual pattern could probably be described more accurately as subsistence interspersed with feasts. The women were primarily responsible for the subsistence component (plant foods, small animals, reptiles, fish, insects, honey, and so on), and although the men did assist with these activities, they were primarily hunters and provided the less regular, but highly valued, "feasts."

Food was usually consumed at the time it was available, and wastage was rare. There are numerous reports of Aborigines eating 2–3 kg of meat at one long sitting, taking maximum advantage of an abundant food supply on those irregular occasions when it was available *(10,22)*. It can be argued that the feasts were critical to the survival of Aborigines as hunter-gatherers as they provided excess energy that could be converted into fat and deposited as adipose tissue, thereby providing an energy reserve to tide an individual over periodic food shortages *(6)*.

4.4. Food Preferences of Aborigines
as Hunter-Gatherers

The most highly prized components of the Aboriginal hunter-gatherer diet were the relatively few energy-dense foods: depot fat, organ meats, fatty insects, and honey. In general, muscle provided the bulk of the energy from a carcass. Although many animals were actively hunted at those times of the year when their fat depots were largest (with Aborigines following the "fat cycle"), the fat depots on most animals were usually small through most of the year and needed to be shared among many people. Thus, high fat foods were either only available seasonally, or were present in small quantities on the animal. Similarly, wild honey was available only at certain times of the year and its procurement was often associated with high energy expenditure.

Nevertheless, it is significant that in a diet that was generally characterized by its low energy density, the foods that were most actively sought and most highly valued were those that had a high energy density. Two important components of the survival strategy of Aborigines as hunter-gatherers were to maximize energy intake and minimize energy output. The reality of the lifestyle, however, resulted in a generally low energy intake (subsistence) in combination with a relatively high energy expenditure. The scenario changed dramatically with Westernization.

5. Westernization

After westernization, these hunter-gatherer behaviors and food preferences were retained by Aborigines in an environment where there is essentially unlimited availability of energy-dense equivalents of these survival foods (sucrose, fatty meats, processed foods), and the energy output involved in their procurement is minimal. Evidence is accumulating that the diet selected by contemporary Aborigines in many parts of Australia is particularly rich in sucrose and fat and very high in energy (9). The contrast is striking: The traditional diet was low in energy, high in protein, low in carbohydrate and fat, with most of the carbohydrate derived from fiber-rich, slowly digested foods (32) and the fat was polyunsaturated (27,28); the Western diet is much higher in energy, lower in protein, higher in carbohydrate and fat, with the carbohydrate derived almost exclusively from fiber-depleted, rapidly digested foods and the fat is highly saturated (Table 6).

Table 6
Comparison of Lifestyle and Diet of Australian Aborigines
as Hunter-Gatherers and After Westernization

	Hunter-gatherer	Western
Physical activity level	high	low
Principal characteristics of diet		
Energy density	low	high
Energy intake	usually adequate	excessive
Nutrient density	high	low
Nutrient composition of diet		
Protein	high	low-moderate
Animal	high	moderate
Vegetable	low	low
Carbohydrate	moderate	high
	(slowly digested)	(rapidly digested)
Complex carbohydrate	moderate	moderate
Simple carbohydrate	usually low	high
	(honey)	(sucrose)
Fiber	high	low
Fat	low	high
Vegetable	low	low
Animal	low	high
	(polyunsaturated)	(saturated)
Sodium:potassium ratio	low	high

6. How Does the Hunter-Gatherer Lifestyle Protect Against the Chronic Diseases of Western Lifestyle?

6.1. Obesity

The high bulk and low energy-density of most components of the hunter-gatherer diet, combined with the energy intensity of their procurement provided inbuilt constraints against overconsumption for much of the time.

The irregular feasts were too infrequent to result in long-term excess fat accumulation.

6.2. Noninsulin-Dependent Diabetes Mellitus

The regular physical activity and relatively low energy intake combined to maintain a lean body weight and high insulin sensitivity. In addition, other qualities of the diet, such as its low fat and high fiber contents, and its slowly digested carbohydrate, are also likely to have contributed independently to protection against NIDDM.

6.3. Hypertriglyceridemia

Regular physical activity and relatively low energy intake combined to maximize insulin sensitivity. In addition, the low sucrose/fructose (honey) and low saturated fat content of the diet, and the absence of alcohol would also have protected against elevated triglyceride levels.

6.4. Atherosclerosis

The low saturated fat and high fiber intake were probably critical to maintaining low plasma cholesterol levels, despite relatively high dietary cholesterol intakes. Again, the regular physical activity and relatively low (but adequate) energy intake would have maximized insulin sensitivity—now recognized to be important in modifying the risk of CHD (syndrome X). Other factors contributing to the low risk of CHD were the relatively long bleeding time, absence/rarity of hypertension and absence of smoking.

6.5. Thrombosis

Relative to the Western diet, the intake of saturated fat was very low and that of the long chain PUFA of both the n-3 and n-6 series was high. These long chain PUFA may have mediated the increased bleeding times observed in Aborigines on their traditional diet by selectively favoring the production of the antiaggregatory vasodilatory prostanoid, prostacyclin, at the expense of the proaggre-gatory vascoconstrictor prostanoid, thromboxane. We have recently reported that when white Australians consume low fat diets rich in either kangaroo meat or tropical fish, there is an inhibition of the cold pressor response (33), analogous to the consequences of an infusion of prostacyclin acutely (34). This effect occurred together with increased proportion of arachidonic acid and, to a lesser extent, of the n-3 PUFA in the plasma lipids (33).

6.6. Hypertension

Salt was not used as a condiment by Aborigines—the major source of sodium was extracellular fluids (blood in particular), which were conserved by the cooking meth-ods. Measurements of sodium:potassium ratios in perspi-ration and urine of Aborigines in different stages of acculturation confirmed low sodium and high potassium intakes of the least acculturated groups in association with low blood pressure. The more Westernized groups had higher sodium:potassium ratios in perspiration and urine, and higher blood pressures (35). Other aspects of the hunter-gatherer lifestyle that may have been even more important in maintaining low blood pressure were regu-lar physical activity and low body fat—since both weight loss and regular exercise have been shown to lower blood pressure in hypertensives (36,37).

7. Conclusion

These observations have implications not only for the prevention of obesity, diabetes and cardiovascular disease in Aborigines, but can also be applied more broadly. It has been argued that the hunter-gatherer or paleolithic diet and lifestyle is the one to which we as modern humans are genetically programmed (38), since the human genetic con-stitution has changed little in the past 40,000 years since the appearance of modern humans. Thus, it is to this diet and lifestyle that we should turn when seeking explanations for (and solutions to) the characteristic pattern of chronic diseases that emerges in all populations when they become more "affluent" economically and adopt a sedentary, Westernized way of life.

Acknowledgments

Much of the work described would not have been possible without the help and excellent cooperation of the Mowanjum Aboriginal Community (Derby, Western Australia). Their support, patience, and good humor was an inspiration. The many other people who have contributed so generously to the research over the years have been acknowledged (6–10, 24–33). The work was supported by grants from the Australian Institute of Aboriginal Studies (now the Australian Institute of Aboriginal and Torres Strait Islander Studies) and the National Health and Medical Research Council of Australia.

References

1. Flood J. *Archeology of the Dreamtime*. Sydney: Collins, 1983; 67–33.
2. Kirk RL. *Aboriginal Man Adapting: The Human Biology of Australian Aborigines*. Melbourne: Clarendon Press, 1981; 39–62.
3. Wise PH, Edwards FM, Thomas DW, Eliott RB, Hatcher L, Craig R. Diabetes and associated variables in the South Australian Aboriginal. *Aust NZ J Med* 1976; **6**:191–196.
4. Bastian P. Coronary heart disease in tribal Aborigines—the West Kimberley Survey. *Aust NZ J Med* 1979; **9**:284–292.
5. Williams DDR, Moffitt PS, Fisher JS, Bashir, HV. Diabetes and glucose tolerance in New South Wales coastal Aborigines: possible effects of non-Aboriginal genetic admixture. *Diabetologia* 1987; **30**:72–77.
6. O'Dea K. Westernization, insulin resistance, and diabetes in Australian Aborigines. *Med J Aust* 1991; **155**:258–264.

7. O'Dea K, Spargo RM, Nestel PJ. Impact of westernization on carbohydrate and lipid metabolism in Australian Aborigines. *Diabetologia* 1982; **22**:148–153.

8. O'Dea K, Traianedes K, Hopper JL, Larkins RG. Impaired glucose tolerance, hyperinsulinemia and hypertriglyceridemia in Australian Aborigines from the desert. *Diab Care* 1988; **11**:23–29.

9. O'Dea K, Lion RJ, Lee A, Traianedes K, Hopper JL, Rae C. Diabetes, hyperinsulinemia, and hyperlipidemia in small Aboriginal community in Northern Australia. *Diab Care* 1990; **13**:830–835.

10. O'Dea K, White NG, Sinclair AJ. An investigation of nutrition-related risk factors in an isolated Aboriginal community in northern Australia: advantages of a traditionally-orientated lifestyle. *Med J Aust* 1988; **148**:177–180.

11. Knowler WC, Pettit DJ, Saad MF, Bennett PH. Diabetes mellitus in the Pima Indians: incidence, risk factors and pathogenesis. *Diab/Metab Rev* 1990; **6**:1–27.

12. Zimmet P, Dowse G, Finch C, Sargentson S, King H. The epidemiology and natural history of NIDDM: lessons from the South Pacific. *Diab/Metab Rev* 1990; **6**:91–124.

13. Dowse GK, Gareeboo H, Zimmet PZ, Albert KGMM, Tuomilehto J, Fareed D, Brissonnette LG, Finch CF. For the Mauritius Noncommunicable Disease Study Group: High prevalence of NIDDM and impaired glucose tolerance in Indian, Creole, and Chinese Mauritians. *Diabetes* 1990; **39**:390–396.

14. Maher HM, Keen H. The Southall Diabetes Survey: prevalence of known diabetes in Asians and Europeans. *Brit Med J* 1985; **291**:1081–1084.

15. Stern MP. Primary prevention of type II diabetes mellitus. *Diab Care* 1991; **14**:399–410.

16. Glatthaar C, Welborn TA, Stenhouse NS, Garcia-Webb P. Diabetes and impaired glucose tolerance. A prevalence estimate based on the Busselton. 1981 survey. *Med J Aust* 1985; **143**:435–440.

17. Rutishauser IHE, McKay H. Anthropometric status and body composition in Aboriginal women of the Kimberley region. *Med J Aust* 1986; **144**:S8–S10.

18. Modan M, Halkin H, Almog S, Lusky A, Eshkol A, Shefi M, Shitrit A, Fuchs Z. Hyperinsulinemia. A link between hypertension, obesity and glucose intolerance. *J Clin Invest* 1985; **75**:809–817.

19. Reaven GM. Role of insulin resistance in human disease. *Diabetes* 1988; **37**:1595–1607.

20. Elphinstone JJ. The health of Australian Aborigines with no previous association with Europeans. *Med J Aust* 1971; **2:** 293–301.

21. Lowenstein FW. Blood-pressure in relation to age and sex in the tropics and subtropics—A review of the literature and an investigation in two tribes of Brazil Indians. *Lancet* 1961; **i:**389–392.

22. White, NG. Sex differences in Australian Aboriginal subsistence: possible implications for the biology of hunter-gatherers. In: Ghesquierre J, Martin RD, Newcombe F, eds. *Human Sexual Dimorphism*. London: Taylor & Francis, 1985; 323–361.

23. Day J, Carruthers M, Bailey A, Robinson D. Anthropometric, physiological and biochemical differences between urban and rural Maasai. *Atherosclerosis* 1976; **23:**357–361.

24. O'Dea K, Spargo RM. Metabolic adaptation to a low carbohydrate-high-protein ('traditional') diet in Australian Aborigines. *Diabetologia* 1982; **23:**494–498.

25. O'Dea K, Spargo RM, Akerman K. The effect of transition from traditional to urban life-style on the insulin secretory response in Australian Aborigines. *Diab Care* 1980; **3:**31–37.

26. O'Dea K. Marked improvement in the carbohydrate and lipid metabolism in diabetic Australian Aborigines after temporary reversion to traditional lifestyle. *Diabetes* 1984; **33:** 596–603.

27. O'Dea K, Sinclair AJ. The effects of low fat diets rich in arachidonic acid on the composition of plasma fatty acids and bleeding time in Australian Aborigines. *Int J Nutr Vitaminol* 1985; **31:**441–453.

28. Naughton JM, O'Dea K, Sinclair AJ. Animal foods in traditional Aboriginal diets: polyunsaturated and low in fat. *Lipids* 1986; **21:**684–690.

29. Sinclair AJ, O'Dea K, Naughton JM. Elevated levels of arachidonic acid in fish from northern Australian coastal waters. *Lipids* 1983; **18:**877–881.

30. Sinclair AJ, Slattery WJ, O'Dea K. The analysis of polyunsaturated fatty acids in meat by capillary gas liquid chromatography. *J Sci Fd Agric* 1982; **33:**771–776.

31. Brand JC, Rae C, McDonnell J, Lee A, Cherikoff V, Truswell AS. The nutritional composition of Australian Aboriginal bushfoods. 1. *Food Technol Aust* 1983: **35:**293–298.

32. Thorburn AW, Brand JC, O'Dea K, Spargo RM, Truswell AS. Plasma glucose and insulin responses to starchy foods in Australian Aborigines: a population now at high risk of diabetes. *Am J Clin Nutr* 1987; **46:**282–285.

33. Butcher LA, O'Dea K, Sinclair AJ, Parkin JD, Smith IL, Allardice J, Blombery P. The effects of very low fat diets enriched with fish or kangaroo meat on cold-induced vasoconstriction and platelet function. *Prostaglandins, Leukotrienes and Essential Fatty Acids* 1990; **39**:221–226.

34. Cowley AJ, Heptinstall S, Hampton JR. Effects of prostacyclin and of the stable prostacyclin analogue ZK36374 on forearm blood flow and blood platelet behaviour in man. *Thromb Haemost* 1985; **53**:90–94.

35. Macfarlane WV. Aboriginal desert hunter-gatherers in transition. In: Hetzel BS, Frith HJ, eds. *The Nutrition of Aborigines in Relation to the Ecosystem of Central Australia*. Melbourne: CSIRO, 1978; 49–62.

36. MacMahon SW, Macdonald GJ, Bernstein L, Andrews G, Blacket RB. Comparison of weight reduction with Metroprolol in treatment of hypertension in young overweight patients. *Lancet* 1985; **i**:1233–1236.

37. Jennings G, Dart A, Meredith I, Korner P, Laufer E, Dewar E. Effects of exercise and other nonpharmacological measures on blood pressure and cardiac hypertrophy. *J Cardiovasc Pharmacol* 1991; **17(Suppl. 2)**:570–574.

38. Eaton SB, Konnor M. Paleolithic nutrition. A consideration of its nature and current implications. *N Engl J Med* 1985; **312**:283–289.

PART IV

PRACTICAL MEANS TO PREVENT
WESTERN DISEASE

Organized Medicine

An Ounce of Prevention or a Pound of Cure

Norman J. Temple

The reasonable man adapts himself to the world: the unreasonable one persists in trying to adapt the world to himself. Therefore all progress depends on the unreasonable man.

George Bernard Shaw in *Reason*

1. The Relative Failure of Therapy

Medicine, which here means medical practice in general, is a prisoner of its past. Medicine has scored tremendous successes in helping to eliminate infectious diseases, notably by immunization and by antibiotics. Similarly, in the latter part of the nineteenth century effective treatments were developed for a range of noninfectious diseases, such as gallstones, appendicitis, and various cancers. One hundred years later medicine is continuing its relentless search for new therapies. But success stories are increasingly rare. The golden age of therapeutics is dead.

For decades medicine has put all its eggs in one basket. Great efforts have been made to comprehend the inner mysteries of disease processes and by this means to seek out effective new cures. Hand in hand with this, medicine has enthusiastically embraced modern technology. The

From: *Western Diseases: Their Dietary Prevention and Reversibility*
Edited by: N. J. Temple and D. P. Burkitt Copyright ©1994 Humana Press, Totowa, NJ

modern doctor has an arsenal of sophisticated techniques at his/her disposal: CAT scans, transplantation, microsurgery, and so forth. Yet, a sober appraisal of the facts reveals that over the last four decades little progress has been made in finding effective new cures for Western disease.

This is illustrated by cancer. There are two versions of the current state of play concerning medicine's war against that family of diseases. In one version, which is regularly trumpeted by the cancer establishment, especially the major cancer charities, and is echoed in the press, cancer is slowly but surely being conquered. Certainly, in some areas important advances have been made, most notably for certain cancers of children and young adults. However, this has only a small impact on the overall picture, since less than 2% of cancer cases occur in those under 25 *(1)*.

When we look at the major cancers, a rather depressing picture emerges. This was discussed in more detail elsewhere *(2)*. One way to view the situation is to consider the 5-yr survival rates. It is regularly claimed that in the United States these have been steadily improving over the last few decades *(3,4)*. However, the trend toward cancer detection at an earlier stage has created an artificial rise in the 5-yr survival rate *(2)*. This occurs for two reasons. First, there is lead time bias. By having cancer diagnosed at an earlier stage, the average time gap between diagnosis and death is bound to lengthen, even with no improvements in treatment. Second, screening widens the net of cancer diagnosis to include many slow-growing tumors that, even if untreated, would not produce clinical symptoms of a cancer within the lifetime of the subject. Today, they can be "cured." For instance, Cairns *(5)* pointed out that in recent decades the incidence of prostate cancer supposedly increased greatly, as did survival times, but without any corresponding important advance in treatment. This is because there is now much overdiagnosis of the cancer. Needless to say, treatment of these spurious cases has a remarkably high success rate. This has also occurred with lung and breast cancer, at least to some degree. After these errors induced by early detection are allowed for, we are left with very little true improvement in recent decades in the 5-yr survival rates for the major cancers.

An alternative approach to the question of whether medicine is winning the war against cancer is to look at changes in the age-adjusted mortality figures. This leads to much the same conclusion: studies from the United States (6) and West Germany (7) found no real evidence that between 1950–1952 and 1985 improvements in treatment had any significant effect on overall cancer mortality. The situation with cancer can be summarized thus: "If things are so good, why are they so bad?"

Cancer is far from the only disease where medicine's achievements since 1950 have been unspectacular. For instance, it is well known that most overweight and obese patients on supposed slimming diets fail to achieve and maintain their desired weight. The $33 billion in sales made by the US diet industry in 1989 (8) attests to the failure of medicine to offer effective remedies.

Of course, some successes have been achieved in the therapeutic war against Western disease. Improved drug treatments are now available for peptic ulcers, a partially Western disease, as well as for hypertension. Yet even here success has been patchy. A recent study of hypertension in Scotland supported the rule of halves: In half of those with hypertension it is not detected, in half where it is detected, it is not treated, and in half where it is treated, it is not controlled (9).

Another area in which significant therapeutic advances have been made is in diet therapy. As discussed in earlier chapters by Diehl, Anderson, and O'Dea several Western disorders, notably hypertension, coronary heart disease, obesity, and type II diabetes can be at least partially cured by nutritional means. Paradoxically, this very success is symptomatic of medicine's failure. Increasingly over the last four decades medicine has become synonymous with drugs, surgery, and technological advance. Diet therapy is the reverse: It is the ultimate in nontechnological treatment, the use of diets that are closer to the Stone Age than to the computer age.

Overall, then, medicine has had remarkably little success in developing new and effective treatments for Western disease. This is certainly not caused by lack of resources. The National Institutes of Health in the United States spent approx $8.3 billion in 1991 on research. Of this, a billion and a half dollars were devoted just to cancer. Nor is there a lack of resources devoted to treatment. The percentage of the GNP in the United States consumed by medicine grew from 8 to 12% between 1970–1990. The overwhelming majority of this vast expenditure was consumed by treatment.

Even if the money spent on health care in the United States were substantially increased, it is highly unlikely that the health of the American people would significantly improve. A major reason for this is that Western diseases are nontransmissible and their incidence cannot therefore be reduced by improvements in therapy. Another major obstacle to therapeutic success is that the first symptom of coronary thrombosis and stroke is often sudden death, which brings the story to an abrupt end. The limited impact of health expenditure on people's health is shown by the fact that although the United States has the world's highest per capita health costs, the health of its people is no better, and in several respects is worse than that of other nations. For instance, Walker *(10)* has pointed out that the United States has over double the per capita health care expenditure of Japan yet has double the infant mortality and has a life expectancy at birth some four years shorter. This illustrates the overwhelming importance of lifestyle factors on the national health as opposed to health care expenditure.

On the basis of the above arguments some cautious future predictions can be made. First, medicine will continue its relentless pursuit of high technology treatments. In consequence, medicine will continue its rapid cost escalation. Second, new treatments for Western diseases will regularly appear but will have relatively little impact on the curability of these diseases, and even less on their incidence.

An obvious question is why does medicine have such an infatuation with a treatment-based approach? One reason is historical: In the past great progress was made, as noted earlier. A second reason is financial and organizational: In some countries, particularly the United States, the rules of the game force doctors to focus most of their efforts on therapy. However, this does not explain why British doctors are as infatuated with therapeutic medicine as are American doctors. A major reason for this appears to be that medicine suffers from technology worship. Most aspects of high technology medicine can only be used in the context of diagnosis and treatment. This includes drugs, surgery, and complex diagnostic procedures. On the other hand prevention is low technology: basic lifestyle changes, particularly diet and exercise.

Technology worship also explains why medicine is generally so slow to adopt treatments centered on changes in lifestyle. In particular, medicine has been slow off the mark in utilizing diet therapy, particularly when drug alternatives are available. For instance, a symposium published in the *American Journal of Medicine* (November, 1984) concerned the treatment of peptic ulcers: The 15 articles ran to over 120 pages but contained scarcely a line on nutrition. Two later symposia published in the same journal (December, 1986 and January, 1989) dealt with therapy for hypertension. Of the combined total of 167 pages only one related to diet. A recent editorial in the *British Medical Journal* on the treatment of hypertension did not even see fit to mention diet *(11)*.

2. The Need for a Prevention-Based Strategy

The critical need for medicine to reform is based on two sets of facts. First, as described above, medicine has over the period since 1950 been largely unsuccessful in discovering effective new cures for Western disease, the dominating medical problem in Western countries. Second, Western disease is preventable. These facts compel the conclusion that medicine must concentrate on prevention.

This argument is reinforced by the fact that prevention is essentially painless. On the other hand, a strategy based on crisis intervention (waiting for people to become sick before taking action) can never be more than partially successful. Such a strategy will always entail a combination of both the pain and the emotional trauma associated with any form of disease. Even with conditions amenable to treatment, side effects are all too common. For instance, with drug treatment of hypertension, one of medicine' relative success stories, approx one-fifth of patients discontinue taking the drug because of adverse effects *(12–14)*.

There is a story that in World War II the US Pacific fleet was steaming toward the enemy when they picked up an object on their radar. They repeatedly ordered it to get out of the way but to no avail. Finally, the reply came back, "We're a lighthouse." Like the US fleet, medicine will have to chart a new course.

The paucity of resources currently devoted to prevention was highlighted in 1987 by Joseph Califano, former US Secretary of Health, Education and Welfare *(15)*, who pointed out that less than three-tenths of 1% of money spent by the United States on health is for health promotion and disease prevention. Another illustration of the low priority that medicine gives to prevention was revealed in a document published in 1985 *(16)*. The following British organizations have large holdings in tobacco stocks: the Imperial Cancer Research Fund, the Institute of Cancer Research, the Royal College of Pathologists, the Royal College of Surgeons of England, the British Heart Foundation, and the Medical Research Council.

Let us now examine the extent to which physicians dispense preventive advice. A recent survey carried out in California reported that only half of current smokers had been advised by their physician to quit *(17)*. Young people are the group most in need of such advice yet were the group least likely to have received it.

In 1983 Boulton and Williams reported on the everyday work of 16 general practitioners in southeast England *(18)*. They observed that:

> Only one doctor routinely inquired about smoking habits and this was confined to patients newly registering with the practice. There was no indication that any of the other doctors routinely raised the issues of smoking or diet and alcohol with patients. … G.P.s now generally accept that they have a significant part to play in helping their patients to curb their smoking habits. Nevertheless, only a small fraction of the opportunities available for doing so were in fact used and on no occasion was advice supplemented by printed material … advice on health behaviors was viewed almost entirely as a means of dealing with current problems. The doctors seemed to have little notion of health promotion or the role of health education for currently healthy individuals. … The fact that dietary factors implicated in the etiology of cardiovascular disease were not discussed at all could also be seen as a major lost opportunity for health education.

A recent study confirmed the low rate with which lifestyle advice is given to patients. The OXCHECK trial studied patients living in the Bedfordshire area (north of London) and who had visited their general practitioner within the previous year *(19)*. Among smokers, 27% had received advice on the habit. Only 7% of the inactive subjects and only 8% of those with a heavy alcohol intake had received appropriate advice. Among those with several risk factors for cardiovascular disease, only 24% had received dietary advice.

Studies carried out around 1984–1986 also documented the low rates with which doctors in Britain give lifestyle advice to their patients, particularly in the area of diet, alcohol, and exercise *(20–22)*.

A survey of US physicians indicated that only 15% counsel most of their patients with poor exercise habits about regular exercise *(23)*.

In the remainder of this chapter specific means by which medicine can adopt a prevention-based strategy are discussed.

3. A New System of Health

What is being proposed is not a fine tuning of medicine or some shift in orientation. Rather, it is a fundamental change of strategy. In essence, medicine must make the prevention of disease a major part of its day-to-day work.

The situation of modern medicine can be likened to that of the former Soviet Union. A combination of history and philosophy led both down a road paved with good intentions but with little hope of achieving the desired goals. The question is: Can medicine find a Gorbachev?

The key element in disease prevention is persuading people to follow a healthy lifestyle. Medicine should play a pivotal role in this in order to counter decades of its own misleading "public education." Most people have an attitude to health that can be summed up as: Forget about keeping healthy, wait until you're sick, then have a doctor write a prescription. This mind set owes much to doctors largely ignoring the lifestyles of their patients, concentrating the vast majority of their efforts on crisis intervention in the sick, and making regular pronouncements on the latest breakthrough. As we have seen, the accumulated effect of decades of these "breakthroughs" is that most Western diseases are still incurable. Since the medical profession still has widespread public confidence, it has the credibility, as well as the responsibility, to re-educate the public. In plain English it must say: "If you want good health, lead a healthy lifestyle. Don't come to us looking for wonder cures; you'll likely be disappointed."

In a broader sense the above is but one component of what must be the overriding aim: to change people's thinking so that the dominating attitude to health is based on prevention rather than on pouring pills down the esophagus. A useful model for this is the dramatic change in public attitudes seen in recent years on the question of the environment. Thanks to wide media coverage there is now a general understanding that our environment is endangered and that radical action must be taken.

A World Health Organization Study Group *(24)* considered this question. Their report states:

> The professional organizations should be asked to encourage community physicians and general practitioners, in particular, to take a leadership role in developing a community-based program for the prevention of chronic diseases. A professional organization can also be asked to see how best to initiate the changes in attitude needed if ideas on medical care are dominated by a demand in the community for effective therapeutic advice and management. If physicians can play a leading role, this will help to promote the active participation of the whole health care team in a multifaceted community-based

prevention program. The wide range of personnel involved should be recognized, e.g., nurses, pharmacists, dentists, health educators, public health personnel in factories, offices, and the retail trade, as well as health personnel involved in schools and in the community care of the sick, infirm, and elderly.

This brings us to the question of how public attitudes are to be shaped. In essence, the medical profession must seize every opportunity to deliver its message.

The obvious starting place is the doctor's office. By visiting his or her doctor, a patient demonstrates a certain confidence in the doctor's medical knowledge and a willingness to follow prescribed advice. Consequently, health education in the doctor's office is likely to fall on receptive ears. This is supported by the fact that patients advised by their doctor to quit smoking often do so (25–27). Potentially, if this could be achieved on a mass scale, the benefits to the national health would be immense and of far more value than, for instance, all advances in cancer therapy over the last 40 years. Moreover, the financial and human cost of disease would be massively reduced.

To be successful the prevention message must permeate all sections of the community. The most efficient means to achieve this is via the mass media. Medicine therefore needs to study and apply modern techniques of public education.

Schools are a prime target for health education. Accordingly, medical planners should cooperate with education authorities so that all aspects of health education are integrated into the curriculum.

The place of work should also be targeted. Many employers have already created their own programs but there remains enormous scope for further development. If medical personnel went to employers and offered antismoking clinics, nutrition counseling, advice on exercise programs, and so forth, the lifestyles of large numbers of people would, no doubt, be benefited. An excellent example of a worksite health promotion scheme was implemented by the Johnson & Johnson Company *(28)*.

Much of this activity does not require physicians but could be undertaken by nurses, dieticians, or other health care professionals. The time is ripe for creating a new type of health care worker to specialize in prevention. This health promotion specialist could be trained in several relevant disciplines, particularly nutrition, physical fitness, the role of lifestyle factors in disease, and techniques of behavior modification. The health promotion specialist could be employed by school boards, homes for the elderly, local government, or by large employers in the private sector. At risk of straying outside the scope of this book we can note that the health promotion specialist could also dispense advice in such areas as sexually transmitted diseases, alcohol abuse, and accidents.

The next question is: Who would control the health promotion specialist? So far I have referred to medicine as being the midwife of the health revolution. There is every possibility that the medical profession is too set in its attitudes to play such a role. Indeed, it has been little more than a spectator in those advances in preventive medicine that have occurred in recent years. It is therefore possible that medicine is incapable of being galvanized into action. If so, it may well fall on central or local government to create the new system.

But what about the cost of these initiatives? In actuality, they are likely to be remarkably low. This was demonstrated by Farquhar et al. *(29)* in their Five-City Project. Each adult in the target cities received an average of 527 educational episodes over 5 years (about 26 h/adult). This was achieved at a per capita cost of a mere $4/yr.

It must be stressed that medicine alone cannot bring about the new system discussed here. Much is beyond its power and demands concerted action by government and industry. These aspects have been discussed elsewhere *(30)*.

4. The Problem of Medical Schools

Many of the problems of medicine can be traced back to the medical school curriculum. Students learn huge tracts of biochemistry, physiology, and anatomy. This then forms the basis of diagnosing ailments and using drugs and surgery to fix things. The weak link in this strategy is the limited effectiveness of treatments for Western disease.

In contrast to its massive emphasis on therapeutic medicine, medical education pays little attention to the prevention of Western diseases. If doctors are to be competent to practice preventive medicine, then medical education must be substantially reformed. In short, medical students must receive in depth study of the role of lifestyle factors in causing Western disease. Yet, according to one estimate, US medical schools devote less than 1.5% of their teaching time to health promotion and disease prevention (15).

Central to this problem are the gross deficiencies in the teaching of nutrition. In the United States the Food and Nutrition Board of the National Research Council set up the Committee on Nutrition in Medical Education. After carrying out a detailed study of medical schools it reported in 1985 that only about 20% teach nutrition as a required course (31,32). Perhaps even more alarming, despite numerous appeals, there is no evidence that US medical schools increased their teaching of nutrition between 1984–1989 (33). Where nutrition is taught, the scope of the coverage leaves much to be desired. For instance, the role of nutrition in health promotion and disease prevention receives scant attention. It was also noted that faculty positions specifically devoted to nutrition are rare. An earlier British study reached similar conclusions (34).

In 1981 and 1982 I carried out a survey of final year medical students. This was done at the medical school in Cayey, Puerto Rico (UCCEM), which is accredited by the

American Medical Association. The results revealed an appalling ignorance of nutrition as shown by the following examples:

- Few students had any idea what percentage of calories are obtained from fat or protein in the average diet.
- Although most knew that saturated fats raise the blood cholesterol level, almost no one knew that polyunsaturated fats can lower it.
- About half did not know that the fat in beef steak is mainly saturated and that the fat in sunflower seed oil is mainly polyunsaturated.
- When presented with a list of diet supplements (lecithin, pineapple, gelatin, iron, and vitamin C), seven in ten believed that at least one of them can stimulate weight loss.

Studies of medical students in Britain have painted a similarly poor picture of their nutrition knowledge (35–37). These medical students from Britain, the United States, and elsewhere now constitute today's younger generation of doctors.

It is predictable, therefore, that practicing physicians have a poor knowledge of nutrition. This was indicated in a survey of physicians at Southampton University. Most rated their nutrition knowledge as "poor" or "very poor" (37). Similarly, in a study of physicians in Miami very few realized that diet is a major factor in cancer causation (38). Commenting on their results Schapira and Pozo state:

(The physicians) attribute the causes of cancer to uncontrollable factors such as family history, pollution and radiation. Under these circumstances they are unlikely to modify their own lifestyle and to imbue in their patients the importance of dietary change.

Clearly, there is a tremendous need to hugely expand the teaching of nutrition. Others have proposed how this might be done *(39,40)*. One important aspect of nutrition teaching is to present it in a meaningful and practical way; the student must develop an understanding of the foods people eat and not merely the nutrients present in the food. The teaching of nutrition in this manner also applies to the training of nurses.

Although disease prevention should be the primary aim of nutrition education, the possibility of disease cure by dietary change is also an essential subject for medical students. Thus the role of diet in disease reversal, as described in earlier chapters by Diehl, Anderson, and O'Dea, should be taught.

Apart from nutrition another aspect of medical education that is severely neglected concerns how to persuade people to practice prevention. According to a World Health Organization study group *(24)*:

> Medical education will need to change in many countries to include instruction, demonstration projects, and the development of skills in disease prevention. Education should include some knowledge of the behavioral sciences, in particular information relevant to both individual and community change.

Many valuable ideas have been expressed on this topic *(41,42)*.

Expanding nutrition and health promotion means cutting into established courses. Unfortunately, existing departments, both preclinical and clinical, are more likely to favor expanding their own teaching to include new advances rather than sacrificing teaching hours for nutrition and health promotion. In this regard, it is significant that in the 4-yr period up to 1991–1992 the full-time faculty at US medical schools increased by 20% but faculty positions in public health and preventive medicine rose by only 9% *(43)*. One is tempted to conclude that medical schools, like many other parts of the medical establishment, need to be firmly jolted into new attitudes.

5. Conclusion

The mind set of most doctors is to see therapy as the key to dealing with disease. To a large extent this can be explained by technology worship. This attitude pervades the whole medical system, including medical schools. Medicine is therefore in crying need for radical reform. The day is surely coming, though it is probably still a long way off, that major concerns for a physician will be the patient's lifestyle, namely, diet, smoking, and exercise.

References

1. Canadian Cancer Society. *Canadian Cancer Statistics 1988*. Toronto: Canadian Cancer Society, 1988.
2. Temple NJ, Burkitt DB. The war on cancer: the failure of therapy and research. *J R Soc Med* 1991; **84**:95–98.
3. Silverberg E, Lubera JA. Cancer statistics, 1988. *Ca* 1988; **38**:5–22.
4. Schottenfeld D. The epidemiology of cancer. *Ca* 1981; **47**: 1095–1108.
5. Cairns J. The treatment of diseases and the war against cancer. *Sci Am* 1985; **253(3)**:51–59.
6. Bailar JC, Smith EM. Progress against cancer? *N Engl J Med* 1986; **314**:1226–1232.
7. Becker N, Smith EM, Wahrendorf J. Time trends in cancer mortality in the Federal Republic of Germany: progress against cancer? *Int J Cancer* 1989; **43**:245–249.
8. McBride G. U.S. diet industry under fire. *Br Med J* 1990; **300**:1481,1482.
9. Smith WCS, Lee AJ, Crombie IK, Tunstall-Pedoe H. Control of blood pressure in Scotland: the rule of halves. *Br Med J* 1990; **300**:981–983.
10. Walker ARP, Labadarios D. What are the prospects for improved health and increased longevity? *S Afr Med J* 1990; **78**:383–385.
11. Swales JD. First line treatment in hypertension. *Br Med J* 1990; **301**:1172,1173.

12. Curb JD, Borhani NO, Blaszkowski TP, Zimbaldi N, Fotiu S, Williams W. Long-term surveillance for adverse effects of antihypertensive drugs. *JAMA* 1985; **253:**3236–3268.

13. Medical Research Council Working Party. MRC trial of treatment of mild hypertension: principal results. *Br Med J* 1985; **291:**97–104.

14. Croog SH, Levine S, Testa MA, Brown B, Bulpitt CJ, Jenkins CD, Klerman GL, Williams GH. The effects of antihypertensive therapy on the quality of life. *N Engl J Med* 1986; **314:**1657–1664.

15. Califano JA. America's health care revolution: health promotion and disease prevention. *JAMA* 1987; **87:**437–440.

16. Medawar C. Report on investment in the UK tobacco industry. London: British Medical Organization/Social Audit, 1985; 8,9.

17. Frank E, Winkleby MA, Altman DG, Rockhill B, Fortmann SP. Predictors of physicians' smoking cessation advice. *JAMA* 1991; **266:**3139–3144.

18. Boulton MG, Williams A. Health education in the general practice consultation: doctors' advice on diet, alcohol and smoking. *Health Ed J* 1983; **42:**57–63.

19. Silagy C, Muir J, Coulter A, Thorogood M, Yudkin P, Roe L. Lifestyle advice in general practice: rates recalled by patients. *Br Med J* 1992; **305:**871–874.

20. Davies L, Anderson JP, Holdsworth MD. Nutrition education at the age of retirement from work. *Health Ed J* 1985; **44:**187–192.

21. Coulter A. Lifestyles and social class: implications for primary care. *J R Coll Gen Pract* 1987; **37:**533–536.

22. Wallace PG, Brennan PJ, Haines AP. Are general practitioners doing enough to promote healthy lifestyle? Findings of the Medical Research Council's general practice research framework study on lifestyle and health. *Br Med J* 1987; **294:**940–942.

23. Wells KB, Lewis CE, Leake B, Schleiter MK, Brook RH. The practices of general and subspecialty internists in counselling about smoking and exercise. *Am J Public Health* 1986; **76:**1009–1013.

24. World Health Organization Study Group. *Diet, Nutrition and the Prevention of Chronic Disease.* Geneva: WHO, 1990; 154,155.

25. Multiple Risk Factor Intervention Trial Research Group. Multiple risk factor intervention trial. Risk factor changes and mortality results. *JAMA* 1982; **248:**1465–1477.

26. Hjermann I, Velve Byre K, Holme I, Leren P. Effect of diet and smoking intervention on the incidence of coronary heart disease. Report from the Oslo Study Group of a randomised trial in healthy men. *Lancet* 1981; **ii**:1303–1310.

27. Russell MAH, Wilson C, Taylor C, Baker CD. Effect of general practitioners' advice against smoking. *Br Med J* 1979; **2**:231–235.

28. Breslow L, Fielding J, Herrman AA, Wilbur CS. Worksite health promotion: its evolution and the Johnson & Johnson experience. *Prev Med* 1990; **19**:13–21.

29. Farquhar JW, Fortmann SP, Flora JA, Taylor B, Haskell WL, Williams PT, Maccoby N, Wood PD. Effects of communitywide education on cardiovascular disease risk factors. The Stanford Five-City Project. *JAMA* 1990; **264**:359–365.

30. Temple NJ, Burkitt DP. Towards a new system of health: the challenge of Western disease. *J Comm Health* 1993; **18**: 37–47.

31. Berkow SE, Committee on Nutrition in Medical Education. Nutritional education in U.S. medical schools. *Med J Aust* 1989; **151**:S27–S29.

32. National Research Council Food and Nutrition Board Committee on Nutrition in Medical Education (M. Winick, chairman). *Nutrition Education in U.S. Medical Schools*. Washington, DC: National Academy Press, 1985.

33. Swanson AG. Nutrition sciences in medical-student education. *Am J Clin Nutr* 1991; **53**:587,588.

34. Gray J. *Nutrition in Medical Education*. London: British Nutrition Foundation, 1983.

35. Brett A, Godden M, Keenan RA. When and how should nutrition be taught to medical students? *Proc Nutr Soc* 1985; **45**:13A.

36. Parker D, Emmett PM, Heaton KW. Final year medical students' knowledge of practical nutrition. *J R Soc Med* 1992; **85**:338.

37. Heywood P, Wootton SA. Nutritional knowledge and attitudes towards nutrition education in medical students at Southampton University Medical School. *Proc Nutr Soc* 1992; **51**:67A.

38. Schapira D, Pozo C. Physicians, nurses and medical students knowledge of cancer prevention and nutrition. *J Cancer Ed* 1986; **1**:201.

39. Kushner RF, Thorp FK, Edwards J, Weinsier RL, Brooks CM. Implementing nutrition into the medical curriculum: a user's guide. *Am J Clin Nutr* 1990; **52**:401–403.

40. Lopez SA, Read MS, Feldman EB. 1987 ASCN workshop on nutrition education for medical/dental students and residents—integration of nutrition and medical education: strategies and techniques. *Am J Clin Nutr* 1988; **47:**534–550.
41. Weare K. *Developing Health Promotion in Undergraduate Medical Education*. London: Health Education Authority, 1988. (Available from Departments of Education and Community Medicine, University of Southampton, Southampton, UK).
42. Albright CL, Farquhar JW, Fortmann SP, Sachs DPL, Owens DK, Gottlieb L, Stratos GA, Bergen MR, Skeff KM. Impact of a clinical preventive medicine curriculum for primary care faculty: results of a dissemination model. *Prev Med* 1992; **21:**419–435.
43. Jonas HS, Etzel SI, Barzansky B. Educational programs in United States medical schools. *JAMA* 1992; **268:**1083–1089.

Changes for Health

Marjorie Gott

1. New Perspectives on Health and Well-Being

In this chapter the author addresses the question of health promotion with particular reference to ill health caused by a Western lifestyle. Health promotion requires a combined social as well as a biomedical strategy. The author describes several examples in which health issues, teaching, and choice were part of a larger strategy in health promotion and risk reduction programs, resulting in behavioral change at an individual and community level ("empowerment"). Shortcomings in existing health services and public health policy that impede these changes are also discussed.

During the last two decades public and professional views on health and health care have begun to shift. This has occurred as a result of recognition of the following:

- Heavy investment, both financial and ideological, in curative medicine has not paid off; it is neither efficient nor effective *(1,2)*.
- Because health is a complex phenomenon no single professional group can deliver "health." This also begs the question as to what professionals corporately can, or should, seek to deliver.
- The individualistic approach (one person, one disease) is inappropriate in terms of both disease prevention

From: *Western Diseases: Their Dietary Prevention and Reversibility*
Edited by: N. J. Temple and D. P. Burkitt Copyright ©1994 Humana Press, Totowa, NJ

and health promotion. Disease prevention requires mass, as well as "high risk," interventions; health promotion requires a social as well as a biomedical strategy *(3)*.

- The health of people is a valuable asset and has, in the past, received insufficient health (as opposed to disease) investment; the budget for disease prevention and health promotion is very small compared to that for treatment and care.
- New theories of health have been generated that call for major changes in the professional practice of health and other public sector workers *(4,5)*.

Within this context of a general shift from therapeutic to preventive medicine, the promotion of health and well-being is a field of rapidly growing importance. Among both the medical community and the general public, people are increasingly recognizing that tackling the major health threats we face, from cancers and coronary heart disease (CHD) in the West to malnutrition and water borne diseases in the Third World, demands changes in lifestyle and policy, rather than an increase in the provision of high technology medicine. At the same time concepts of health are shifting away from a focus on "absence of illness" models to those that stress positive and holistic well-being.

Against this backdrop the nature of disease itself has also changed. Until the turn of the century the major "killer" diseases in Western countries were the infectious ones (e.g., measles, poliomyelitis, and diptheria in children, and tuberculosis in adults). Today the major "killer" diseases, notably cancer, CHD, and stroke, are ones that are associated with individual lifestyle *(6,7)*.

2. The New Public Health

Yet this shift to "lifestyle" to account for health and ill health is, in itself, an account in transition. Recognizing the influence of public health measures in reducing the incidence of infectious diseases in the last century, McKeown *(8)* perceived that these efforts far outweighed the contribution of medicine. In the latter part of the present century the lessons learned a century ago have been rediscovered and reborn as "The New Public Health." For instance, Ashton and Seymour *(9)* advise,

> The rational approach would appear to be to tackle those behavioral and environmental factors and their determinants which affect health in the twentieth century with the same vigor and determination that was applied to their predecessors in the nineteenth century.

They go on to advocate the need for widespread coordinated social policy measures.

The major forces for the development and growth of the New Public Health movement have been the European Office of the World Health Organization (WHO) in Europe, and Health and Welfare, Canada. Drawing on the thinking of such people as Lalonde in Canada *(10)* a WHO Declaration was formulated in 1977 *(11)*. It recognized the rights of all citizens to a level of health that would permit them to lead a socially and economically productive life. In 1977 the WHO "Alma Ata" Declaration was signed. This recognized Primary Health Care (PHC) as the key to reducing health inequalities and sees PHC as the way to achieve the goal of Health For All (HFA).

By the mid-1980s it was recognized that if there was to be any significant progress toward reducing health inequalities, some action strategies needed to be developed. This was the purpose of the Ottawa Conference in 1986 *(12)*. Following the conference, a Charter for Action was published; it addressed five key issues:

- Building "healthy" public policy;
- Creating supportive environments;
- Strengthening community action;

- Developing personal skills;
- Reorientating health services.

3. Western Diseases: The UK Response

UK government policy in relation to reduction of Western diseases was identified in the 1987 White paper *Promoting Better Health (13)* and most recently in *The Health of the Nation (14)*. Like WHO policy makers, the UK government identified the importance of PHC in health promotion (disease reduction) work. Responsibilities for health promotion are set out in both the National Health Service (NHS) reforms *(14)* and related papers on the revised General Practitioner (GP) Contract.

The UK government does not share the vision of health espoused in the Ottawa Charter. Current UK government policy is based on a medical reductionist ("symptom swatting") and commodities view of health (healthy bodies, healthy people fit to work). Equity is taken to mean access to a range of physical measures and professional advice about how to adapt individual behavior. Choice is based on whether one chooses to take the health care measures provided. Responsibility for ill health thus becomes lodged with health workers who are required to meet "health" targets, and individuals who are required to help them do this by being the willing recipients of their interventions.

UK health promotion practice is confused and inconsistent. Lip service is paid toward WHO ideologies of health, but practice remains medically dominated and "risk focused." The principles of the Ottawa Charter require the formulation of public policy, the creation of supportive environments, strengthening community action, developing personal skills, and the reorientation of health services. But this conflicts with vested interests in health provision and organization in ways that subvert the coherence of the Health For All philosophy. Instead of generating alternative relationships between individuals and the social structures with which their health and health care interact, health promotion philosophies and the professional bodies responsible for their implementation in the United Kingdom are incorporated into traditional rationales and practices. The focus is exclusively on the individual. Meanwhile, the role of social contexts in shaping obstacles and opportunities is ignored. Recent UK policies relating to health provision exacerbate the division between health workers' awareness of socioeconomic dimensions and their ability to address these dimensions in practice *(15)*.

This division, together with escalating role conflict, is set to sharpen in the immediate future for the major caregiving group, namely nurses. Nurse curricular reform (Project 2000) *(16)* runs alongside NHS reform, but they draw from quite different ideologies; one, P2000, is concerned with redressing inequalities in health experiences and expectations, the other, NHS, with efficient and effective ways of promoting individual self-care. The role of nurses in the latter is to screen people. Recording physiological and behavioral data about people, as ends in themselves, is unlikely to affect the incidence of disease. Recognition of this act is, in the longer term, also likely to be demoralizing for nurses. This is not to imply that screening and data collection are unnecessary activities, rather it is to signal that they are not ends in themselves but part of a much broader approach to improving health. Mapping patterns of disease by aggregating individual characteristics gives important information that helps in recognition and planning for reduction of regional and local morbidity and mortality inequalities. To successfully do this, however, social data need to be matched with medical data. This needs to be followed up by the implementation of the social policies that the data suggest are necessary.

4. The Myth of "Lifestyle"

The issue of beliefs and behavior in relation to health is controversial, but the field becomes less politically charged by focusing on the characteristics of population members, rather than on the characteristics of the population as a whole. Because of this, health promotion policy and thinking in general, and especially in the area of CHD, has erected the concept of "lifestyle" as the major vehicle for health intervention. The Stanford Heart Disease Prevention Program (17), the North Karelia Project (18), and the Minnesota Heart Health Program (19), to name but three, focus their activities on lifestyle with particular emphasis on risk factor analysis and change. After more than two decades of research, experimentation, and evaluation there exists no conclusive evidence that the emphasis on risk factors alone actually achieves the fundamental aims of health promotion (20,21).

With no clear evidence to demonstrate the benefits of the lifestyle focus in isolation from its social context, it must be concluded that its major impact operates at the ideological level, serving to reinforce the power base of medicine. Professionals use "lifestyle" as an interventionist approach to legitimize continuation of their traditional disease related activities. Policy makers use it as an excuse to evade responsibility for the social causes of ill health. Ross (22) notes that societies that we call primitive usually took environmental influence on the human condition for granted and understood how individual health depends on it. In contrast, Ross charts the rise of a savagely individualistic ethos in the heyday of capitalist industrialization that interprets high rates of illness, accidents, and death among the poor as being mainly owing to personal inadequacy.

5. Disease Prevention and Motivation

In response to the recognition of common "casual" factors in Western diseases there are increasing calls for preventive interventions that are global rather than specific in their focus: healthy citizens as opposed to healthy hearts. Global intervention recognizes the complexity of social behavior and the competing rationales and imperatives for social action. Global interventions are more likely to be successful in terms of sustained behavior change than are victim blaming interventions that focus on single diseases (23–25).

Temple and Burkitt (26) have recently suggested that a crucial area of research that has been neglected is behavior modification. They call for greater investment of effort and finance in preventive activities, particularly in relation to understanding the factors that cause Western disease. Referring to cancer prevention they comment:

> It would be myopic to view cancer in isolation from other degenerative, particularly Western, diseases. There is a great overlap between the factor which cause or prevent all Western diseases. Moreover, a prevention campaign is likely to be far more successful in motivating people if the goal is prevention of the diseases that cause 80 per cent mortality (all Western disease) rather than those that cause 21 per cent (cancer only).

These authors recognize the crucial role that motivation plays in behavior change.

Similarly, Kelly et al. (27), reporting on recent research on behavior change into six lifestyle areas, found that:

> Behavior change was poorly predicted by beliefs, support and self-efficacy for most lifestyle areas ... findings strongly suggest that motivation is a very important intervening variable when evaluating health promotion and resulting behavior change.

Motivation to change is likely to be influenced by the real opportunities afforded by a supporting social system *(28)*. This can be created by the generation of healthy public policy.

6. Promoting Health and Reducing Disease

There is now growing agreement that an improvement in public health requires simultaneous action on a broad front, and at a number of levels. The need is for multiple prevention strategies, not just case finding of risk factors (traditional "prevention"). Interventions need to be context sensitive and the gap between knowledge and behavior change needs to be addressed by enabling social policy actions and primary health care interventions at a number of levels as detailed below.

6.1. Health: International Aspects

The international agenda on action to reduce Western diseases has been set by WHO in Europe, and by people such as Marc Lalonde and Irving Rootman in Canada *(29)*. In addition to recognizing the effect of social influences on health and disease, the necessity for a different approach on the part of health care practitioners has been identified. What is called for is greater teamwork among all concerned, and better intersectoral collaboration in such areas as health, housing, and employment *(12)*. The need now is for professionals, as well as the public, to change, and for national governments to support the international collaboration that is required. But collaboration should not mean imperialism. The great danger for the new democratic central European countries is that of Western cultural imperialism, as new markets are created for glamorous Western products, many of which are health damaging.

One way forward is through genuine partnership with, and collaboration between, coalitions of people and practitioners in both parts of Europe. The Tipping The Balance Towards Primary Health Care (TTB) is an example of just such a collaboration (Yugoslavia was a founding partner, the Baltic countries are now joining). The TTB Project is a unique, multinational initiative involving primary health care workers, researchers, and local populations in seven European countries. Its underlying rationale is that of the Health For All (HFA) movement *(11)*. HFA philosophy underpins local action research projects in the countries described. The aim of the program is to implement the HFA intention of making primary health care (PHC) the focus of health protection and promotion activities. TTB is a European Community (EC) funded program, run in collaboration with WHO. For a fuller description *see* Godinho, 1990 *(30)*. Project objectives are to:

- Establish methods of promoting healthier lifestyles in the community;
- Improve intersectoral collaboration;
- Reorient health care from the secondary and tertiary care sector to the primary care sector;
- Improve the health promotion skills of health practitioners.

An initial task for the team was to identify what partners understood by the term Primary Health Care. The fact that the organization of PHC varies widely between "developed" countries has been noted by WHO *(31)*:

> In some countries well coordinated teams form the established and recognized first point of contact with the official health system. In others, access to health care may be through general practitioners, specialists or nurses, working alone. In many countries, people belonging to particular social groups or living in particular geographical areas are still without good access to primary health care services, while in other instances preventive and rehabilitative services are less generally available than curative services.

In addition to differences in levels of, and access to, PHC services between countries, there are also differences of interpretation. Vuori *(32)* has remarked on the distinction between primary *health* care, and primary *medical* care; a distinction that he believes few, particularly health care workers, recognize.

The difficulty in achieving mutual understanding about the role and remit of PHC was recognized at the outset of the TTB project; it was unanimously agreed that the operational concept of PHC would be based on the Alma Ata Declaration *(11)*. Four ways of interpreting PHC were identified:

1. As a philosophy, reinforcing the notions of justice and equity;
2. As a principle, incorporating the notion of free, open, and relevant access to all who need it;
3. As the first point of health care contact;
4. As a set of activities concerned with disease prevention and health promotion.

Some TTB projects are worth special mention in relation to the themes of this chapter. The Dudley (UK) project concentrated on traditional preventive work but introduced an additional PHC worker, namely a facilitator. Prior to commencing work the facilitator provided training to the health promotion staff. In addition, a referral support structure was developed that included the health promotion staff and the community dietician. A pilot study was carried out to develop a protocol for nurses to use (with doctor support) when measuring and managing CHD risk factors. This protocol was then used with other teams, although they were encouraged to develop their own protocols. Following screening, doctors and nurses gave verbal and written information about healthy behavior changes that should be made. The health promotion staff set up group education sessions to reinforce messages that had been given and to provide support for behavior change.

Results show that advising patients individually at screening sessions was more effective than attempting to persuade them to attend groups. It is believed that this is because the population did not see group education as part of their culture and were particularly resistant to "quit smoking" groups as they saw smoking as helping them to cope. Dietary advice was much more successful; PHC teams reported a significant fall in cholesterol levels following dietary advice.

In Blekinge, Sweden, two CHD reduction approaches were used; high risk and mass population. It is believed that this dual approach to behavior change is more likely to be successful than a single approach (33). PHC staff in Blekinge recognized that prevention of CHD demands a community strategy in combination with a medical one. The TTB initiative therefore concentrated on:

- Health For All in the local community through CHD focussed community development work;
- Primary prevention of cardiac and arterial disease by opportunistic screening in a PHC district.

Results of the PHC high risk intervention gave a lowering of serum cholesterol after two years for men aged between 40–59 years. Unlike the Dudley study an extra PHC worker was not introduced, and screening was integrated into regular case finding. In addition to the above project, Blekinge County Council utilized mass population strategies in health education. Local study circles have been formed and are run by people in the community. They are provided with specially prepared educational material. To support their work, local groups can request assistance from any of 150 "resource persons" from a list of available politicians and experts. It is believed that discussions will increase awareness about CHD risk factor protection and management and therefore promote healthier lifestyles.

Another important international collaborative venture is also worthy of note: a common European Masters Degree in Public Health (EMPH). The first module of the European course, on Lifestyles For Health For All, was taught in Valencia, Spain, during the summer of 1991 *(34)*. The module can be seen as a "free standing" continuing education course for senior health promoters. More importantly, however, it is the first of four modules in the Association of Schools of Public Health in Europe (ASPHER)/WHO Masters Degree in Public Health program. The EMPH program is seen by Kohler *(35)* as an important landmark in the new "public health renaissance" and "a new step taken in order to bridge the sometimes wide and difficult implementation gap between ideas and reality." WHO and ASPHER have started a joint project to develop curricula based on WHO's targets, which could, in due course, build up to a European Masters in Public Health program. This would then serve as a framework for the training of public health and related professionals for leadership positions in the field. The Valencia course was designed with this possibility in mind. It was repeated in Goteborg, Sweden, in the summer of 1992. The overall aim of the course is to enable students to put into practice the philosophy and ethics of HFA in a European context. Course goals are:

- To develop the skills of participants so as to enable them to promote healthy lifestyles within their communities;
- To establish a European register of health promotion projects and a live network for future collaboration.

The international collaborative course team approach to teaching is seen as vital. Course team members should successfully model the principles of HFA; in other words practice what is preached. The first (Valencia) Course Team comprised Spanish, Yugoslavian, British, and Finnish teachers, who worked together exceptionally well, and were drawn from the four schools in the consortium (Valencia, Zagreb, Liverpool, and Goteborg).

Eighteen students from eight different countries attended the first course. They came from Spain, Portugal, Belgium, England, Germany, Croatia, The Netherlands, and Bulgaria. They were doctors, nurses, educators, sociologists, and managers. All are in senior or middle rank positions in their own country and working in health promotion related work; some directly (for example, running health promotion programs), others less directly (for example, training health professionals).

Evaluation of the first course was very positive. A major goal, the foundation of a live European Register of health promotion projects, was achieved. Three linked areas of project activity were established:

- Reorientating professionals;
- Reorientating environments;
- Reorientating lifestyles.

Projects are established based on common concepts and criteria; all are multinational and multidisciplinary, and model the principles of "The New Public Health."

6.2. Health: UK Aspects

Policy in relation to the reduction of Western diseases in the United Kingdom was described earlier. Within the United Kingdom, however, two quite different approaches are occurring: narrow disease reductionism in England, and broader health promotion in Wales.

In addition to the extremely narrow approach in England is the paradox that the UK government, because of its reluctance to adopt fiscal restraints on the legal drugs (alcohol and tobacco), is actually promoting ill health. O'Neill (6) in 1984 remarked:

> What folly for a government to reap vast taxes for its exchequer from tobacco and alcohol only to have to spend as much, or more, on coping with their effects on health!

Interestingly, exactly the same view was put to the author by a GP in a personal communication during the summer of 1991:

> The government could do a lot more. A lot of these things attach to lobbying the food industry. They are reluctant to do this, one can only speculate as to why. The government needs to do something in some fiscal way about high cholesterol foods ... and to subsidize healthy food. Millions are spent treating people with heart disease; having no campaign to eat healthy food is crazy ... it's schizophrenic.

A much broader approach to health promotion and disease prevention is evident in Wales. That country has a single national agency, the Health Promotion Authority, which coordinates all health promotion activity in the country. Its major focus has been the reduction of CHD rates. "Heartbeat Wales," launched in 1985, works with all sectors of the community *(36)*. It has joined forces with food retailers and restaurants (to provide healthier eating choices), industry (to promote healthier workplaces), and environmental health officers. There is an annual Industrial Award for those employers who do most to improve the health of their workforce.

The Welsh approach, as opposed to the English one, recognizes the centrality of healthy contexts if healthy choices are to be made by individuals. The emphasis is on teamwork, collaboration, and community participation. The agenda is also broad. What began as a CHD agenda now encompasses health damaging lifestyles per se. Early indications are that this approach is successful *(36)*.

6.3. Health: Local Aspects

In much of the foregoing discussion, stress has been placed on the importance of involving people at a grassroots level in social change strategies. This is vital if the rhetoric of health promotion and disease reduction is to become

meaningful in terms of health experiences and expectations. Local action requires local (context) sensitivity and
so cannot be based on an inflexible plan. However, common principles for action can be identified, as has been
done in the WHO Healthy Cities movement *(9,37)*.

The intention of "Healthy Cities," of which there are
over 30 throughout Europe, is to make healthy choices the
easy choices for local populations. Under the Healthy Cities umbrella important collaborative initiatives are occurring. One, focusing on nutrition, is the SUPER project *(38)*.
It is designed to exchange ideas and use available information and experience about shopping and food choice in
different European cities. Its objectives are:

1. To review food patterns, nutritional problems, and food
   policy issues in each of the cities, taking account of
   cultural differences;
2. To identify "at risk" groups and problem areas in each
   city with respect to nutrition and retail activities;
3. To develop a nutrition promotion program for the
   supermarkets in the different areas by bringing
   together supermarket managers, schools, and people
   from local communities;
4. To evaluate these programs by assessing dietary
   changes, changes in knowledge and attitudes within
   target groups, and changes in retail activities in the
   different areas;
5. To investigate similarities and differences between
   cities with respect to nutrition and retail activities;
6. To use survey evaluations to influence central and
   local food policies in Europe; and
7. To develop a resource pack that can be used by non-
   professional groups.

The SUPER project models many of the principles for
successful change identified throughout this chapter. It
works on contexts rather than individuals, on a broad, rather
than a narrow, front, and is "bottom up" and therefore

locally relevant. It is also underpinned by a network of collaborative support at a number of decision making levels.

7. Conclusions

Changing for health, instead of working for illness, requires an ideological shift and a pragmatic commitment on the part of health professionals and politicians alike. The general public is, in the main, aware of the causes of Western diseases, and also some possible strategies for their reduction *(39)*. Professionals and policy makers need to acknowledge this and to form coalitions with them to reduce the incidence of "killer" Western diseases. Some possible ways forward have been identified in this chapter. Foremost among these is the need for the individual to be abandoned as a focus for social policy interventions, and the community (homes, schools, shops, workplaces) to be embraced. It needs to be officially acknowledged that the sickest in society are also the poorest *(40,41)* and redress to be made accordingly. It is not only economically inefficient to keep people unemployed and reduce their opportunity for health, it is morally indefensible.

References

1. Mohan J, Killoran A, Johnson K, McKenzie J. Reducing coronary heart disease in England: targets and implications. *Health Ed J* 1990; **49:**176–180.
2. Creese A. Primary health care—what still needs to be done? Round table discussion. *World Health Forum* 1990; **11:**359–366.
3. Whitehead M. *Swimming Upstream—Trends and Prospects in Education for Health*. London: Kings Fund, 1989.
4. Nutbeam D. Health promotion glossary. *Health Prom* 1986; **1:** 113–127.
5. Siler-Wells G. Public participation in community health. *Health Prom (Canada)* 1988; **27(1):** 7–23.

6. O'Neill P. *Health Crisis* 2000. London: Heinemann Medical Books, 1983.
7. Smith A, Jacobsen B. *The Nation's Health. A Strategy for the 1990's*. London: Kings Fund, 1988.
8. McKeown T. *The Role of Medicine: Dream, Mirage or Nemesis?* London: Nuffield Trust, 1976.
9. Ashton J, Seymour H. *The New Public Health*. Milton Keynes: Open University Press, 1988.
10. Lalonde M. *A New Perspective on the Health of Canadians: Working Document*. Ottawa: Government of Canada, 1974.
11. World Health Organization. Regional Office for Europe. *Alma Ata Declaration*. Copenhagen: World Health Organization, Regional Office For Europe, 1978.
12. *Ottawa Charter for Health Promotion*. Ottawa. Copenhagen: World Health Organization, Regional Office For Europe, 1986.
13. Department of Health and Social Security. *Promoting Better Health: The Government's Programme for Improving Primary Health Care*. London: HMSO, 1987.
14. Department of Health and Social Security. *The Health of the Nation. A Consultative Document for Health in England*. London: HMSO, 1991.
15. Gott M, O'Brien M. *The Role of the Nurse in Health Promotion: Policies, Perspectives and Practice*. London: Department of Health and Social Security, 1990.
16. *Project 2000, UKCC: A New Preparation for Practice*. London: United Kingdom Central Council For Nursing, Midwifery and Health Visiting, 1986.
17. Farquar JW. The community-based model of life style intervention trials. *Am J Epidemiol* 1978; **108:** 103–111.
18. Puska P, Nissinen A, Tuomilehto J, et al. The community based strategy to prevent coronary heart disease: conclusions from ten years of the North Karelia project. *Ann Rev Public Health* 1985; **6:**147–193.
19. Blackburn H, Luepker R, Kline FG, et al. The Minnesota heart health program: a research and development project in cardiovascular disease prevention. In: Mattarazzo JD, Weiss SM, Herd JA, Miller NE, Weiss SM, eds. *Behavioral Health. A Handbook of Health Enhancement and Disease Prevention*. New York: John Wiley, 1984; 1171–1178.
20. Turpeinen O, Karvonen MJ, Pekkarinen M, Miettinen M, Elosuo R, Paavilainen E. Dietary prevention of coronary heart disease: The Finnish Mental Hospital Study. *Int J Epidemiol* 1979; **8:**99–118.

21. Mitchell JRA. Diet and arterial disease—the myths and the realities. *Proc Nutr Soc* 1985; **44:** 363–370.
22. Ross E. The origins of public health: concepts and contradictions. In: Draper P, ed. *Health Through Public Policy.* London: Green Print, 1991; 26–40.
23. Tannahill A. Health education and health promotion: planning for the 1990's. *Health Ed J* 1990; **49:**194–198.
24. Allison KR. Theoretical issues concerning the relationship between perceived control and preventative health behaviour. *Health Ed Res* 1991; **6:**141–151.
25. Ben-Sira Z. Eclectic incentives for health protective behavior: an additional perspective on health oriented behavior change. *Health Ed Res* 1991; **6:** 211–229.
26. Temple NJ, Burkitt DP. The war on cancer: failure of therapy and research. *J R Soc Med* 1991; **84:** 95–98.
27. Kelly RB, Zyzanski SJ, Alemagno SA. Prediction of motivation and behavior change following health promotion: role of health beliefs, social support, and self-efficacy. *Soc Sci Med* 1991; **32:**311–320.
28. Hellman R, Cunnings KM, Haughey BP, Zielezny MA, O'Shea RM. Predictors of attempting and succeeding at smoking cessation. *Health Ed Res* 1991; **6:**77–86.
29. Rootman I. Knowledge development: a challenge for health promotion. *Health Prom (Canada)* 1988; **27(2):** 26–30.
30. Godinho J. Tipping the balance towards primary health care; managing change at the local level. *Int J Health Plan Mgnt* 1990; **5:**41–52.
31. World Health Organization. *Evaluation of the Strategy for Health for All by the Year 2000.* Seventh report on the world health situation. Vol 5. Copenhagen: World Health Organization, Regional Office for Europe, 1986.
32. Vuori H. Primary health care in Europe: problems and solutions. *Commun Med* 1984; **6:**221–231.
33. Corson J. Heartbeat Wales: a challenge for change. *World Health Forum* 1990; **11:**405–411.
34. Gott M. *Report of the First European Health Promotion Training Course.* Valencia, Spain: Institut Valencia D'Estudis En Salut Publica, 1991.
35. Kohler L. Public health renaissance and the role of schools of public health. *Eur J Public Health* 1991; **1:**2–9.
36. *Health for All in Wales. Strategies for Action.* Policy document. Health Promotion Authority For Wales, 1990.

37. World Health Organization. Regional Office for Europe. *Healthy Cities: Action Strategies for Health Promotion*. First project brochure. Copenhagen: World Health Organization, Regional Office For Europe, 1990.

38. Vaandrager HW, Colomer C, and Ashton J. Inequalities in nutritional choice: a baseline study from Valencia. *Health Promotion Int.* 1992; **7**:109–118.

39. Farrant W, Taft A. Building healthy public policy in an unhealthy political climate: a case study from Paddington and North Kensington. *Health Prom* 1988; **3**: 287–292.

40. Smith GD, Bartley M, Blane D. The Black Report on socio-economic inequalities in health 10 years on. *Br Med J* 1990; **301**: 373–377.

41. Black D. Inequalities in health. *Public Health* 1991; **105**: 23–27.

PART V

MEDICAL RESEARCH

Medical Research

A Complex Problem

Norman J. Temple

On the big issues the experts are very rarely right.

Peter Wright in *Spycatcher* (1987)

1. Introduction

In 1974 I was working on my PhD in biochemistry. At that time I firmly believed that biochemistry was the leading route to successful medical research. Over the previous year I had read two of Cleave's books on the concept of refined carbohydrates as the cause of what we now call Western disease *(1,2)*. His arguments greatly impressed me and I started to speculate on how refined carbohydrates might cause coronary heart disease and other maladies. In a letter to Cleave I pointed out that the poor vitamin and mineral content of refined carbohydrates might be an important factor in these diseases and suggested a few biochemical mechanisms.

The reply from Cleave was anything but what I had expected:

I fear it will take you many years to think things out in terms of simplicity, but if you could analyze even one case of feminine intuition, you would find

From: *Western Diseases: Their Dietary Prevention and Reversibility*
Edited by: N. J. Temple and D. P. Burkitt Copyright ©1994 Humana Press, Totowa, NJ

there some indication of what is involved. Perhaps one day you'll be married and your wife will give you such a case! I have referred in the preface (of my book) to your great-grandmother—your great-grandfather would be relatively useless! ... You will, I fear, regard my present letter as utterly incomprehensible, and the art of seeing the wood for the trees will be the more difficult for you the more you know.

I am terribly sorry, but if you want to achieve salvation you will have to think far more simply than you do at present.

Cleave maintained his position that the only effects of foods on the body he would consider were those that were simple to understand. His litmus test for a particular concept was whether a barmaid or a grandmother could understand it. Even such areas as vitamins, minerals, and blood cholesterol were out of bounds.

Slowly I came round to Cleave's way of thinking. Five years after I received the above letter things were brought to a head while I was working in the Department of Biochemistry at the University of Surrey, England. My colleagues and I were investigating how carcinogens in cigarette smoke cause cancer. The more I attended research seminars, the more I realized how extraordinarily complex the whole subject is and the more I became convinced that Cleave was essentially right. Real progress on cancer depends on finding simple answers to simple questions.

But why stop at cancer? The same arguments apply equally well to all other areas of medical research.

2. What Is Simple and Complex Research?

On the basis of the above experiences I developed the concept of simple research as the most fruitful means of medical research (3).

Medical research is divided into two types, "simple" and "complex." Major forms of simple research are studying the relationship between lifestyle and subsequent disease, investigating whether alterations in lifestyle will prevent or cure disease, and carrying out similar studies on animals. Simple research, as defined here, includes measuring body parameters that may be closely correlated with health (e.g., the gastrointestinal transit time or the blood level of cholesterol, glucose, insulin, or nutrients). (It should be noted that my definition of "simple" is less restrictive than Cleave's.) The key feature of simple research is that the results provide clear clues as to how disease may be prevented or cured. Complex research, on the other hand, deals with the intricate mechanisms of disease that, generally speaking, are complex in the extreme.

Denis Burkitt (4) has used the analogy of a chemical plant to illustrate this (Fig. 1). The inlet pipes supply the raw materials (lifestyle factors). Within the processing unit multistage and exquisitely complex processes take place. Eventually, the product (health or disease) leaves through the outlet pipe. It is not necessary to comprehend the inner workings of the chemical plant to see the relationship between what goes in and what comes out; one need only study how changes in the initial ingredients affect the product. So with disease. We can learn the relationship between lifestyle factors and health without understanding the mechanisms by which disease comes about. That, in a nutshell, is the rationale for simple research.

Complex research, by contrast, studies the detailed mechanisms of disease etiology. Using the above analogy it attempts to comprehend the inner workings of the chemical plant. Because the body is of such extreme complexity, it is generally a multidecade process to gain a sufficient understanding of how disease comes about and to be able to apply this knowledge in a practical manner.

RAW MATERIALS
(lifestyle factors)

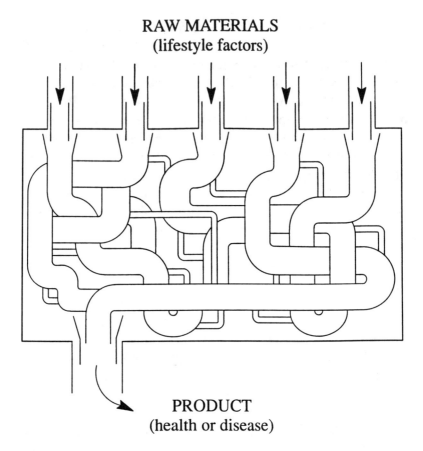

PRODUCT
(health or disease)

Fig. 1. Analogy between a chemical plant and the human body.

It is impossible to be precise but it is certainly true that complex research has swallowed the bulk of the research budget in most medical areas. Nevertheless, most of our useful information has come from simple research. It is not difficult to see why. Complex research produces vast amounts of information but by the very nature of the human body we never see a complete picture and are seldom able to put these discoveries to good use. Simple research, on the other hand, time and time again, points to which lifestyle factors (diet, exercise, and so on) cause or prevent which diseases and therefore what we can do about it. Even if all the facts concerning disease mechanisms were somehow assembled, there is no guarantee that we would actually understand disease. This is because we would need to integrate more information than the human mind can cope with. Thus trying to gain a full understanding of disease etiology is likely to be self-defeating; it is today's search for the Holy Grail.

Since disease mechanisms are in the great majority of cases highly complex, simple research (by definition) does not concern itself with them. Sometimes, however, disease mechanisms are simple. Examples are acid production from food in the mouth causing tooth decay and lack of dietary fiber causing diverticular disease. In such cases studying mechanisms become a valuable part of simple research as we get the best of each world: seeing the whole picture but without entering a quagmire of complexity.

The usefulness of simple research is mainly in the realm of disease prevention. This is because simple research is mostly concerned with determining which lifestyle factors cause or prevent particular diseases. Occasionally, though, the health rules that tell us how to prevent disease also apply to the treatment of that disease. Examples include CHD, obesity, type II diabetes, and hypertension and were discussed in earlier chapters by Diehl, Anderson, and O'Dea. Complex research, by contrast, deals mainly with therapy. This is particularly the case with pharmacology; studies of the detailed mechanisms of disease sometimes lead to new drug treatments.

Let us now summarize simple research. First and foremost, it efficiently leads us to the causes of disease (and sometimes to their cure), whereas with complex research, more often than not, we sink into a quicksand of disease mechanisms. It was stressed in the earlier chapter "Organized Medicine," by Temple, that the focus of our efforts against Western disease must be based on prevention, which, in turn, demands that research concentrates its efforts into discovering how this can be achieved. This points to the second advantage of simple research: It is largely relevant to prevention, whereas complex research is mainly relevant to therapy.

3. Simple and Complex Research in Practice

In previous papers various examples were given of the superiority of simple research (3,5). Further examples are given here. A classic example of simplicity at work is the case of vitamin C. In 1747 James Lind, a surgeon's mate in the Royal Navy on HMS Salisbury, carried out what may well have been the world's first clinical trial. He demonstrated that oranges and lemons cure scurvy. As a result Gilbert Blane in 1796 had lemon juice officially introduced in the British navy to prevent scurvy. This simple and inexpensive measure prevented more medical problems than had previously been attributed to all other causes, all battle casualties, and all shipwrecks, and virtually doubled the proportion of fit men in the British navy. The alternative approach to conquering scurvy would have been to search for its biochemical basis. This would have entailed a delay of two centuries until Albert Szent-Gyorgyi isolated ascorbic acid followed later by an elucidation of its biochemical role. Had the world waited for the detailed mechanism to be revealed rather than being satisfied with simple observations, untold numbers would have perished.

In cancer research, as in most other areas, complex research grabs the lion's share of resources. This is illustrated by the 1991 budget of the (US) National Cancer Institute (6). With respect to research into the causes and nature of cancer, simple research is mainly confined to two programs, namely, epidemiology and nutrition. Together they receive a scant 8.6% of the total budget. Complex research, by contrast, receives about 36% of the budget; the money going to such area as tumor biology, virology, immunology, and, to a major extent these days, molecular genetics. With treatment oriented research it is much less clear how to differentiate between expenditures on the two types of research. Simple research, however, probably gets less than half. Of the total National Cancer Institute Budget, therefore, simple research receives only about 21–25%.

The situation is no better in the United Kingdom and Canada and may be even more lopsided. One index of where the money goes is advertisements for positions in cancer research. The overwhelming majority of such advertisements refer to complex research. For instance, *Nature* (January 10, 1991) carried an advertisement for graduate studentships at the Imperial Cancer Research Fund in London. The positions were in 35 different projects. Each and every one was for some type of complex research; molecular genetics alone accounted for over half of them.

Another indication of the extent to which molecular genetics is dominating medical research in general—not just cancer research—is to look at the number of papers being published. In the 1991 edition of *Index Medicus* it required 118 pages to list all the papers under the subject heading DNA. Several more pages of papers come under related headings, such as oncogenes and nucleic acids. By contrast the major nutritional areas (vitamins, ascorbic acid, iron, and diet, including dietary fats, dietary fiber, dietary proteins, and dietetics) required only 51 pages. The imbalance is probably even greater than these figures indicate, since many of the nutrition-related papers are actually complex research (mainly studies of the biochemistry of dietary components).

As indicated above, molecular genetics has become of considerable importance in cancer research. The objective of molecular genetics is to gain a total comprehension of how genes express themselves. The reasoning is that if we can only understand oncogenes, then we can learn how to control them and thereby make tumors regress. The race is on to turn oncogenetics into a "magic bullet" against cancer. It is, of course, impossible to predict the future with any certainty, but one can, however, try to learn from the past. What we learn from complex research (and molecular genetics is as complex as research gets) is that a gulf separates the reality from the promise. Indeed, the snail's pace of advances in molecular genetics can be gaged from the fact that the double helix was first described in 1953, yet it was not until the early 1980s that oncogenes emerged as a distinct research field.

In stark contrast to the study of oncogenes an excellent illustration of the superiority of simple research is provided by smoking and lung cancer. The original demonstration in 1950 that smoking is the likely cause of the disease needed little more than a supply of lung cancer patients and a questionnaire *(7,8)*. Refinement of this research tool over the years has given us the prospective and case-control studies. These immensely valuable techniques have facilitated the demonstration that smoking causes several cancers and other diseases.

Researchers, needless to say, have asked the question: "Why does smoking cause cancer?" Analysis of cigaret smoke has revealed at least 2000 chemicals. A huge effort has gone into unraveling how these chemicals cause cancer. Like so many other areas of disease mechanisms, the more one looks, the more complex things get. Numerous obstacles block the road to the answer to the above question *(9–11)*. These include:

1. These chemicals can interact (e.g., one will induce enzymes that alter the metabolism of others);
2. They can often be metabolized by several competing pathways (including pathways that activate carcinogens and pathways which detoxify them);

3. The metabolism varies depending on the species, the tissue, whether it is whole tissue or a subcellular fraction and on the incubation conditions;
4. Cancer is a multistage process with each stage requiring different chemicals; and
5. The whole process is influenced by such extraneous factors as hormones, diet, and personality.

There can be little doubt that the task of properly understanding why smoking causes cancer is one that will take many decades. In the meantime, of course, we can take the straightforward approach to the problem of lung cancer—we can utilize the solution that has been available for fully 40 years—we can avoid smoking.

Research into the dietary control of the blood cholesterol level has parallels with research into smoking and lung cancer. Based on simple feeding studies it was established in the 1950s and 1960s that dietary saturated fat and cholesterol raise the blood cholesterol level. That the biochemical basis of this was unknown caused few problems for dieticians. In the early 1980s Brown and Goldstein demonstrated the crucial role of the low density lipoprotein receptor in the liver. This discovery, the fruit of years of complex research, was technically brilliant. But it is difficult to see what advances it has generated for the control of hypercholesterolemia. Nevertheless, in 1985 Brown and Goldstein received the Nobel Prize in physiology and medicine.

Much the same applies to coronary heart disease research in general. Most of our useful information has come from simple research. In particular, prospective and case-control studies have demonstrated the importance of the blood cholesterol level, blood pressure, smoking, and exercise. On the other hand, countless studies based on complex research—such as studies of the biochemistry of the liver, the platelet, or the arterial wall—do not really tell us what causes the disease or how to prevent it.

The superiority of simple research is also seen in the history of diabetes. More than six decades have passed since Banting and Best discovered insulin. During that time many thousands of studies have been reported concerning the mode of action of insulin, the metabolism of carbohydrate and fat, the biochemical and physiological abnormalities present in diabetics, and so forth. Yet this has not told us how to prevent or cure the disease. On the other hand, the radical change in the dietary treatment of type II diabetes that was introduced in the 1970s, namely, the introduction of the high carbohydrate/high fiber/low fat diet, is a testament to simple research. Clinical trials demonstrated that the new treatment led to an improvement in several key parameters of the diabetic patient *(12,13)*. These papers indicate that the rationale for conducting the trials was the effect of diets high in carbohydrate and fiber and low in fat on such parameters as the blood glucose and triglyceride level. I strongly suspect that an additional reason was the realization that populations that consume a high carbohydrate/high fiber/low fat diet rarely suffer from diabetes *(14)*. It is apparent, then, that simple research on diabetes, mainly by clinical studies, has led to a rapid advance in our understanding of how to treat the disease. Moreover, the new knowledge is equally relevant to the prevention of diabetes.

Obesity teaches us the same lesson. Library shelves groan under the weight of our accumulated knowledge in such forms of complex research as intermediary metabolism and endocrinology. But how many useful advances in treatment have such studies produced? Most useful treatments for obesity, whether based on diet, exercise, or surgery, owe far more to common sense (which is simple research at its simplest) than to complex research. Likewise, the introduction of the high fiber diet represents a valuable step forward in the treatment of obesity. The evidence in its favor comes from simple research similar to that used in support of the high carbohydrate/high fiber/low fat diet in diabetes.

The chapter "Vitamins and Minerals in Cancer, Hypertension, and Other Diseases," by Temple, describes the key role of salt in the causation of hypertension. This immensely important piece of information is based on the following simple evidence: The condition is most common in individuals and populations with a high salt intake and the blood pressure tends to fall when salt intake is reduced.

We shall now look at some examples outside the confines of Western disease. An interesting example of the power of simple research concerns food intolerance. A number of pioneering doctors observed how particular foods induce various symptoms. This led to the development of the exclusion diet, which is now a valuable tool in the diagnosis and treatment of conditions related to food intolerance, such as migraine, childhood hyperactivity, and the irritable bowel syndrome (15–20).

The relationship between neural tube defects (NTD) and folate status is a discovery of considerable importance. Studies were done in England (21,22) and Northern Ireland (23,24) on women who had previously given birth to a child with the condition. It was demonstrated that supplements containing folate and, in some studies, with other nutrients, dramatically reduce the incidence of NTD in subsequent births. The rationale given for carrying out these trials relates only to simple research:

1. The presence of malformations in offspring of animals fed a deficient diet;
2. The poor folate status of women who had given birth to a baby with NTD; and
3. The possibility of diet as an explanation for social class differences in the incidence of NTD (21,22,24).

In theory, this discovery could have come from fundamental research into the biochemistry of neural tube development. But this would have necessitated a detailed understanding of how suboptimal levels of folate (or of other nutrients) affect the neural tube. In reality, of course, there is exceedingly little chance that complex research could yield the crucially important information that was furnished by the above clinical trial.

Another dimension of simple research is to take a general biological perspective. This approach, like the concept of simplicity itself, was strongly advocated by Cleave *(1,2)*. Guiding principles for medical researchers should be:

1. An appreciation that humans are just one animal species among thousands;
2. A consideration of human evolution; and
3. A recognition that the great majority of human disease is caused by deviating from the environment to which we are genetically adapted.

The development of the concept of Western disease is a testimony to these principles. This came from the realization that Western disease arises where the diet and other aspects of lifestyle are far removed from what had existed in evolutionary times.

4. Conclusions

There are several types of research that qualify as simple and all can be endorsed. These are:

1. Population studies (the relationship between environmental, particularly lifestyle, factors and disease incidence);
2. Prospective and case-control studies (similar to population studies but dealing with individuals within a population);

3. Experimental studies on animals (investigating the effect of environmental variables in animal models of human disease); and

4. Controlled trials (testing hypotheses developed in the above studies as to their efficacy for the prevention and treatment of disease in humans).

In all four types of study not only is disease itself recorded but so are parameters closely associated with disease, such as the blood cholesterol level, glucose tolerance and the gastrointestinal transit time.

Where does the investigation of disease mechanisms fit into this scheme? Complex research is almost synonymous with the study of mechanisms, which, as we have seen, is a recipe for failure to make real progress. On occasion, though, a disease mechanism may be studied without loss of simplicity. When that situation arises, the picture becomes clearer. An excellent example of this is that of dietary fiber and colon cancer: Burkitt (25) speculated that the increase in fecal bulk and the speeding of the transit time induced by fiber work to prevent colon cancer. The role of calcium in that disease provides another example. Based on simple chemistry Newmark et al. (26) reasoned that calcium may be protective. A high fat diet increases the concentration of free ionized fatty acids and bile acids in the colon contents and this increases the likelihood of cancer. Calcium can counteract this by converting the free acids into insoluble soaps.

In situations such as these the disease mechanism can be studied and the research can still be categorized as simple. As a result our understanding of the cause of the disease and how it may be prevented is much improved. When a disease mechanism can be studied utilizing simple research, this complements the previous four types of investigation.

A strong advantage of simple research is that its major relevance is in the realm of prevention. With cancer, for instance, we learn a great deal about prevention but almost nothing about cure. Only in some limited areas does simple research pertain to therapy. It is precisely those areas— particularly coronary heart disease, obesity, diabetes, and hypertension—that have been covered in the previous chapters on disease reversal.

Complex research, in the main, is targeted at finding new treatments. Despite receiving the majority of resources expended on research, it has only produced a minor share of the useful information (i.e., that information that can help reduce the burden of human disease).

If complex research is such an inferior strategy, then why are most medical scientists seemingly mesmerized by it? Two explanations present themselves. First, medicine is oriented towards therapy rather than prevention (*see* earlier chapter, "Organized Medicine," by Temple). Accordingly, the primary job given to research is to find new treatments rather than to discover how to prevent disease. This in turn leads to an emphasis on complex research. (Complex research is certainly an inferior strategy, but there is no denying that it is capable of producing therapeutic advances.) Second, medicine has for decades had a fatal attraction for high tech science. (Indeed, this also helps explain why medicine concentrates on therapy and shuns prevention; it is therapy not prevention that utilizes high tech science.) As a result, medical researchers are intrigued with discovering what makes the body tick and using high tech science to achieve this. All of us have been fascinated by medical discoveries. Unfortunately, this fascination has become the *raison d'etre* for much research. We live in a high tech world; medicine has greatly profited by the use of CAT scans, computers, and the like. It is only to be expected, therefore, that medicine should look to high tech science (the higher the better) as the key to successful medical research. Unfortunately, there is little attempt to critically evaluate the grand strategy.

5. Some Suggestions

Medical research should be refocused so that the major share of resources goes to simple research, in particular the study of lifestyle factors in disease. The primary objective is to greatly expand our knowledge of how disease may be prevented.

Currently under way is an enormous collaborative study in molecular genetics—the genome project. It strikes me that a far more useful project would be a major international collaborative study on the role of lifestyle factors in disease. This would take the form of a prospective study several times larger than any similar study yet done. One or two hundred thousand apparently normal men and women would be recruited, blood samples collected, and diet histories and other aspects of lifestyle recorded. The relationship between these parameters and disease incidence over the following 20 or more years would provide a gold mine of invaluable information. Currently under way in Europe is a rather more modest version of this type of study but with a focus on cancer (27). Not only would we refine our knowledge of the major factors related to the major diseases but, for the first time, there would be enough statistical power to reveal the role of minor factors in major diseases and of major factors in minor diseases. Examples of the findings we might expect are the role of micronutrients (selenium, β-carotene, and vitamins C and E) in various cancers, the role of different types of dietary fiber in heart disease, and the relationship of different types of fat to a spectrum of diseases. The Chinese study described by Campbell in the chapter "Diet and Chronic Degenerative Diseases," is a fine example of the great value of jumbo projects.

One important problem of research in the past is that many important areas have been relatively neglected. This is because the majority of resources have been given over to complex research. A major aspect of this is the subject of behavior modification. We must study, for instance, how to effectively dissuade young people from smoking and how to motivate the population to exercise and eat a low fat diet. This needs to be researched at both the individual level (e.g., in the doctor's office) as well at the mass media level. Another example of a neglected area is how to help people to quit smoking. Compared to the vast resources the tobacco giants devote to persuading people to smoke, scarcely anything goes to smoking cessation techniques.

Another good example of a neglected area is that of vitamin deficiencies in patients in nursing homes. Drinka and Goodwin *(28)* recently reviewed this subject and concluded that there is evidence that vitamin supplements lead to improvements in health. Moreover, such supplements are both safe and cheap. They then added,

> Perhaps the most striking conclusion one can draw from this review is the remarkably small number of good studies in this very important area.

They then suggested that one explanation for this is that

> It is difficult to envision a proposal for such an intervention study being greeted enthusiastically by an N.I.H. study section. It is descriptive, non-mechanistic, and altogether too pedestrian to elicit much scientific interest.

Acknowledgment

The figure was produced by Margaret Anderson, whose assistance is gratefully acknowledged.

References

1. Cleave TL, Campbell GD. *Diabetes, Coronary Thrombosis, and the Saccharine Disease*. Bristol: Wright, 1966.
2. Cleave TL. *The Saccharine Disease*. Bristol: Wright, 1974.

3. Temple NJ. Simplicity—the key to fruitful medical research. *Med Hypotheses* 1985; **17:**139–145.
4. Burkitt DP. Dietary fiber and cancer. *J Nutr* 1988; **118:**531–533.
5. Temple NJ, Burkitt DB. The war on cancer—failure of therapy and research. *J R Soc Med* 1991; **84:**95–98.
6. National Cancer Institute. *1991 Budget Estimate*. Bethesda: National Cancer Institute, 1989.
7. Doll R, Hill AB. Smoking and carcinoma of the lung; preliminary report. *Br Med J* 1950; **2:**739–748.
8. Wynder EL, Graham EA. Tobacco smoking as a possible etiologic factor in bronchiogenic carcinoma. *JAMA* 1950; **143:**329–336.
9. Miller EC. Some current perspectives on chemical carcinogenesis in humans and experimental animals. *Cancer Res* 1978; **38:**1479–1496.
10. Slaga TJ, Fischer SM, Weeks CE, Klein-Szanto AJP, Reiners J. Studies on the mechanisms involved in multistage carcinogenesis in mouse skin. *J Cell Biochem* 1982; **18:**99–119.
11. Conney AH. Induction of microsomal enzymes by foreign chemicals and carcinogenesis by polycyclic aromatic hydrocarbon. *Cancer Res* 1982; **42:**4875–4917.
12. Kiehm TG, Anderson JW, Ward K. Beneficial effects of a high carbohydrate, high fiber diet on hyperglycemic diabetic men. *Am J Clin Nutr* 1976; **29:**895–899.
13. Anderson JW, Ward K. High-carbohydrate, high-fiber diets for insulin-treated men with diabetes mellitus. *Am J Clin Nutr* 1979; **32:**2312–2321.
14. Trowell HC, Burkitt DP. *Western Diseases: Their Emergence and Prevention*. London: Edward Arnold, 1981.
15. Mackarness R. *Not All in the Mind*. London: Pan Books, 1976.
16. Egger J, Carter CM, Graham PJ, Gumley D, Soothill JF. Controlled trial of oligoantigenic treatment in the hyperkinetic syndrome. *Lancet* 1985; **i:**540–545.
17. Egger J, Carter CM, Wilson J, Turner MW, Soothill JF. Is migraine food allergy? *Lancet* 1983; **ii:**865–868.
18. Grant ECG. Food allergies and migraine. *Lancet* 1979; **i:**966–969.
19. Mansfield J. *The Migraine Revolution*. Wellingborough, Northants: Thorsons, 1986.
20. Nanda R, James R, Smith H, Dudley CRK, Jewell DP. Food intolerance and the irritable bowel syndrome. *Gut* 1989; **30:**1099–1104.

21. Laurence KM, James N, Miller MH, Tennant GB, Campbell
    H. Double-blind randomised controlled trial of folate treat-
    ment before conception to prevent recurrence of neural-tube
    defects. *Br Med J* 1981; **282**:1509–1511.
22. Laurence KM, James N, Miller MH, Campbell H. Increased
    risk of recurrence of pregnancies complicated by fetal neural
    tube defects in mothers receiving poor diets, and possible ben-
    efit of dietary counselling. *Br Med J* 1980; **281**:1592–1594.
23. Smithells RW, Nevin NC, Seller MJ, Sheppard S, Harris R,
    Read AP, Fielding DW, Walker S, Schorah CJ, Wild J. Fur-
    ther experience of vitamin supplementation for prevention of
    neural tube recurrences. *Lancet* 1983; **i**:1027–1031.
24. Smithells RW, Shepphard S, Schorah CJ, Seller MJ, Nevin
    NC, Harris R, Read AP, Fielding DW. Apparent prevention
    of neural tube defects by periconceptual vitamin supplemen-
    tation. *Archs Dis Childh* 1981; **56**:911–918.
25. Burkitt DP. Epidemiology of cancer of the colon and rectum.
    Cancer 1971; **28**:3–13.
26. Newmark HL, Wargovich MW, Bruce WR. Colon cancer and
    dietary fat, phosphate, and calcium: a hypothesis. *J Natl
    Cancer Inst* 1984; **72**:1323–1325.
27. Cherfas J. Europeans launch diet, cancer study. *Science* 1991;
    254:1448,1449.
28. Drinka PJ, Goodwin JS. Prevalence and consequences of vita-
    min deficiencies in the nursing home: a critical review. *J Am
    Geriatr Soc* 1991; **39**:1008–1017.

Western Disease

End of the Beginning

Norman J. Temple and Denis P. Burkitt

Two full decades have now elapsed since the concept of "Western disease" was proposed. During that time it has gained widespread acceptance. Often this has been in the form of adopting the philosophy underlying the concept—that many of our common noninfective diseases are caused by our environment—but without actually using the term "Western disease." Be that as it may, few can doubt the enormous impact of the Western disease concept.

We can illustrate this by the following example. A recent book produced by the World Health Organization Study Group explored the relationship between diet and the prevention of chronic disease *(1)*. The introductory chapters clearly reveal that the concept of Western disease has strongly influenced the thinking of this group of experts.

By no stretch of the imagination can we conclude that complete success has been attained. The concept of Western disease is not a mere intellectual exercise intended only to persuade the medical profession and medical scientists of a particular viewpoint. Inherent within the concept is the possibility of prevention. We are reminded of the words of Confucius: "The essence of knowledge is having acquired it to apply it." The more one considers the twin facts of the preventability of Western disease and the

From: *Western Diseases: Their Dietary Prevention and Reversibility*
Edited by: N. J. Temple and D. P. Burkitt Copyright ©1994 Humana Press, Totowa, NJ

serious limitations of medicine at curing these diseases, the more one is driven to conclude that the only sensible strategy is one based on prevention. Not only is such a strategy potentially far more effective, but it also avoids the suffering inherent in so much disease. We must bear in mind that even when diseases are successfully treated, pain and anguish are often unavoidable.

This conclusion is reinforced by a consideration of the financial state of medicine. It is a classic situation of an irresistible force (increased demand created by the combined effect of medical advances and an aging population) pushing against an immovable object (financial constraints). The results of this are reduced availability of medical treatment: In the United States tens of millions of the less well off can no longer afford medical insurance, whereas in the United Kingdom there is the more subtle method of rationing by use of a million-person waiting list. Like our worsening environment, the problem is not going to conveniently cure itself. Bold initiatives are required.

We believe the time has now come to commence a long-term campaign on several fronts aimed at making prevention a reality. Many of the component parts of such a campaign have been described in previous chapters. What we need now is action.

The proposals made in this book are, of course, intended only as guidelines, not as a definitive plan. We fully expect major developments on the prevention and health promotion fronts in the next few years. Indeed, this is undoubtedly the new frontier, offering as much opportunity for the prevention of disease as that offered by the discoveries made in the last century concerning infectious disease. Louis Pasteur would feel at home. Not only would he recognize the tremendous opportunity for eliminating disease, but he would also recognize the great inertia of the conservative forces in medicine and government. The phrase *déjà vu* applies perfectly.

To defeat disease, a war must be waged on it. At present the armies have not even been mobilized. Not only must the armies be sent into action, but they must be coordinated. Perhaps we should follow the Allied strategy in 1944 at D-Day, "combined ops"—the highly integrated use of the army, navy, and air force. Against disease we must totally integrate medicine, government, industry, and research. Unlike D-Day, however, they should prevent the disease, not wait until it has almost overwhelmed us.

It is the authors' profound belief that in half a century people will look with incredulity at today's world, where narrow commercial interests and government laissez-faire have free reign while the national health is left to founder.

Reference

1. World Health Organization Study Group. *Diet, Nutrition, and the Prevention of Chronic Disease*. Geneva: WHO, 1990.

Index